THE MILKY WAY AS A GALAXY

THE MILKY WAY AS A GALAXY

Gerard Gilmore
UNIVERSITY OF CAMBRIDGE

Ivan R. King
UNIVERSITY OF CALIFORNIA, BERKELEY

Pieter C. van der Kruit
UNIVERSITY OF GRONINGEN

Edited by
Roland Buser
UNIVERSITY OF BASEL

Ivan R. King
UNIVERSITY OF CALIFORNIA, BERKELEY

Saas-Fee Advanced Course No. 19
Lecture Notes
Swiss Society of Astrophysics and Astronomy
1989

UNIVERSITY SCIENCE BOOKS
Mill Valley, California

University Science Books
20 Edgehill Road
Mill Valley, CA 94941

Production manager: Mary Miller
Manuscript editors: Roland Buser and Ivan R. King
Copy editor: Aidan Kelly
Proofreader: Sylvia Stein Wright
Text and jacket designer: Robert Ishi
TEX formatter: Ed Sznyter
Indexer: Roland Buser
Printer and binder: Maple-Vail Book Manufacturing Group

Library of Congress Catalog Number: 90-070276

ISBN 0-935702-62-8

Printed in the United States of America
10 9 8 7 6 5 4 3 2 1

CONTENTS

FOREWORD

The 19th Advanced Course of the Swiss Society of Astrophysics and Astronomy was held in Leysin, Vaud, Switzerland, from March 13–18, 1989. It was attended by more than 80 participants from 14 countries, including Israel and the United States. Although this is the largest audience ever admitted to a "Saas-Fee" course, a very large number of applicants still had to be refused, unfortunately. It is our hope, however, that the present volume will be welcomed by them as well as by an even larger community of astronomers interested in the subject.

In response to the society's call for proposals, we put forward the concept of *The Milky Way As a Galaxy*, which is intended as a timely response to recent developments in Galactic and extragalactic research. In this era of rapid advancements in both theory and observation of the formation, structure, and evolution of galaxies, the course was to provide young astronomers with a coherent picture of our Galaxy. The "inside" and "outside" views were to be consolidated into a concept of the Galaxy which may eventually prove to be as robust as the Galaxy itself, and which may yet be comprehensive enough to be useful in the more general study of galaxies.

With these demanding goals in mind, the course was designed as a coherent flow of lectures. This in turn required close collaboration among the lecturers at all stages from the planning through the final preparation of the printed text. We are very much indebted to the authors for enthusiastically joining in this formidable task. We are grateful to them for providing 28 highly illuminating and stimulating lectures at Leysin, and for supplying lecture notes written in a common mathematical notation and, largely, also in a common terminology. Particular appreciation is due to Ivan King, who, as an author and editor, coordinated the whole process and composed the manuscripts into the 16 chapters of the present volume.

We hope that the volume can serve as a textbook of Galactic astronomy. A parallel edition of this book will be published and distributed by University Science Books of Mill Valley, California. This will greatly enhance the accessibility of the 1989 "Saas-Fee Lecture Notes" to our American colleagues.

We would like to thank Yves Chmielewski of the Geneva Observatory and Andreas Spaenhauer of the Astronomical Institute of the University of Basel for their assistance during the course at Leysin.

Finally, we gratefully acknowledge financial support by the Swiss Academy of Sciences.

Roland Buser
Gustav A. Tammann

Astronomical Institute
University of Basel

Organizers

PREFACE

This book is the outgrowth of a series of lectures given by the three authors at the 19th "Saas-Fee" course, which was held at Leysin, Vaud, Switzerland, March 13–18, 1989.

The annual "Saas-Fee" courses have been organized by the Swiss Society of Astrophysics and Astronomy and co-sponsored by the Swiss National Academy of Sciences since 1971. They are open to an international audience of astronomers and physicists on the pre- or post-doctoral level, as well as to young researchers. Each year's subject is selected from the field of current astronomical research, and is covered in complementary lectures given by three invited experts. The lectures are designed to expose the fundamental physical principles behind the subject and to discuss the research methods in the context of the major results in the field. Their text is published in the annual volume of the Swiss Society's well-known series of "Saas-Fee Advanced Course Lecture Notes."

The topic of the 19th "Saas-Fee"course had been chosen by Roland Buser and Gustav Tammann, of the University of Basel, who contacted the lecturers and planned the course with them. The lecturers and organizers were able to keep in close touch with each other thanks to the availability of electronic mail, and as a result the details of the course were planned so as to cover its subject matter in a fairly comprehensive way, with a logical flow from one lecture to the next.

This year's Saas-Fee volume is structured somewhat differently from those published in the past. Because of the close collaboration that had taken place, it has seemed appropriate to depart from the traditional Saas-Fee format of three monographic articles. We have therefore interleaved the lectures in the logical order referred to, merely identifying the author of each chapter. In the editing, successive lectures could often be consolidated into a single chapter, so that the material of 28 lectures appears here as 16 chapters.

The unification of three separate contributions was facilitated by the fact that the authors had agreed beforehand on a common mathematical notation and, to a large extent, on a common terminology. Each was free to pursue the details of his topics in his own way, however; and a small amount of overlap has inevitably resulted. Some cross-references have been inserted where it seemed appropriate, and Roland Buser prepared a single index for the whole book. Ivan King saw the manuscript through the various stages of preparation and editing.

We are grateful to many of our colleagues for their valuable help. G.G. acknowledges the hospitality of the Canadian Institute of Theoretical Astrophysics during a period when much of this material was written and notes that a substantial part of his contribution to this book derives from work done in collaboration with Rosemary Wyse. I.R.K. thanks Carl Heiles, Cedric Lacey, John Carlstrom, and Frank Shu for comments on various parts of his writing. P.C.K. thanks Gineke Alberts and Theo Jurriens for help and advice in producing the texts in TeX. We would also like to thank Bruce Armbruster, Publisher, and Harry Nussbaumer, President of the Swiss Society of Astrophysics and Astronomy, for their consent to the agreement which led to the publication of this book. In particular, we are indebted to Bruce Armbruster for his enthusiastic readiness to produce an American co-edition of the Swiss Society's "Saas-Fee Lecture Notes."

Roland Buser
Astronomical Institute
University of Basel

Ivan R. King
Astronomy Department
University of California, Berkeley

Editors

THE MILKY WAY AS A GALAXY

CHAPTER 1

INTRODUCTION

Ivan R. King

The aim of this book is to present the broadest possible overview of the Milky Way, our home galaxy. Our own night sky gives only the barest inkling of its structure: we see the encircling band from which the Galaxy[1] takes its name, and the brightening in the Sagittarius region that indicates in which direction the center lies. And indeed, except for these two features, the superficial appearance of the night sky gives little information about the larger structure in which we are imbedded. The stars that make up our constellations are generally within one or two hundred parsecs of us—whereas the disk of the Milky Way extends hundreds of times farther—and there is no clear indication at all that such a thing as a halo exists.

Clearly the study of the Milky Way as a galaxy needs to deal with far wider questions. It will require, in fact, reaching out to nearly all the subrealms of astronomy. Among these areas, however, two stand out as especially relevant to our problem. One, of course, is extragalactic astronomy. By a careful examination of the properties of other galaxies, we see what to expect in the Milky Way— particularly if we know our own Galaxy well enough to recognize its position within the morphological range of other types of galaxies that we observe. Yet we do not rely merely on expectation; we have a great deal of direct information about the properties and the structure of the Milky Way. Our approach will thus be to develop both points of view in parallel, sketching out what we know about Galactic structure and interweaving that with what we know about the structure of other galaxies.

The other area of special relevance for us is stellar populations. The products of different eras and events of star formation differ from each other physically in striking ways; also, many of them have quite different spatial distributions, as well as different kinematic properties that maintain these distributions. During the nearly half a century in which such differences have been prominently discussed, however, our perceptions of their nature have evolved; and with our changing perceptions our taxonomy and our terminology for such populations have also

---

[1] Note that it is customary to use the word "Galaxy," capitalized, to refer specifically to the Milky Way.

1

evolved. It is thus appropriate to begin with a survey of stellar populations, with particular emphasis on the need to choose a consistent and coherent terminology.

1.1   THE TERMINOLOGY OF STELLAR POPULATIONS

Baade's (1944) initial distinction was between "Population I" in the solar neighborhood and "Population II" in the globular clusters (and in the center of M31, which had enabled him to pass from already recognized differences to a generalization). Within a decade it became clear that Baade's Population II was an old one, in which star formation had ceased a very long time ago (Sandage and Schwarzschild 1952; Hoyle and Schwarzschild 1955), but that in Population I stars are still forming now. And then another variable appeared: the high-velocity halo stars and the globular clusters were found to have a lower abundance of heavy elements than the Sun and its neighbors (Chamberlain and Aller 1952; Helfer *et al.* 1959). It turned out, though, that there was an old component of Population I, as old as the globular clusters (or nearly so), but with solar heavy-element abundance.[2] Morgan and Mayall (1957) showed that the center of M31 had this sort of population, in agreement with Baum and Schwarzschild's (1955) observation that Baade's globular-cluster-type stars were only a minor contributor to the light there. Populations were becoming a confusing subject.

Moreover, in the solar neighborhood, the older stars of Population I showed spatial and kinematic properties somewhat different from those of the younger stars, and the confusion became greater. At the Vatican conference of 1957, Oort outlined a scheme of five population types that spanned the range between Baade's original two extremes (O'Connell 1958, p. 419). But this classification, too, has proven unequal to the more recent recognition that there are young stars of low metal abundance in the Magellanic Clouds, in other galaxies, and even in the outer parts of the Milky Way.

The traditional terminology of stellar populations has thus become outmoded. In this book we shall abandon it as much as possible, and return to simple descriptors—mainly age (or age range) and abundance.[3] These in turn may in the future become insufficient, as mass-function slopes and abundances of different element groups gain in importance; but perhaps they can suffice for now.

---

[2]Note that the abundances referred to here are those of all elements heavier than hydrogen and helium. In addition, we shall use here the loose but usual parlance of "metal abundance."

[3]There is one possible exception to this rule. It is presumed that there was an early generation of stars, now disappeared, which made the heavy elements that are found in the stars of the globular clusters and the stellar halo. These are often referred to as "Population III."

1.2   THE COMPONENTS OF THE MILKY WAY

To consider the components of the Milky Way we must again begin with terminology. The recognizable spatial and population components that make up the
Milky Way have acquired a set of names that have become alarmingly obscure
and confusing. It can be argued that the misnaming began when flat rotation
curves in galaxies were discovered, since these showed that galaxies must contain
an extended component that emits little or no light. This "massive halo" con-
flicted in name with the classical stellar halo, since the two differ both in physical
nature and in spatial distribution. Some misguided individuals then chose to distinguish the stellar halo by calling it the "spheroid." Unfortunately, that word
was already being used to designate the central bulge of external galaxies that
resemble the Milky Way. It has taken only a few years of carelessness to extend
the term "spheroid" to include in a single mélange the dark massive halo, the
low-metal-abundance stellar halo, and the central bulge, which is dominated by
stars of high metal abundance. In this book we shall avoid the term "spheroid"
(except in geometric descriptions where it is literally needed). The central bulge
will be referred to explicitly by its own name. The dark component that makes
the outer rotation curve flat will be called the dark halo (*faute de mieux*), and
the stellar halo will be referred to as such (and often the word "stellar" will be
omitted when the meaning is perfectly clear).

What then are the major components of the Milky Way? Most obvious to
us, of course, is the thin disk, which contains most of the stars of our immediate neighborhood. Its distribution is approximately exponential both in the
radial direction and in z, with scale heights of a few kiloparsecs and a few hundred parsecs, respectively. Its stars move in orbits around the Galactic center
that differ only little from circles. Its abundances are very close to those of the
Sun, although there is almost certainly a modest radial gradient toward lower
abundances with increasing distance from the center. Ages of stars, and of open
clusters, range from nearly that of the Milky Way down to practically zero.

Also well known is the stellar halo, consisting of the globular clusters and
about 100 times as many field stars. The globular clusters, at least, can be
divided into two subgroups, the so-called halo globulars and disk globulars. The
halo globulars have a nearly round distribution that is approximated by a radial
-3.5-power law; their system is pressure-supported (an arcane way of saying
that the motions are random) and has little or no rotation around the Galactic
center. The disk globulars, on the other hand, have a flatter spatial distribution
and a mean rotation around the Galactic center that is about half the circular
speed. The metal abundances of the halo globulars are very low, and those of the
disk globulars are moderately low. The globular clusters are the oldest objects
in the Milky Way; whether there are age differences among them (particularly
between the halo group and the disk group) is controversial.

The status of the field stars of the halo is in many ways puzzling. On the
whole, they seem to be the counterpart of the halo globular clusters, but some
of the correspondences are not as close as they might be. And the counterpart of

the disk globulars is even more puzzling. It is tempting, but speculative and in no way compelling, to identify them with the thick disk that is described below.

In the central part of the disk is the bulge—perhaps a part of the disk, perhaps a separate component, but certainly not merely the center of the halo, from which it differs drastically in metal abundance. The bulge is, moreover, difficult to observe; and insofar as we observe it, it is a mixture. Metal abundances of the old stars range from low to far above solar. The bulge is dominated by old stars, but we do not know if they are all equally old. Its flattening (on the average, in what after all may be a mixture) is modest, with comparable amounts of rotational and pressure support. And in the Galactic plane, in the neighborhood of the center, there exist some young stars.

The most recently recognized stellar component of the Milky Way is the thick disk. Its scale height in the solar neighborhood is between 1 and 2 kpc, and its population is chiefly, or even totally, old. As already indicated, it strongly resembles the disk globular clusters both in abundances and in kinematics, but we do not yet know whether the two are manifestations of the same component or not.

These stellar components of the Milky Way are distinguishable; but it is a separate question whether they are discrete, or whether in some way there exists a continuum of properties in which our instruments of perception happen to distinguish these points. We shall need to address this question directly, whether we can answer it or not.

And finally there are the two non-stellar components. The interstellar medium, most of it thinner even than the thin disk, is the key to present and future star formation. And the dark halo, known only by its gravitational effect, is still very much a mystery. There are indications that it is rather round, and it is almost certainly not made of stars.

One dynamical effect complicates the separation of these components; it might be said, in fact, that this effect forcibly mixes them. This is the fact that in any well-behaved distribution, the density is greater at points that are deeper in the potential well. Thus the stars of the thick disk and the halo have a higher spatial density in the central layer of the thin disk than they do outside it, even though they contribute only a small part of the star density in that central layer. More strikingly, the stars of the halo must have their highest spatial density at the Galactic center, even though they are so heavily outnumbered there by the high-metal-abundance stars of the bulge.

1.3   POPULATION CHARACTERISTICS AND INDICES

As we have noted above, we shall distinguish populations by their ages and their metal abundances. How is each of these measured?

Ages come from the theory of stellar evolution, applied to observations of colors and magnitudes of stars. Here the study of star clusters offers a distinct advantage, since all the stars of a cluster (with a few possible exceptions) have the

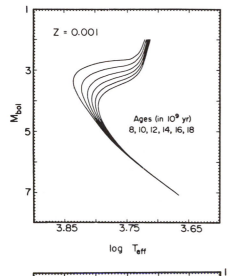

FIGURE 1.1
Isochrones for different ages, at the same
metal abundance (adapted from
VandenBerg and Bell 1985).

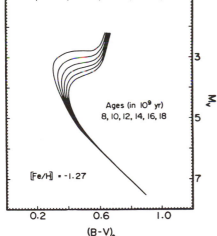

FIGURE 1.2
The isochrones of Figure 1.1, converted to
observational coordinates (from
VandenBerg and Bell 1985).

same age and chemical composition, so that they should lie along an *isochrone* (Fig. 1.1). The fitting of isochrones to observations is simple in principle, but in practice extreme care must be taken (1) to use isochrones calculated for the correct chemical composition and (2) to make a correct conversion from the theoretical HR-diagram coordinates $M_{\rm bol}$ and $\log T_{\rm e}$ to the color and magnitude system of the actual observations (cf. Fig. 1.2).

Abundances pose a complicated observational problem. The only true measures of abundances are made from high-dispersion spectra. These are expensive of telescope time, and are unable to reach very faint stars. The reason that high dispersion is needed is that reliable abundances can be derived only from weak spectral lines, which are formed in the photospheric layers whose temperatures can be calculated reliably. Strong lines not only suffer from curve-of-growth

effects, but are formed in a boundary layer for which the temperature (and therefore the atomic excitation) is uncertain.

Since direct measures of abundance are so difficult (or at least expensive),
many secondary methods are used. One is to study, at lower spectroscopic dispersion, the strong lines in the spectra of individual stars. Other methods rely
on the integrated spectrum of a stellar system (for example, a globular cluster or the bright central region of a galaxy). Even though the spectrum is a
mixture of the light of many stars of different types, higher-abundance systems
tend to have stronger lines. For globular clusters there is a complication here,
though, because the strengths of the lines can also be influenced by the amount
of blue-horizontal-branch light with which they are diluted.

Even more crude is the measurement of colors. Usually interstellar reddening
must be corrected for, so that at least three wavelength bands are needed. An
example of an abundance criterion is ultraviolet excess, measured for the integrated light of a cluster rather than for an individual star. For this purpose, any
color system will serve if it can produce a reddening-free index of the amount of
line blanketing at the shorter wavelengths.

It is appropriate to refer to the indirect ways of estimating abundances as
criteria rather than as measurements. Each of these methods is like a pointer on
a dial that has no numbers on it. In order to be useful, the criterion must be
calibrated by means of observations of stars or clusters whose abundances have
been measured directly by high-dispersion spectroscopy.

To specify abundances, it is almost universal practice to give the logarithmic
abundance of iron, related to hydrogen, with the ratio related in turn to that in
the Sun:

$$[\text{Fe/H}] = \log \frac{N_{\text{Fe}}}{N_{\text{H}}} - \log \left(\frac{N_{\text{Fe}}}{N_{\text{H}}} \right)_{\odot}.$$

Observed values of [Fe/H] range from near $+1$ for some stars in the central bulge
of the Milky Way down to about -2.3 for the most metal-poor globular clusters,
and field stars are known with even lower metal abundances.

1.4   THE MILKY WAY AND OTHER GALAXIES

The study of the Milky Way and that of other galaxies are in many ways complementary. Here in the midst of an outer part of the disk of our Galaxy, we
can study our own region in great detail. In particular, we can see the stars of
low luminosity that are unobservable at the distance of other galaxies. On the
other hand, we suffer from our imbedded position, being unable to see the overall
structure, and in many directions being unable to see much at all on account
of interstellar absorption. In observing other galaxies, by contrast, we grasp the
grand structure easily, but the fine details are out of our reach. The aim of this
book is to combine the advantages of each of these points of view to reach a
picture of *The Milky Way As a Galaxy*.

REFERENCES

Baade, W. 1944. *Astrophys. J.*, **100**, 137.

Baum, W. A., and Schwarzschild, M. 1955. *Astron. J.*, **60**, 247.

Chamberlain, J. W., and Aller, L. H. 1952. *Astrophys. J.*, **114**, 52.

Helfer, H. L., Wallerstein, G., and Greenstein, J. L. 1959. *Astrophys. J.*, **129**, 700.

Hoyle, F., and Schwarzschild, M. 1955. *Astrophys. J. Supp.*, **2**, 1.

Morgan W. W., and Mayall, N. U. 1957. *Publ. Astron. Soc. Pacific*, **69**, 291.

O'Connell, D. J. K., ed. 1958. *Stellar Populations*. Amsterdam: North Holland.

Sandage, A. R., and Schwarzschild, M. 1952. *Astrophys. J.*, **116**, 463.

VandenBerg, D. A., and Bell, R. 1985. *Astrophys. J. Supp.*, **58**, 561.

CHAPTER 2

THE DISTRIBUTION OF STARS IN SPACE

Gerard Gilmore

The spatial distribution of stars in the Galaxy is related, through the gravitational potential ψ, to their kinematics. Specifically, the scale length of the spatial distribution is determined by the combination of the total energy of the stellar orbits and the properties, primarily the gradient, of the gravitational potential. The shape of the spatial distribution allows us to calculate the relative amounts of angular momentum (rotation) and pressure (stellar velocity dispersion) by which the potential gradients are balanced. These observables are interesting in part because they help us understand the Galaxy in its own right, in part because they provide a fossil record of the physical processes that took place during the formation and early evolution of the Galaxy, and in part because they give us a way to measure the Galactic potential. Since the potential is generated by *all* the mass in the Galaxy, comparison of the total potential with that generated by the (known) stellar and gas distribution allows us to determine the spatial distribution of any dark matter, and perhaps to discern its nature.

To be more explicit, the total orbital energy and angular momentum of the gas that will become a star depend on the maximum distance from the center of the proto-Galaxy that it ever reached, the angular momentum of its orbit at that time, the depth of the potential well (generated by both dark and luminous mass) through which it fell, the fraction of the total orbital energy that was dissipated, the amount of viscous transport of angular momentum before star formation, and the subsequent dynamical evolution of the stellar orbit. That is, the present kinematic properties of old stars in the solar neighborhood are determined in part by the initial conditions in the proto-Galaxy, and in part by the physics of galaxy formation. Hence, measurement of the large-scale structural and kinematic properties of the Galaxy can help reveal both the detailed physics of galaxy formation and the distribution of gravitating mass, both identified and dark, in the Galaxy. As well as satisfying our inherent curiosity about our neighborhood, such studies provide the most promising opportunity to understand the fundamental features of the formation, evolution, and structure of galaxies in general.

2.1   A DIDACTIC RAMBLE

Every child who lives away from public lighting is aware of the basic features of the structure of the Milky Way, and quite likely wonders how it got like that. The generally flat distribution, with dark bands (of obscuration) covering large arcs, and the noticeable asymmetry, with a substantial excess of brightness toward Sagittarius, are clearly apparent. Attempts to provide a more specific description, based primarily on star counts, have been a major part of observational astronomy since the invention of the telescope. Although these efforts have produced a reasonably coherent and (we hope) reliable picture of the large-scale structure of the Galaxy, there is also an important lesson to be learned from the failure of some careful attempts to model star-count data. The most widely known of these erroneous models is that devised by Kapteyn, whose analysis of star-count data from photographic plates led to the "Kapteyn Universe" (Kapteyn and van Rhijn 1920), in which the Sun was near the center of the Milky Way.

Kapteyn's conclusion was based on star counts in small areas toward what we now know as the Galactic center and the Galactic anticenter, which showed equal numbers of stars in these two directions. Looking at the data at face value, and accepting the widely believed assumption of the time that interstellar obscuration was small, any rational person must conclude that the observer is at a point of symmetry, i.e., near the center. Yet the canonical child mentioned above could tell that such a model is false from casual observation, provided, of course, that the child lived sufficiently far south to be able to see Sagittarius. (Kapteyn could have told this also if very much larger data sets had been available to him.) One of the more illustrious of Kapteyn's predecessors, John Herschel, who had seen the southern sky, had written (1849, art. 788)

> our situation as spectators is separated on all sides by a considerable interval from the dense body of stars composing the Galaxy, which in this view of the subject would come to be considered as a flat ring of immense and irregular breadth and thickness, *within which we are excentrically situated,* nearer to the southern than the northern part of its circuit   (my emphasis).

A glance at a wide-angle picture of the Milky Way, or perhaps the excellent Lund Observatory map, is an easier way for a modern inhabitant of northern latitudes or of a city to appreciate the same point, though Herschel (1847, Plate 13) had himself published an excellent map making the same point very clearly. In view of the importance of this result, and the difficulty in obtaining access to it today, this map is reproduced here as Fig. 2.1.

This little example makes an important and generally valid point: the interpretation of star-count data *in isolation* is unreliable, and prone to provide erroneous results. Reliable interpretation of Kapteyn's data required an unavailable appreciation of the large-scale structure of the Galaxy, and of the extreme point-to-point variations in the observed stellar surface density in the Galactic

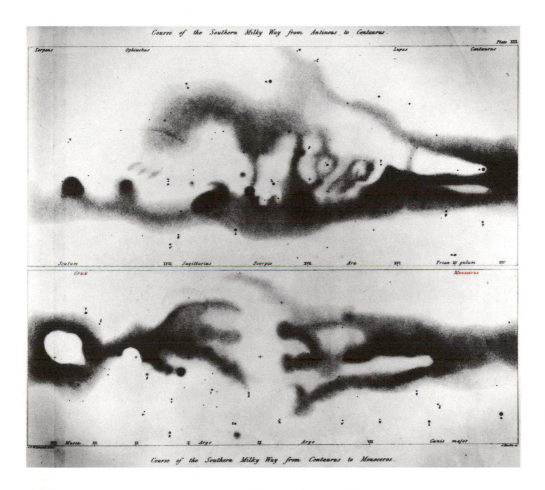

FIGURE 2.1

The surface-brightness distribution of the central regions of the Milky Way, as drawn by
John Herschel (1847) based on his star-count survey of the southern skies. The features
relevant to later studies are the general broad high-surface-brightness distribution through
Sagittarius and Scorpio, and the perceptible minimum in this distribution near the
Galactic plane. Thus optical star counts at low Galactic latitude were (and are)
inappropriate for discerning the true distribution of stars in the Galaxy.

plane. Reliable interpretation of modern data similarly requires knowledge of these points if we are working at low Galactic latitudes, and in addition consideration of chemical abundance, age, and kinematic data. Star-count data *in isolation* are incapable of providing a reliable, let alone a unique, description of the spatial structure of the Milky Way Galaxy.

Nevertheless, substantial progress has been made in determining the spatial distribution of stars in the Galaxy. By the time of the 1957 Vatican conference on stellar populations (O'Connell 1958), the combination of kinematic, spectroscopic, and star-count data had provided a picture of the distribution of the several stellar populations in the Galaxy that differed only in trivial ways from our current picture. The extensive data set and understanding available at that time are reviewed in many excellent articles in Blaauw and Schmidt (1965). Unfortunately, few are now familiar with this very substantial body of information.

The next period of rapid progress in studies of Galactic structure occurred when the new deep high-quality data of King and collaborators at Berkeley in the late 1970's became available. The application of computer modeling to these data by van den Bergh (1979) led to a considerable resurgence of interest, which is still continuing. Presumably because of the growth of extragalactic studies in the 1960's and 1970's, however, most efforts to understand these new data were based on the assumption that the Galaxy is rather like other galaxies, rather than on the available, detailed studies of the Milky Way itself. Thus considerable weight was given to the expectation that spiral galaxies were well described by the sum of a ("Population I") exponential (both radial and vertical) disk and a roughly spherical ("Population II") halo, whose luminosity profile in projection followed the $R^{1/4}$ law.[1] Some of these assumptions have since been discovered to have been unjustified. For example, recent photometric analyses have shown that the preceding description almost always fits the high-quality surface-photometric data of other spiral galaxies (Schombert and Bothun 1987) much too poorly to be acceptable. Rather, the luminosity profile of NGC 891, the galaxy often said to be most like the Milky Way, shows no evidence for a detectable $R^{1/4}$ component in its luminosity distribution (Shaw and Gilmore 1989); so we should not necessarily expect a star-count model based on an $R^{1/4}$ halo to be an adequate description of the Milky Way Galaxy. In fact, an exponential is a better approximation of the vertical structure of the Galaxy near the Sun, even well away from the (exponential) thin disk. Some other problems in interpreting photometric, kinematic, and spectroscopic data for other galaxies in terms of stellar populations are discussed in later sections of this book.

It is an interesting historical lesson that the vertical profile of the Galactic disk which we now adopt, on the basis of an enormous investment of large-telescope time, is in excellent agreement with Figs. 1 and 2 of the review by Elvius (1965) and with Fig. 7 of the review by Plaut (1965), and in quite adequate agreement with Table 2 of Oort (1958; see Table 3 of Blaauw 1965).

---

[1] In this chapter, contrary to usage elsewhere in the book, R will denote a spherical rather than a cylindrical radial distance. In projection it will denote a polar coordinate.

2.2 STAR COUNTS AND GALACTIC STRUCTURE

The number of stars N countable in a given solid angle ω to a given magnitude limit m is given by a simple linear integral equation often known as the fundamental equation of stellar statistics. It is

$$N(m) = \omega \int \Phi(M, \vec{x}) \, D(M, \vec{x}) \, d^3\vec{x}, \tag{2.1}$$

where $\Phi(M, \vec{x})$ is the distribution function of absolute magnitude M, which will also be a function of position $\vec{x}$; $D(M, \vec{x})$ is the stellar space-density distribution, which will also be a function of absolute magnitude; and $d^3\vec{x}$ is a volume element. This is a Fredholm equation, which is rarely invertable, being ill-conditioned. A detailed discussion of its use and the approximate methods for its solution are presented by Trumpler and Weaver (1953). The essential feature of this equation is that the number of stars counted depends on the *product* of the luminosity function and the space-density distribution. Consequently, unless independent supplementary information is available to define either the stellar luminosity function or the spatial-density distribution, no unique inversion of Eq. (2.1) to derive $\Phi(M, \vec{x})$ or $D(M, \vec{x})$ independently is possible.

The basic problem with use of the fundamental equation is that both the stellar luminosity function and the stellar density law are functions of many parameters, with most of these parameters affecting both Φ and D. The most important of these include the mean value and the range of age, chemical abundance, and kinematics in each subset of stars of interest. Few of these parameters are sufficiently well known to be fixed on external criteria. Consequently, a wide variety of combinations of Φ and D is allowed mathematically as solutions of Eq. (2.1). Other astrophysical constraints are necessary, whose choice remains subjective unless very large amounts of observational data have been obtained, hence my emphasis on the importance of familiarity with the very substantial data set available in the literature.

There are three solutions to the indeterminacy inherent in Eq. (2.1). The most straightforward is to identify precisely defined subsets of stars whose absolute magnitude, and hence distance, is known independently or can be determined from independent observations. Examples of such subsets include pulsating variable stars (RR Lyrae stars and Cepheids are particularly important here), readily identifiable stars of fixed absolute magnitude (*e.g.*, blue-horizontal-branch stars), or objects whose distance can be established in some other way (*e.g.*, open and globular clusters, with distances from main-sequence fitting). Many such analyses have been carried out with conspicuous success. A second solution is to use as much astrophysical data as is available to limit the solution space of Eq. (2.1) for general star-count data. This latter approach is clearly more model-dependent and more work, but potentially provides more information.

The third solution is to restrict the region of study in such a way that only one of the variable functions in Eq. (2.1) need be considered. This approach is the basis of surface photometry, where we assume that the integral over the luminosity function is a constant, and the resulting variation in integrated surface brightness corresponds to a variation in the stellar spatial density distribution. We will now discuss each of these approaches to determination of the spatial distribution of stars in the Galaxy.

2.3   INTEGRATED SURFACE-BRIGHTNESS MEASUREMENTS

Since the Galaxy is not extragalactic, and occupies rather a large solid angle on the sky, the interpretation of surface photometry must proceed in a model-dependent way. Surface photometry at low Galactic latitudes ($b \lesssim 20°$) is, of course, very severely affected by interstellar obscuration, which requires careful consideration. This is especially important for studies of the Galactic disk. At higher latitudes obscuration is less serious, though local obscuration near the Sun still requires careful treatment. We proceed by adopting some model for the spatial luminosity distribution of the halo, calculating the line-of-sight integral of this distribution, subtracting from that model a model for the disk contribution, and then correcting for a model of the distribution of interstellar extinction. We then compare this resulting model to observations at a range of Galactic latitudes to find values for the several parameters (central surface brightnesses, density profiles, characteristic scale lengths, axial ratios) of the halo and disk luminosity distributions. In practice the good data are limited to a narrow range in surface brightness, because of the dominance of the disk and extinction problems at low latitudes, and the faintness of the halo at high latitudes. Thus, as we might imagine, the resulting parameters have not been evaluated to high precision.

The most careful attempt to carry out an analysis of this type is that by de Vaucouleurs and Pence (1978). They showed that their surface-brightness data over a range in surface brightness of 1.5 magnitudes between Galactic latitudes 15° and 30° were consistent with a luminosity distribution that follows the $R^{1/4}$ law, with an assumed effective radius $r_e = 2.7$ kpc. The c/a axial ratio of the halo was equally well fit by models with $c/a = 1$ and models with $c/a = 0.6$. Other combinations of r_e and c/a were not tried.

The difficulties imposed on models of this type by interstellar extinction may be mitigated by using observations in the near infrared (~ 2 to $2.5\,\mu$m), where the effects of reddening are substantially reduced. Suitable $12\,\mu$m data from the IRAS satellite, and $2.4\,\mu$m observations by Hiromoto et al. (1984) and Hayakawa et al. (1981), were analyzed for this reason by Garwood and Jones (1987). They investigated a variety of $R^{1/4}$ models, and deduced an axial ratio $c/a = 0.25$. They also showed that an $R^{1/4}$-law halo, with the normalization in the central regions of the Galaxy fixed by the infrared data, was inconsistent with the observed space density of Extreme Population II subdwarfs near the Sun.

Thus the most likely interpretation of available surface photometric data is that the Galactic halo is consistent with, but does not require, an $R^{1/4}$-law distribution with effective radius $R_e \sim 2.5$ kpc and axial ratio $c/a \lesssim 1$. In the central few kiloparsecs of the Galaxy, an additional, flatter distribution is seen, with a shorter spatial scale length. The relationship between this central component and the rest of the Galaxy remains problematic, and will be discussed further.

2.4 SPECIFIC TRACERS OF GALACTIC STRUCTURE

Determination of the spatial density distribution of a complete (or incomplete in some well-understood way) sample of tracers whose intrinsic luminosity can be found, and hence whose distance from the Sun can be calculated, is clearly a simple exercise. One simply counts the number of tracers in each of several distance intervals, divides by the corresponding volume elements, and has the answer. The luminosity function then degenerates to a constant (delta) function, and the relation between number counts and the density law is trivial. Fortunately for the intellectual interest of observers, reality is less simple. The most important complication in principle, apart from the technical problems in finding reliable absolute magnitudes and/or distances, involves selection effects that are correlated with distance. When these exist, the luminosity function Φ and the density function D are no longer independent, and cannot be separated. The most likely way for such a correlation to occur comes from the joint dependence of Φ and D on a common parameter.

An important real example is the use of RR Lyrae stars to trace the density profile of the halo. The principal advantages of RR Lyraes as tracers of the Galactic density distribution are that they are intrinsically luminous, and so can be seen at large distances; they are large-amplitude variables, and so can be identified easily; and their absolute magnitudes can be calculated from relatively low-resolution data, so that distances can be found without too much difficulty. However, corresponding complications exist. The origin of RR Lyrae stars is not well understood. Examination of open- and globular-cluster color–magnitude diagrams shows that RR Lyraes are not found in systems younger than about 12 Gyr old. Similarly, although high-metallicity field RR Lyraes certainly exist, RR Lyraes are not found in apparently old globular clusters that are sufficiently metal-rich to have red-horizontal-branch stars. Finally, not all metal-poor globular clusters have blue horizontal branches and RR Lyraes (the second-parameter problem). Hence the radial distribution of RR Lyrae stars in the Galactic halo depends on the radial density profile of all stars that are potentially their progenitors, on the radial chemical-abundance profile, on the radial age profile, and on the radial profile of the "second parameter," which may or may not be the same as one of the other parameters. Thus we have an exact analog of the situation in Eq. (2.1): there are too many correlated and interdependent parameters to allow a unique solution for the density profile. Nevertheless, insofar as we believe that the selection effects are understood and that the various parameters listed above are or are not a problem, RR Lyraes have provided invaluable information about the shape and the kinematics of the Galactic halo.

There are many other types of object that are intrinsically luminous, and for which we can, given appropriate data, find distances. The more important of these include early-type stars, late-type giants, long-period variables, novae, H_{II} regions, Cepheids, open clusters, and globular clusters. The detailed results of careful study of these tracers are well reviewed in Blaauw and Schmidt (1965), and need not be repeated here. The general picture resulting is, however, important, and we will illustrate how single tracers reveal the spatial structure of the Galaxy, by discussing the two most important—RR Lyraes and globular clusters—in some more detail.

RR Lyrae Stars as Probes of the Galaxy

Early work based on RR Lyrae stars in the Palomar–Groningen and Lick surveys toward the Galactic center (Kinman *et al.* 1966; Oort and Plaut 1975) concluded that these stars were distributed in a nearly spherical system. These results have now been superseded by better photometric data (Wesselink *et al.* 1987); the more modern analysis finds in contrast that the RR Lyrae stars toward the Galactic center have a rather flattened distribution, with axis ratio $\lesssim 0.6$, in tolerable agreement with the integrated-light result above.

The method required to analyze the RR Lyrae data is very straightforward. Since the absolute magnitudes of the stars are known *a priori*, the luminosity-function term in Eq. (2.1) is a delta function, and can be ignored. A reliable distance is therefore known for each star. We note that finding out how *complete* the sample is is of considerable importance, and that the stellar absolute magnitude may well be a function of metallicity, so that supplementary spectroscopic data are also necessary. Similarly, at low Galactic latitudes there can be much interstellar extinction, and it can vary on very short angular scales. The several effects of extinction must be treated carefully, but they present problems merely in practice, not in principle. Hence, only the density function remains to be evaluated. The density function is, however, a function of several parameters, including the (possibly variable) shape of the stellar distribution, the distance to the Galactic center, and the numerical function describing the density distribution. We therefore proceed by calculating the expected distribution of tracer stars in a variety of models, convolving these predictions with the observational error and incompleteness functions, and comparing the model predictions with observation, to deduce the most appropriate set of numerical parameters.

An important example of this is the use of RR Lyrae stars to map the density distribution in the Galactic bulge. This was first studied in detail by Oort and Plaut (1975), based on Plaut's photometric studies. The geometry here depends on R_0, the solar Galactocentric distance, the major-axis/minor-axis shape of the bulge isodensity contours a/c, the Galactic latitude b (their fields were all near the $\ell = 0°$ meridian), and the distance r to a star. Then

$$\frac{a}{R_0} = \left\{ \left(\frac{r}{R_0} \right)^2 \left[\cos^2 b + \left(\frac{a}{c} \right)^2 \sin^2 b \right] - 2 \frac{r}{R_0} \cos b \cos \ell + 1 \right\}^{1/2}. \qquad (2.2)$$

The number of stars expected in a volume element at specified latitude b is easily calculated, convolved with the error distribution, and compared with the data. The Oort and Plaut analysis included data from five fields with $4° \lesssim b \lesssim 29°$, and concluded that the solar Galactocentric distance $R_0 = 8.7$ kpc, that the halo stellar distribution was well described by a power-law (R^{-n}) decrease with exponent $n = 3$, and that the equidensity contours were rounder than $c/a = 0.8$. A more recent recalibration of the photometric data analyzed by Oort and Plaut has revised these values somewhat; the best available current values are $c/a = 0.6 \pm 0.1$ (Wesselink *et al.* 1987) and $R_0 = 7.8$ kpc (Feast 1987). At large distances from the Galactic center, many RR Lyrae surveys have shown that the apparent halo-density profile is adequately described by a roughly round power law R^{-3} to distances of at least 20 kpc, beyond which some evidence for a cutoff is apparent. It remains to be confirmed if this is a real feature of the density distribution, a difficulty with completeness in the surveys, or a consequence of one of the several other relevant parameters listed above. It also remains to be investigated from independent tracer samples to what extent the radial distribution of the halo defined by RR Lyrae stars is affected by radial gradients in such subsidiary parameters as age, metallicity, and the elusive "second parameter," and to what extent it corresponds to the density profile followed by all stars that exist in the halo.

A complication in this picture of the halo having a simple power-law profile was introduced by the results of a kinematic analysis by Strugnell *et al.* (1986), who showed that the c-type (low amplitude of variability) and the small ΔS (*i.e.*, those with [Fe/H] $\gtrsim -1$) RR Lyraes have kinematics significantly different from those of the more metal-poor RR Lyraes. These subgroups of the RR Lyraes belong to a (thick) disk system, with dynamically significant rotation and a considerable flattening of their spatial distribution, as deduced from their kinematics. (The techniques leading to these deductions are discussed in Chapter 11.) Thus the RR Lyraes apparently form a two-component system.

Yet another complication arises from the analysis of Hartwick (1987). Hartwick analyzed all the available data for *metal-poor* RR Lyrae stars, in the same way as the Oort/Plaut analysis outlined above, and concluded that the axis ratio of the metal-poor RR Lyrae system varies with Galactocentric radius, being flattened (axis ratio $c/a \sim 0.6$, and scale height ~ 1.5 kpc; *i.e.*, rather similar to the parameters of the more metal-rich thick-disk RR Lyraes) interior to the solar circle. At very large Galactocentric distances the RR Lyrae data somewhat favor a more spherical distribution. Hartwick thus suggested that the *metal-poor* RR Lyrae stellar system is itself two-component, in addition to the (third) metal-rich RR Lyrae system.

Globular Clusters as Probes of the Galaxy

High intrinsic luminosity is an obvious observational advantage for a tracer of the halo. Globular clusters also have calibrated distances, metallicities, and age estimates, though finding values for each of these is a major research effort in its own right, and not without its difficulties. The major drawback of these clusters is simply that they are few in number. It has been suspected for many years that the globular clusters form a two-component system, with the more metal-rich (G-type) clusters forming a thick-disk system, and the more metal-poor (F-type) clusters forming a more extended, roughly round system (Becker 1950; Baade 1958). Conclusive evidence in favor of this suggestion has resulted from finding reliable and internally consistent values for chemical abundance, usually [Fe/H], for almost all observable clusters (Zinn 1985). These data show the globular cluster system to have a bimodal distribution in [Fe/H], with a natural division near [Fe/H] $= -1$. Clusters more metal-rich than this value form a disk-like distribution, with equivalent exponential scale height ~ 1 kpc, and disk-like kinematics. The more metal-poor clusters (the majority) could have a roughly round distribution described by $R^{-3.5}$ for 5 kpc $\lesssim R \lesssim 100$ kpc, though with a marginally significant deficit, in that no clusters are known with $35 \lesssim R \lesssim 60$ kpc. A similar disk–halo two-component distribution is apparent in the globular cluster system in M31 (Elson and Walterbos 1988).

This picture, however, remains uncertain, in part because of problems with distance errors, and in part because of the analysis by Hartwick (1987). Hartwick's analysis of the distribution of the projected positions on the sky of the metal-poor ([Fe/H] < -1) globular clusters found a two-component structure similar to that which he found for the metal-poor RR Lyraes. Again, this two-component structure is additional to the well-established distinction between metal-rich and metal-poor clusters. However, the small number of clusters involved gives Hartwick's multi-component model small statistical weight. It must also be remembered that globular-cluster formation may not be intimately related to the formation of field stars, and that empirically globular-cluster systems associated with external galaxies have different spatial distributions (the cluster systems tend to be more extended) and different colors (the clusters tend to be bluer) than the background starlight. We should use considerable caution if adopting some measured property of a globular-cluster system as an indicator of the likely value of that property for one of the field-star systems in the Galaxy, and *vice versa*.

IRAS Sources as Probes of the Galaxy

The large-scale distribution of sources which are luminous at $12\,\mu$m and $25\,\mu$m was mapped by the IRAS satellite, and forms a striking disk–bulge picture which is complementary and almost completely orthogonal to optical stellar studies (Habing 1987). From the IRAS color data we can select those sources that are

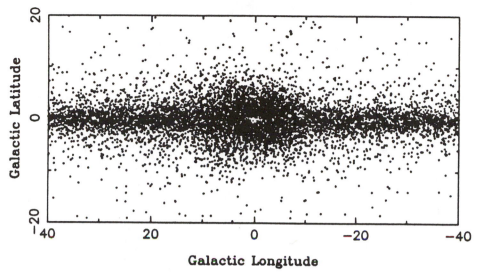

Galactic Longitude

FIGURE 2.2

The distribution of those IRAS sources with $0.4 \lesssim f_{12\,\mu m}/f_{25\,\mu m} \lesssim 1.6$ and 1 Jy$\lesssim f_{12\,\mu m} \lesssim 8$ Jy, where $f_{x\,\mu m}$ is the IRAS flux at the stated wavelength. This color and flux range have been chosen to maximize the contrast between the bulge and the disk (from Harmon and Gilmore 1988, Royal Astronomical Society).

luminous (dust-enshrouded) stars, and exclude those hotter stars that are simply nearby and highly reddened. The resulting spatial distribution is shown in Fig. 2.2. The sources outlining this distribution can be shown to be a mixture of two types of late-type giants, with those in the bulge being intermediate-mass AGB stars ($1.2 \lesssim \mathcal{M}/\mathcal{M}_\odot \lesssim 3$; Harmon and Gilmore 1988) and those in the disk being a mixture of intermediate-mass and low-mass stars with dust shells. The parameters of the large-scale spatial distribution have been derived by Habing (1989), using evaluations of Eq. (2.1), to show that the extended distribution forms two disks. About 80 percent of the stars form a thin disk (scale height $\lesssim 200$ pc) with radial exponential scale length 4.5 kpc, and with a cutoff at ~ 9 kpc, or about at the solar circle. The remaining 20 percent of stars form a thick disk, with scale height ~ 2 kpc, radial exponential scale length ~ 6 kpc, and no evident cutoff.

These disk parameters may be understood in terms of the two types of star detectable by IRAS. IRAS could see low-mass stars with high-optical-depth dust shells, and higher-mass young AGB stars with high mass-loss rates. For low-mass stars there is a correlation between the optical depth of the dust shell in the late stages of evolution (the Mira variable stage) and metallicity, with those stars that have [Fe/H] $\gtrsim -1$ having the highest-optical-depth shells (Harmon and Gilmore 1988). These same Mira variables have pulsation periods from ~ 150 days to ~ 200 days, and have been known for many years to outline a thick disk. In fact, the Vatican-conference stellar-population classification scheme used these variables to define the Intermediate Population II.

The thin-disk IRAS sources are predominantly higher-mass AGB stars: Miras and OH/IR stars.These stars are young, and hence their distribution reflects that of the young disk and the molecular gas, which corresponds to regions of current and recent star formation. The distribution of molecular clouds is now well defined (Solomon and Rivolo 1987) and also shows a cutoff at about the solar circle, in excellent agreement with Habing's analysis of the IRAS stellar data.

The IRAS sources in the central few degrees of the Galaxy outline the central bulge, which is hidden from optical study by interstellar obscuration. The detectable outer edge of the IRAS bulge in Fig. 2.2 in fact is near Baade's Window, where optical studies are first possible. IRAS and optical data are therefore nicely complementary. The spatial distribution of the central IRAS bulge ($4° \lesssim |b| \lesssim 10°$, where the lower latitude limit is set by satellite confusion) is well described by a roughly spherical exponential with scale height 375 pc, corresponding to a half-light radius (~ 1.68 exponential scale lengths) of ~ 600 pc. That is, the central bulge contains at least some young stars (those seen by IRAS; Harmon and Gilmore 1988) and has a scale length that is about five times smaller than that followed by more metal-poor halo stars. Lacking adequate age data for the sources in the central few degrees of the Galaxy, we cannot yet tell if the small scale length is due to a steep abundance gradient in an old stellar population, or to the presence of a discrete young central component of the Galaxy.

2.5   THE ANALYSIS OF STAR-COUNT DATA

The type of data that could provide the most information about Galactic structure is obviously data for a large number of field stars. We could then derive from Eq. (2.1) both the stellar luminosity function, as a function of several other parameters, such as age, chemical abundance, and position, and the stellar density distribution, again as a function of several parameters. In the remainder of this section we will present the available data, discuss their reliability, and then discuss available alternative approaches to their analysis.

Modern Star-Count Data

The availability of high-efficiency, linear, two-dimensional detectors (CCDs) and fast automated photographic-plate-scanning microdensitometers (PDS, COS-MOS, APM) has revolutionized stellar statistics. Complete samples of stars can be measured to useful precision in several wavebands over sufficiently large areas of sky that random errors due to counting statistics are unimportant. The minimization of systematic errors still requires an enormous effort, however. The relevant photometric techniques are detailed elsewhere (Gilmore 1984a), and need not be described here. Table 1 of Gilmore and Wyse (1987) summarizes those recent high-Galactic-latitude studies in which the magnitude calibration

colour–magnitude diagram

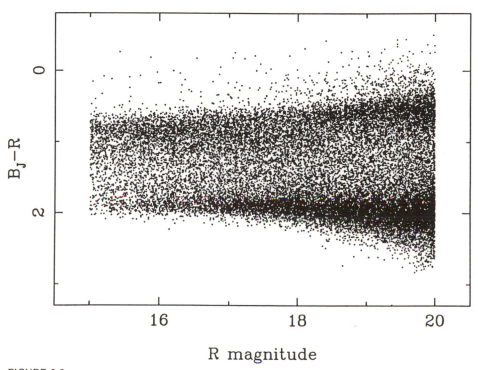

FIGURE 2.3
A representative color–magnitude diagram of the stellar distribution in the high-latitude sky, derived from APM scans of UK-Schmidt-telescope plates.

was derived directly from photoelectric or CCD standards. With the conspicuous exception of the Basel group (Fenkart 1989a,b,c,d; see below), no substantial new data sets have been published since that table was prepared.

The general features of the high-latitude sky are illustrated in Fig. 2.3, which shows stellar photometric data derived from APM scans of high-Galactic-latitude plates from the UK Schmidt Telescope. The important aspects of these data are as follows.

1. The sharp edge to the distribution near the equivalent of $B - V \approx 0.4$, with very few stars being seen significantly bluer than this limit. This corresponds to the main-sequence turnoff color of an old, metal-poor population. The absence of a younger turnoff shows that no significant number of stars have continued to form in the halo, though note the existence of some apparently young high-latitude metal-rich A stars whose place of formation remains an extremely important mystery (Lance 1988), and some distant B stars that cannot have traveled from the thin disk in their main-sequence lifetimes (Keenan *et al.* 1986; Conlon *et al.* 1988). There is also

some evidence that there are fewer blue-horizontal-branch stars than would be expected from the field subdwarf population. If so, the field stars near the Sun may suffer from the second-parameter problem noted above.

2.   A peak in the distribution near the equivalent of $B - V \approx 0.6$. This is similar to the main-sequence turnoff color of metal-rich globular clusters, and shows the mean abundance of the field halo sampled here to be approximately -0.5 to -1 dex.

3.   A peak in the distribution near the equivalent of $B - V \approx 1.5$. These stars are normal disk K and M dwarfs, typically a few hundred parsecs from the Sun. This feature arises from the insensitivity of blue colors to effective temperature in cool main-sequence stars. It is not seen in similar diagrams that use a redder passband, in which sensitivity to stellar color temperature is maintained.

4.   The absence of a large number of stars at the red edge of the distribution at faint apparent magnitudes. Such very red stars are cool low-mass M dwarfs, and have often been hypothesized as candidates for the local missing mass. Their absence in this diagram was the first direct evidence that low-mass luminous stars do not contribute significantly to the mass density in either the disk (Gilmore and Reid 1983) or the halo (Gilmore and Hewett 1983) of the Galaxy.

The appearance of Fig. 2.3 does not change significantly to $V \sim 22$ (Kron 1980). At fainter magnitudes it is expected that the blue edge will move to progressively redder colors, as intrinsically fainter subdwarfs dominate the counts.

The most useful external estimates of the accuracy of data of this type come from observations of the Galactic poles, as only there do truly independent duplicating data exist. For each Galactic pole observations exist from three independent surveys. Comparison of the detailed star counts from these sources shows that the various authors agree to within ~ 5 percent for apparent magnitudes brighter than about $V = 18$. At apparent magnitudes fainter than $V = 18$ and in other directions only one other such comparison is possible. Similar precision is suggested to $V \lesssim 21$. At fainter magnitudes the precision of the star-count data becomes dominated by the difficulties of reliable star–galaxy separation, leading to *systematic* errors whose amplitude is impossible to quantify from extant data (this is discussed further below). The best available estimate of the external accuracy of modern automated photographic data that are calibrated by a large number of photoelectric and/or CCD standards is about 10 percent in both apparent magnitude and color for $V \lesssim 21$. The internal precision is typically a factor of two better.

Photometric-Parallax Analysis of Star-Count Data

We may proceed to analyze general star-count data much as we did for specific tracers, provided that some information is available to limit the luminosity function. Straightforward number counts, with no color, spectral-type, or similar information, cannot be interpreted uniquely, because of the very large number of parameters involved in the fitting (see below), and are of relatively little use. If color data are available, however, analysis is substantially improved.

The most straightforward analysis technique for stellar number–magnitude–color data is photometric parallax. This involves use of the absolute magnitude–color relation for a galactic or globular cluster of appropriate chemical abundance. The absolute magnitude for a field star of measured (dereddened) color and assumed luminosity class is read directly from this diagram, and combined with the apparent magnitude to give a photometric distance. From a large set of distances, with appropriate Malmquist corrections, a density law is derivable directly. A representative set of color–magnitude diagrams is shown in Fig. 2.4. It is evident from this figure that both chemical-abundance and luminosity-class

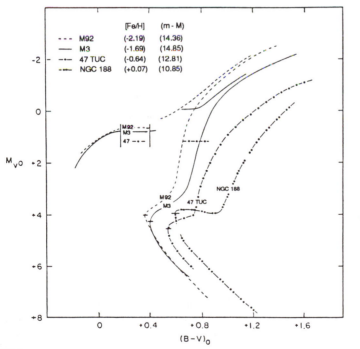

FIGURE 2.4

Color–absolute magnitude diagrams for several globulars and an open cluster. For old stars the main sequences are seen to shift systematically to the blue and to lower luminosities at low metallicities, whereas the giant branches move to the blue and to higher luminosities. For younger stars, particularly near the subgiant region, there is no simple monotonic behavior, because age and metallicity produce compensating effects.

estimates must be reliable, or a very substantial error will ensue in the derived distance. For stars sufficiently blue to be near the main-sequence turnoff and for giants, we really also need an age. In practice we know none of these things; so we must proceed in a model-dependent way.

The model-dependence is a less severe difficulty than might be supposed. The effects of different chemical abundances are readily included by assuming a range of different chemical-abundance gradients in analyzing the data, and then seeing how sensitive the derived density profiles are to the assumption. In practice we have adequate spectroscopic measurements of abundance gradients for most situations that interest us. Sometimes abundance-sensitive photometry (*i.e.*, U band) is also available, so that we can derive internally consistent density and abundance gradients simultaneously. The effects of luminosity-class uncertainties (*i.e.*, the reliability of dwarf–giant classification) have often been misunderstood. There are two limiting situations of interest. For stars that are red enough to correspond to the color of the nearly vertical part of the HR diagram (the giant branch, though the term "giant" is occasionally used to describe any evolved star; here we use "subgiant" for the roughly horizontal part of the HR diagram after the main-sequence turnoff), and sufficiently bright or at sufficiently low Galactic latitude that significant numbers of giants are expected, we cannot simply use photometric parallax. In practice, for high-latitude data, we must exclude K stars brighter than $V \sim 15$ from consideration, or obtain spectroscopic data for a suitable subset of the stars. For stars near the main-sequence turnoff, a star's evolution is nearly vertical in the HR diagram. Thus, age and metallicity differences between stars introduce a "cosmic scatter" into the absolute magnitude–color relation. This may be compensated for to adequate precision by choice of an appropriate scatter in the Malmquist correction, for stars more than ~ 0.05 magnitude redder than the turnoff.

The specific techniques for the consideration of the metallicity dependence of stellar absolute magnitudes and for the Malmquist correction are as follows.

(1) Vertical metallicity gradients

The color–absolute magnitude (CM) relation of Fig. 2.4 changes with metallicity, so that metal-poor dwarf stars are fainter than metal-rich stars of the same temperature or color, and the opposite is true for giants. Consequently, if there is a vertical metallicity gradient, the photometric parallaxes used to derive a density law will be systematically in error as we move away from the plane. We can use the metallicity gradients that have been measured using stars (mostly K giants) for which true [Fe/H] values can be derived reliably to calculate the effect that such a gradient will have on a derived density law. In what follows, let $\nu_0(z)$ be the density that would be calculated if there were no abundance gradient.

Suppose that at a height z the mean metallicity is such as to correspond to a mean difference $\Delta M(z)$ between the true absolute magnitude of a star and the absolute magnitude assigned to that same star assuming a solar metallicity calibration. Then a star whose solar-neighborhood parallax places it at a distance z_0 is really at a distance

$$z = z_0 10^{-0.2\,\Delta M(z)}. \tag{2.3}$$

This skew will both move and compress the volume bins that were used to calculate $\nu_0(z)$:

$$\frac{dV}{dV_0} = \frac{\Omega z^2 dz}{\Omega z_0{}^2 dz_0}$$

$$= \frac{z^3}{z_0{}^3} \frac{d\log z}{d\log z_0}$$

$$= 10^{-0.6\,\Delta M(z)} \left(1 + \frac{0.2z}{\log e}\frac{d\Delta M}{dz}\right)^{-1}. \tag{2.4}$$

Thus the density $\nu(z)$ we deduce assuming a metallicity gradient is

$$\nu(z) = \nu_0(z_0)\frac{dV_0}{dV}$$

$$= 10^{0.6\,\Delta M(z)} \left(1 + 0.4605z\frac{d\Delta M}{dz}\right) \nu_0\left(10^{0.2\,\Delta M(z)}z\right). \tag{2.5}$$

The most recent study of the sensitivity of M_V to [Fe/H] is by Laird $et\ al.$ (1988). From a study of local subdwarfs with trigonometric parallaxes, they find the following relation between the metallicity-dependent UV-excess photometric parameter $\delta_{0.6}$ and the magnitude difference ΔM_V, relative to a Hyades main-sequence star of the same $B - V$ color:

$$\Delta M_V = 0.862\left(-0.6888\,\delta_{0.6} + 53.14\,\delta_{0.6}^2 - 97.004\,\delta_{0.6}^3\right). \tag{2.6}$$

The UV excess–metallicity relation was studied by Carney (1979), and can be approximated as

$$\delta_{0.6} = -0.0776 + \sqrt{0.01191 - 0.05353\,[\text{Fe/H}]}. \tag{2.7}$$

Eq. (2.6) and Eq. (2.7) can be combined to give $z(z_0)$, and a new density law calculated using Eq. (2.5). (The resulting density law shown in Fig. 2.6 illustrates the effect of these gradients on the density law.) There is a relatively precise and internally consistent calibration of the dependence of absolute magnitude on metallicity when $B - V$ colors are used. More-reliable effective temperatures for K and M stars can be derived from redder colors than can be derived from $B-V$ values; so we might imagine that $V - I$ colors (say) would be more appropriate for photometric parallax. However, too few metal-poor stars have both adequate parallax distances and good $V - I$ photometry to allow a well-defined empirical definition of $\Delta M_V(V - I)$ as a function of metallicity.

(2) Malmquist bias

There is a bias in the mean absolute magnitude of a sample of stars that is introduced between volume-limited and magnitude-limited samples of objects, and that arises out of uncertainties in the absolute magnitudes (Malmquist 1936; Mihalas and Binney 1981). If the uncertainty in the absolute magnitude M of an individual star is modeled as a Gaussian distribution of dispersion σ_M about the true mean $\overline{M}_{\rm vol}$ present in a volume, then the mean magnitude that would be observed in a magnitude-limited sample is

$$\overline{M}_{\rm mag} = \overline{M}_{\rm vol} - \ln 10 \, \sigma_M{}^2 \frac{d \log A}{dm}. \tag{2.8}$$

Here $A(m)$ is the number of stars in an apparent magnitude bin centered on m. If $A(m)$ increases, stars are on average brighter in a magnitude-limited sample than in a volume-limited sample, because there are more intrinsically brighter (and hence more distant) stars that scatter into the sample than there are intrinsically fainter (and hence nearer) stars that scatter out of it. We must correct to the magnitude-limited magnitude $\overline{M}_{\rm mag}$ of the stars to compensate for this effect, and hence must know σ_M.

The "cosmic" scatter about the solar-neighborhood color–magnitude relation is 0.3 mag for the absolute magnitudes being considered here (Gliese 1971; Reid and Gilmore 1982). This is the intrinsic dispersion in the calibrating color–magnitude relation for nearby field stars with good distances; it arises from age differences, abundance and distance uncertainties, and unrecognized duplicity. Photometric errors in typical modern star-count data increase this dispersion slightly: representative photometry has an internal rms accuracy of 0.05 mag (see above), corresponding to ~ 0.20 in M_V. Thus the cosmic scatter dominates, and the photometric parallax distances we can derive are as good as we could hope to obtain from broad-band photometry. The resulting rms error σ_M on the absolute magnitude of a single star is 0.36 mag, corresponding to an uncertainty in the distance of 18 percent.

The Stellar Distribution Derived from Photometric Parallax

When we apply the procedures outlined in the sections above, and fit exponential density profiles to the resulting data perpendicular to the plane in the distance range $z \lesssim 1000$ pc, the variation of scale height with absolute magnitude shown in Fig. 2.5 results. The sudden increase in the thickness of the Galactic disk for stars with $M_V \gtrsim +4$ is well understood in terms of the main-sequence lifetimes of stars of different absolute magnitudes, the age of the Galactic disk, and the correlation between stellar velocity dispersion and age. It will be discussed in Chapter 9.

The most reliable assessment of the density profile toward the Galactic pole at larger distances comes from a recent study of the Galactic $K_z(z)$ law (see

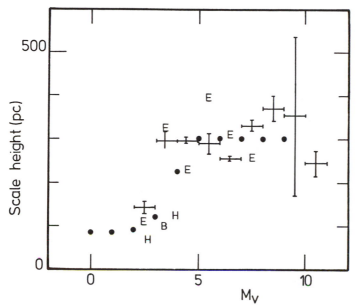

FIGURE 2.5
Variation of the thin-disk exponential scale height for $100\ \mathrm{pc} \lesssim z \lesssim 1000\ \mathrm{pc}$ with absolute magnitude. The solid points are from Schmidt (1963), points with error bars from Gilmore and Reid (1983). Letters are as follows: E, Eriksson (1978); B, Bok and Basinski (1964); H, Hill *et al.* (1979). (Royal Astronomical Society)

Chapter 8). In that study the density profile was derived, including spectroscopic-luminosity-class, metallicity, and Malmquist corrections to the photometric absolute magnitudes, using the techniques outlined above. The resulting density profile is shown in Fig. 2.6, and is adequately described by the double-exponential fit

$$\frac{\nu_0(z)}{\nu_0(0)} = 0.959\,e^{-z/249\,\mathrm{pc}} + 0.041\,e^{-z/1000\,\mathrm{pc}}. \tag{2.9}$$

This fit was made to the star-count data over the z-range 300–4000 pc, where the lower limit was chosen to avoid any possible residual giant contamination of the sample. The individual exponential scale heights and normalizations quoted are not definitive. The numerical values are very highly anticorrelated, and other equally acceptable double-exponential fits exist. A density profile that does not include from one to a few percent of the stars near the Sun in a component of the Galaxy with a characteristic scale height of ~ 1 kpc is, however, quite inconsistent with observation.

The component with a scale height of ~ 1 kpc has become known as the thick disk, to distinguish it from the old-disk population, which is dominant near the Sun, and from the halo subdwarf population, which is presumably dominant at large distance from the Sun. The significance of the thick disk in the general scheme of stellar populations and Galactic evolution will be discussed further in Chapter 11.

DENSITY LAW FOR K DWARFS

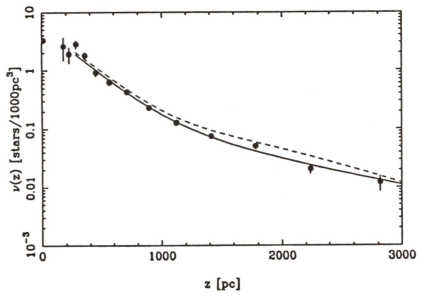

FIGURE 2.6
The stellar density distribution perpendicular to the Galactic plane, derived from K-dwarf stars with spectroscopically determined luminosity classes and metallicities (from Kuijken and Gilmore 1989, Royal Astronomical Society).

Evaluating the local thick-disk normalization by extrapolation down to the Galactic plane of the thick-disk density law derived from star counts—*i.e.*, extrapolation of the density from $z \sim 1$ kpc to $z \sim 0$ kpc—yields values from $\lesssim 1$ percent to typically a few percent of the local total stellar density, although values as high as 10 percent have been derived. In view of the anticorrelation of the scale height and the intercept at $z = 0$, considerable uncertainty is not surprising. The *physical* local normalization can be calculated from the distant data only if we know the Galactic $K_z(z)$ force law and the velocity distribution function over the distance range of interest.

A density profile of steep exponential form at distances of a few kiloparsecs from the Galactic plane was in fact very well established many years ago (see the reviews by Elvius 1965, especially his Figure 2, and by Plaut 1965, especially his Figure 7b, though it evidently has been forgotten. Results similar to those of Fig. 2.6 were also derived from the Basel surveys by Yoshii (1982), though they were not widely appreciated at that time. More recently Yoshii *et al.* (1987) have analyzed new data for the north Galactic pole, and derived a density profile similar to, though not identical with, that of Eq. (2.9).

The *radial* profile of the thick disk is as yet poorly known. Assuming similarity between the Galaxy and external galaxies (see Chapters 11 and 5), Gilmore (1984b) developed a model in which the radial profile of the thick disk was a constant-thickness exponential, with radial scale length the same as that of the old disk. He showed that the color–magnitude relation and luminosity function of the thick disk were similar to those of a metal-rich globular cluster, by using the techniques outlined in the next section, as well as published spectroscopic and kinematic data. Confirmation of the adopted metallicity distributions and color–magnitude relations in this model was derived from new photometric abundance data by Gilmore and Wyse (1985). The most comprehensive tests of this model have resulted from the Basel halo program, for which the data set is analyzed using the photometric-parallax technique described above. The Basel halo program is a major survey of the stellar distribution at high Galactic latitudes, has been in operation for 25 years, and has provided three-color data in fifteen separate fields (see Fenkart 1989a,b,c,d for a recent summary and description). The Basel data are in better agreement with the density profile of Eq. (2.9) and the constant-thickness radial-exponential thick-disk model than with alternative models of the large-scale stellar distribution.

When comparing observations with the predictions of star-count models, however, we must bear in mind the limitations of the available models. Although the thick-disk model is in good agreement with a wealth of star-count, chemical-abundance, spectroscopic-luminosity-classification, and radial-velocity data (see Chapter 11), almost all high-precision data are restricted to the range from a few hundred to a few thousand parsecs from the Sun. The original set of parameters describing the thick disk was derived solely to reproduce the observed stellar distribution from ~ 1 kpc to ~ 4 kpc from the Galactic plane toward the south Galactic pole. Although parameters similar to those originally derived have subsequently been found to provide a good description of the data in several other high-latitude fields, extrapolation of the fitted exponential far outside the range over which it has been verified is certainly not guaranteed to be a complete description of the Galaxy. It is almost certain to be unreliable at low latitudes toward the Galactic center, as very recent data show that the central few kiloparsecs of the Galaxy are characterized by a stellar distribution with scale length ~ 1 kpc, and not the ~ 3 kpc commonly applied to describe the subdwarf system. Development of improved descriptions of the stellar distribution in the central regions of the Galaxy is currently a very active field of research.

Numerical Models of Star-Count Data

The alternative to photometric parallax for analysis of stellar number–color–magnitude data involves the direct calculation of the integral in the fundamental equation of stellar statistics, Eq. (2.1). This is a straightforward computational exercise. Consequently, many attempts have been made recently to explore parameter space, so that the uniqueness of the results from the direct analysis of star-count data can be decided. In relevant form this equation is

$$N(m, \text{color}) = \omega \int \Phi\left(M_v, \left[\frac{\text{A}}{\text{H}}\right], \tau, \vec{x}, \ldots\right) D(M_v, \tau, \vec{x}) \, dV, \qquad (2.10)$$

with τ representing stellar age, [A/H] representing the stellar metallicity, as distinct from the [Fe/H] abundance, and dV the volume element. The luminosity function Φ (stars mag^{-1} pc^{-3}) has been known for many years to be a function of distance from the Galactic plane (*e.g.*, Bok and MacRae 1941; Fig. 2.5 above), because of the finite age of high-mass stars and the existence of an age–velocity dispersion relation. Since the velocity dispersion defines the density distribution, this couples the luminosity function and the density function in Eq. (2.10). The reality of an age–metallicity relation, though not its detailed form, for old thin-disk stars is also well established. Thus there is some form of age–metallicity–velocity dispersion relation. Since the conversion from absolute magnitude, in terms of which the luminosity function Φ is specified, to observed color requires an absolute magnitude–color relation, which is a function of both age and metallicity, whereas the density function depends on age through the age–velocity dispersion relation, there is a further coupling in Eq. (2.10).

This emphasizes the crucial and irreducible constraint on analyses of this type: both the luminosity function and the density law are functions of the same phase-space parameters. A unique inversion of Eq. (2.10) is therefore impossible. Instead, a large number of parameters must be fixed on external astrophysical grounds. Of these parameters, the most important to allow comparison with observational color data is the adoption of an appropriate absolute magnitude–color relation, in exactly the way required in the more direct photometric-parallax analysis technique described above. Number data are sensitive to the product of the adopted luminosity function and the density profile, so that compensating changes in one function can mask an inappropriate choice of the other. It is this non-uniqueness that must be overcome by the use of supplementary chemical-abundance and stellar-kinematic data, and by careful consideration of internal consistency in the models. This last point may be further illustrated by a recent model that adopts a color–magnitude relation appropriate to a metal-rich globular cluster (specifically, 47 Tucanae, with [Fe/H] ~ -0.7) to describe the halo field stars; at the same time it identifies the local high-velocity subdwarfs as the local halo to provide a normalization of the luminosity function. These local subdwarfs are very much more metal-poor than 47 Tuc, having a mean abundance [Fe/H] ~ -1.5, and therefore cannot be described adequately by the color–magnitude relation of a metal-rich cluster. Such a model is therefore internally inconsistent, though it can reproduce published star-count data to adequate precision.

We emphasize that star-count analyses *in isolation* are incapable of providing a unique description of the structure of the Galaxy. The results of any such analysis should be viewed as merely indicating the type of combination of fitting functions that can be used to represent available data. They should not be accepted as a valid description of the stellar populations in the Galaxy until the chemical-abundance, luminosity-class, age, and kinematic assumptions are tested by spectroscopic observations. Nevertheless, computer modeling of the Galaxy can explore parameter space with ease, and is a valuable tool for finding descriptive parameters of the spatial distribution of stars in the Galaxy from suitable data. It is of particular importance in the analysis of data for stars near the main-sequence turnoff, where classic photometric-parallax techniques are unreliable, and in modeling relatively small-area count data, where statistical fluctuations in the data make inversion of the fundamental equation of stellar statistics quite unreliable. We will now discuss the information necessary to allow calculation of solutions of Eq. (2.10).

The first requirement is to decide how many (discrete) stellar components we wish to consider in the model. This choice affects the range of Galactic co-ordinates and apparent magnitudes over which a model may be compared with observations. For now we will consider the Galaxy to be rotationally symmetric and north-south symmetric. That is, structure due to spiral arms and to such prominent asymmetric features of the sky as Gould's belt will not be discussed. In effect, this excludes early-type stars and bright stars from consideration, and restricts the range of validity of the model to intermediate and high Galactic latitudes. The precise numerical values of the latitude limits remain to be de-cided from observation. The gross features of the Galaxy with which we are left include the thin disk, the thick disk, the (subdwarf) halo, the central ($R \lesssim 3$ kpc from the Galactic center) bulge, and a very central ($R \lesssim 1$ kpc from the Galactic center) structure. This last component may or may not be the same as the central bulge, and there may or may not be continuity between some or all of these components. This question is important for studies of Galactic evolution (see Chapters 11 and 13) but not for our purposes here. Any continuum can be modeled at some level as a sum of discrete functions, and the amount and quality of extant data are such that a model with several discrete components has more than enough degrees of freedom to describe available observations.

Given the number of (discrete) components we want to include, we must now specify the density function D, the luminosity function Φ, and the absolute magnitude–color relation $\Im$ for each component.

(1) The central bulge: $R \lesssim 1$ kpc

Existence of this component is deduced in two ways—in a model-dependent way from the inner rotation curve, and directly from optical and IRAS counts of late M giants and from near-IR integrated-light observations. The rotation-curve modeling is somewhat complicated by the possibility of non-circular motions in the gas (plausible, since most galaxies do not show a maximum in the inner rotation curve like that in the Galaxy), and the star-count data are somewhat complicated, since the stars counted are in a short-lived and poorly understood evolutionary state. Thus we cannot reliably deconvolve a density gradient from an abundance gradient and/or an age gradient. All available data are adequately represented by a spatial density profile $\nu(R) \propto R^{-1.8}$ for $R \lesssim 1$ kpc. The ap-propriate luminosity function Φ and color–magnitude relation $\Im$ are generated entirely empirically (see Frogel 1988). Thus available models of the stellar dis-tribution in the central 1 kpc or so of the Galaxy are entirely *post hoc*, and lack predictive power.

(2) The main bulge: $1 \lesssim R \lesssim 3$ kpc

The annulus between $\sim 10°$ and $\sim 30°$ from the Galactic center is one of the least-understood and yet one of the most significant regions in the Galaxy. It corresponds to the only non-disk regions in external galaxies that have enough surface brightness that they can be studied, and yet it has been almost entirely neglected in our Galaxy. Just sufficient star-count data exist to show that the stellar distribution in this annulus is not consistent with models that do not include it as an extra component. Very preliminary indications suggest that a scale length of ~ 1 kpc is appropriate for the density profile, but the form of that density profile is not constrained. Star-count data also suggest a rather blue main-sequence turnoff, consistent with either very low metallicity or intermediate age (Yoshii and Rodgers 1989). The available data are not adequate to define these parameters consistently, because of difficulties with photometry in crowded fields, the complexities of patchy reddening, and the need to obtain quite large amounts of data to define the density and color distributions adequately. Until better data exist, it is not possible to include a reliable description of the dominant stellar population within $\sim 30°$ of the Galactic center in extant star-count models.

(3) The subdwarf halo

There are several alternative density profiles that might be used to describe the stellar halo distribution. To a precision which is adequate for comparison with presently available data, the subdwarf system can be described by the density distribution that corresponds to an $R^{1/4}$ law when seen in projection. This density profile is, for $R/R_e \gtrsim 0.2$ or $R \gtrsim 0.5$ kpc (latitude $\gtrsim 5°$), given by Young (1976):

$$\nu_{\text{halo}}(r) \propto \frac{\exp\left[-7.669(R/R_e)^{1/4}\right]}{(R/R_e)^{7/8}}. \tag{2.11}$$

In this equation R is the distance from the Galactic center, which is related to the distance r from the observer to a star by

$$R^2 = R_0^2 + r^2 \cos^2 b - 2rR_0 \cos b \cos \ell + z^2 \tag{2.12}$$

where

$$z = r \sin b, \tag{2.13}$$

with R_0 the solar Galactocentric distance (~ 7.8 kpc), b and ℓ the Galactic coordinates of the line of sight, and the transformation

$$z \Rightarrow (c/a)z \tag{2.14}$$

allows for an oblate distribution with axis ratio c/a. The unknown quantities in Eq. (2.11) are the scale factor R_e, the radius which encloses 50 percent of the total halo luminosity, and the proportionality constant. We assume that R_0 is known, and the axis ratio c/a is discussed in detail below. The measurement of R_e was discussed briefly above, where a value of $R_e \approx 2.7$ kpc was noted from surface photometry. The proportionality constant may be merged with the corresponding proportionality constant that arises in the normalization of the luminosity function, since only the product of the luminosity and density functions is observable. This constant is evaluated by counting subdwarfs near the Sun, and is discussed further below.

An alternative density distribution, one derived directly from RR Lyrae and globular-cluster studies, is a power law $\nu(R) \propto R^{-n}$, or, allowing for a "core radius" R_c and combining the normalization constants into f_{halo}, the fraction of all stars near the Sun that belong to the halo,

$$\nu_{\text{halo}}(R) = f_{\text{halo}}\, \nu_{R=R_0}\, \frac{(R_c^n + R_0^n)}{(R_c^n + R^n)}. \tag{2.15}$$

A suitable value for R_c is ≈ 1 kpc, though this value is neither well established nor critical. The normalization constant, the fractional number of all stars in the solar neighborhood that follow this density law and are described by this color–magnitude relation, is $\approx 1/800$, with an uncertainty of at least 50 percent in this value.

The luminosity function appropriate for the field subdwarfs can be constrained in a variety of ways. Samples of field subdwarfs with good distances have been analyzed, and show that, within the very large sampling noise, there is no significant difference in the luminosity function of subdwarfs with $[\text{Fe/H}] \lesssim -1$ and that of solar-abundance disk stars for $5 \lesssim M_V \lesssim 15$. (This similarity imposes some constraints on star-formation theories, showing that the low-mass spectrum is almost independent of the metallicity through the range when the effect of metals on the high-temperature cooling function changes from being negligible to being dominant.) Direct modeling of star-count data reaches similar conclusions regarding the similar shapes of the luminosity functions of halo and disk stars. For evolved stars we may use the age data discussed in Chapter 13. Effectively, all stars with $[\text{Fe/H}] \lesssim -0.8$ are as old as the globular clusters. Hence all evolved halo stars have almost identical masses, and the luminosity function is determined entirely by stellar evolution, rather than by the stellar initial mass function. We may therefore use the observed luminosity functions of evolved stars in globular clusters with some confidence.

Some complexities remain, however. The most important of these is the treatment of the horizontal branch. The second-parameter problem alluded to earlier means that we face considerable uncertainty in treating horizontal-branch stars, and that we must be guided entirely by observation. There is suggestive, though not conclusive, evidence that we would expect more blue-horizontal-branch stars than are seen, so that the second-parameter problem afflicts subdwarfs near the Sun. A further complication is photometric evidence that globular clusters show significant differences in their main-sequence luminosity functions, with a possible systematic change in the slope of the luminosity function with metallicity. The significance of this is complicated by the fact that there are corresponding, and not well defined, systematic changes in the stellar mass–luminosity and absolute magnitude–color relations with metallicity, because of the effect of changing opacity on the stellar atmosphere and interior. The significance of any resulting systematic changes in the stellar *mass* function remains to be studied in detail. Since the relationship (if any) of the field stars to globular clusters remains unknown, this adds to star-count modeling a further note of warning, and another interesting parameter that may be measured.

The remaining sensitive parameter is the absolute magnitude–color relation $\mathfrak{F}$. This is actually defined very well, in spite of the fact that most star-count modeling tends to treat it as a free parameter. The metallicity and age distributions of subdwarfs near the Sun are known (see Chapter 11) to be similar to those of the metal-poor globular-cluster system. Thus we have no choice in the specification of an appropriate color–magnitude relation: it must be that of a representative metal-poor globular cluster. At very large distances from the Galactic center we might wish to adopt a very metal-poor cluster as template, in order to test for a radial abundance gradient, whereas very near the Galactic center a more metal-rich cluster might also be relevant. Given the available limits on the amplitude of an abundance gradient in the field RR Lyraes, however, and the absence of any noticeable metallicity–velocity dispersion relation in subdwarfs (which is the kinematic requirement for an abundance gradient), there is no freedom in the choice of the color–magnitude relation when modeling stars within a few kiloparsecs of the Sun. The choice of an appropriate color–magnitude relation is a very important point, because the abundance distribution of high-velocity stars near the Sun is known in adequate detail. The adoption of an inappropriate color–magnitude relation when modeling halo stars is one of the few errors for which star-count models can provide *prima facie* evidence of a serious problem with either the model or the data.

(4) The thick disk

The vertical density profile of the thick disk is shown in Fig. 2.6, and is adequately represented for $1000 \lesssim z \lesssim 3000$ pc by an exponential

$$\nu(z) = \nu_{z=0} \exp(-z/h_{z,\text{td}}). \tag{2.16}$$

The current best estimates for the normalization constant and scale height are given in Eq. (2.9) to be 4 percent and 1000 pc, respectively. The normalization constant $\nu_{z=0}$ is roughly an order of magnitude larger than that for the halo stars, though again with an uncertainty of about 50 percent in this value. Note that the normalization here is *not* the fractional number of thick-disk stars near the Sun, but the numerical value required to model the stellar distribution a few kpc above the plane, assuming the density profile of Eq. (2.16). The relationship of this numerical value to the actual number of thick-disk stars in a volume near the Sun is a steep function of the Galactic $K_z(z)$ law (see Chapter 8) and the stellar velocity distribution function. We note in passing that a noticeably oblate $R^{1/4}$ density profile ($c/a \approx 1/4$) also provides a good description of the data.

The radial profile of the thick disk is still poorly known. As noted above, the Basel star-count surveys suggest that a radial exponential is an adequate description, so that the two-dimensional (rotational symmetry is assumed, remember) density distribution becomes

$$\nu(R, z) \propto \exp(-z/h_{z,\mathrm{td}} - R/h_{r,\mathrm{td}}). \qquad (2.17)$$

The radial scale length $h_{R,\mathrm{td}}$, based on available star-count modeling and by assumption from photometry of other disk galaxies (see Chapter 12) is the same as the radial exponential scale length of the thin disk, or $\sim 4 \pm 1$ kpc. The numerical value of this scale length is discussed further in Chapter 15.

The color–magnitude relation $\Im$ for the thick disk is defined rather well. The mean metallicity of the thick disk is known (see Chapter 11) to be like that of the metal-rich globular-cluster system, and at least the metal-poor part of the thick disk is about the same age as the globular-cluster system. For more metal-rich thick-disk stars, the situation remains unclear, pending further observations. Thus we should adopt the color–magnitude relation of a metal-rich globular as the most suitable choice, but beware of the possibility that it may be inadequate, in being too red, for stars near the turnoff. A similar situation applies for the choice of an appropriate luminosity function for the thick disk. The luminosity function of a metal-rich globular cluster is the most appropriate choice available, but the possibility remains that it will lack an adequate mass range, corresponding to an age range, near the turnoff.

(5) The thin disk

Models of the spatial distribution of stars in the thin disk suffer from two serious difficulties: the system is a complex mixture of ages, velocity dispersions, and metallicities; and there is a lot of observational information available. Subsidiary complications, particularly the effects of large and irregular obscuration, further confuse the issue. To provide a reliable description of the stellar distribution within ~ 200 pc of the Galactic plane, we would have to model the star-formation history, the present distribution of ages, chemical abundances, and velocities, and the reddening distribution, and would also have to know the $K_z(z)$ force law near the plane to high precision. If that information were available, there would be no point in building crude statistical models of the data. Conversely, since such a level of detail is neither known nor the point of this discussion, we will restrict the discussion to distances $\gtrsim 200$ pc from the plane. This effectively restricts the modeling to a description of the distribution of old disk stars with vertical velocity dispersion $\gtrsim 15$ km sec^{-1}. For this restricted case, the observations summarized in Fig. 2.6 show an exponential to be an adequate description of the spatial density distribution, though a vertical distribution described by $\nu(z) \propto \mathrm{sech}^2(z/2z_0)$, with z_0 the equivalent exponential scale height, is an equally satisfactory description.

The radial distribution of the old disk is invariably assumed to be exponential, though the true distribution is almost entirely unconstrained by observation. The strongest evidence in favor of a radial exponential is the argument that other disk galaxies are (more or less; see Chapter 5) exponentials. Some indirect dynamical arguments (Chapter 11) and some limited photometric data (Chapter 15) support a radial scale length of $\sim 4 \pm 1$ kpc for the old disk. As the reader will have deduced from this uncertainty, the radial distribution is poorly constrained, because it cannot be seen: interstellar extinction forms a very effective screen between the real old disk and our models, and *vice versa*. The actual value of this radial profile is therefore not very important for our purposes here.

Stars near the Sun may be discovered with high efficiency, leading to reliable volume-complete samples of old disk stars. Thus the luminosity function for stars fainter than the turnoff, and the color–magnitude relation, are known directly. Similarly, a sufficiently large sample of white dwarfs is known that their observed luminosity function and color–magnitude relation may be adopted. For the locally rarer evolved stars, the giants and subgiants, the situation is more complex. A wide variety of ages and metallicities occurs, and hence a very broad scatter in the color–magnitude relation, and a complex luminosity function. To some extent this may be approximated by imposing a scatter in our models, but in general no new information is obtained by doing so.

The more sensible course, in view of the complexity of the situation and the negligibly small amount of new information to be obtained from star-count models of the distribution of evolved thin-disk stars, is simply not to do it. The inevitable broad-brush effect of statistical models of the stellar distribution in the Galaxy makes them inherently poorly suited to understanding a complex system like the thin disk. For such systems it is more profitable to pose specific questions and develop methods to find specific answers. As a good rule of thumb, star-count models of the type described here are inappropriate for discussing observations of stars with $V \lesssim 12$.

2.6   THE SHAPE OF THE METAL-POOR HALO

An example of the type of analysis for which star counts *are* ideal is finding out the shape of the metal-poor halo, represented in the solar neighborhood by the high-velocity subdwarfs. The shape of the distribution of the non-thin-disk stars has important implications for the early stages of galaxy collapse and star formation, the interpretation of the kinematics of high-velocity stars, and the shape of the underlying dark matter that generates the gravitational potential in which these stars move. Available kinematic data for these stars (see Chapter 11 for a more detailed discussion) suggest that they are distributed in a substantially flattened system, with axial ratio $c/a \sim 1/2$.

In the light of this kinematic evidence, it is mildly puzzling that direct star-count studies suggest that the subdwarf stellar system is approximately round at the solar Galactocentric distance. Note that an earlier suggestion from RR Lyrae studies that the inner halo was round has now been superseded by new data, as discussed above. The best estimate for the shape of the inner halo is $c/a \sim 1/2$. The most-quoted evidence for a spherical distribution of field halo stars at large distances from the Galactic center comes from the modeling by Bahcall and Soneira (1980, 1984) (hereafter BS) of the faint star counts of Koo and Kron (1982) in two fields; BS conclude that the axis ratio of the halo stars is $c/a = 0.80^{+0.20}_{-0.05}$. Their technique is based on the fact that fields in the $\ell = 90°, 270°$ plane are at equal Galactocentric distances if at equal distances from the solar neighborhood, and hence a spherical distribution of stars will contribute equally to all fields in this plane. Thus if we compare magnitude-limited samples in fields at high and low Galactic latitude, we should obtain equal numbers of halo stars in the two fields. A flattened distribution of stars will yield lower counts in the higher-latitude field.

BS complicate their analysis somewhat by adopting different color–magnitude relations for the two fields, and thus they do not predict equal numbers of stars for a spherical distribution; they are forced to do this to obtain an acceptable fit for their model in each of the two fields, because of a combination of in-adequacies in the model, such as lack of the thick-disk component, and in the data, discussed below. There is no physical basis for such a variation of color–magnitude relation—metallicity gradients are not relevant, since the fields are supposed to be at the same Galactocentric distance—and it is a potential source of uncertainty. Adoption of a metal-poor color–magnitude relation has the effect of assigning a low intrinsic luminosity to main-sequence stars of a given color. Hence, in an apparent-magnitude-limited sample, we will be comparing lower-luminosity, less distant stars in the "metal-poor" field with higher-luminosity, more distant stars in the other field. The predictions of relative star counts are therefore sensitive to the shape of the subdwarf luminosity function, as well as to the shape and density profile of the stellar tracer population. It is then possible to produce predictions for the ratio of counts that can exceed unity in a spherical distribution, as BS derived.

A new study of this problem, using a larger data set and a more general model-Galaxy program that requires internally consistent properties for a given stellar population in different fields, and which allows the inclusion of a thick disk, is described by Wyse and Gilmore (1989). Following BS they counted stars blueward of a color limit $(B - V = 0.6)$ chosen to minimize contamination of the tracer sample by nearby old disk stars, with the precise value of this limit not being critical. The two fields used were $(\ell = 0°, b = 90°$; area surveyed $= 0.75$ square degrees) and $(\ell = 272°, b = -44°$; area surveyed $= 0.75$ square degrees). The observed ratio of blue stars in the two fields was 0.59. This disagrees strongly with Koo and Kron's counts in two fields at similar Galactic latitudes but at much fainter magnitudes, which yield a ratio of 1.09 for blue stars with $20 \lesssim V \lesssim 22$; the most recent recalibration of the Koo and Kron data, by Koo *et al.* (1986), gives a ratio of 1.3. Note that these numbers are based on a somewhat uncertain color cut, but this should not matter, provided the color range chosen is blue enough to have isolated the metal-poor halo stars. Analyses of this type are substantially eased by the bimodality of the color–magnitude data shown in Fig. 2.3.

The most likely explanation of the disagreement between these data sets is the difficulty of reliable star–galaxy discrimination at faint magnitudes. Galaxies substantially outnumber stars at these faint magnitudes. Thus a small fractional error in the star–galaxy discrimination rate, which will have only a small effect on the resulting galaxy counts, will have a very large effect on the star counts. This suspicion is based on the detailed comparison of the Koo *et al.* (1986) "subdwarf" category counts for their north-Galactic-pole field, with predictions from the Gilmore–Wyse model and from the BS model, each model with an assumed halo axis ratio of 0.8. There is a substantial disagreement between the data and the models; the number of stars counted fails to increase toward fainter magnitudes, contrary to both of the models, and contrary to intuition. Similarly counter-intuitive changes in the stellar number–magnitude data for $V \gtrsim 21$ are seen in most other published data, with either very flat or very steep stellar number counts being apparent. See Gilmore (1981) and Pritchet (1983) for further discussions of this point.

The BS model predictions (their Table 3) combined with the new low value of the relative observed counts would imply that the halo had an axis ratio $c/a \lesssim 0.5$. When we consider the presence of the thick disk, and also model the observed *total* counts as well as their ratio, the best estimate for the axis ratio of the metal-poor subdwarf stellar population within a few kiloparsecs of the Sun is $c/a \sim 0.6$. Note that this is 4σ below the quoted errors of BS, illustrating the importance of systematic errors in these analyses.

In spite of the difficulties in obtaining and analyzing reliable data at faint magnitudes illustrated by this discussion, modeling stellar number–magnitude–color data offers the best available method to map the large-scale distribution of stars in the Galaxy. When these data are combined with chemical abundance and kinematic data, detailed study of the present distribution of mass and light, and of the evolution of the stellar populations in the Galaxy, becomes feasible. Such analyses are described in other chapters of this book.

REFERENCES

Azzopardi, M., and Matteucci, F., eds. 1987. *Stellar Evolution and Dynamics in the Outer Halo of the Galaxy.* Garching: European Southern Observatory.

Baade, W. 1958. In O'Connell, p. 303.

Bahcall, J. N., and Soneira, R. M. 1980. *Astrophys. J. Supp.*, **44**, 73.

———. 1984. *Astrophys. J. Supp.*, **55**, 67.

Becker, W. 1950. *Sterne und Sternsysteme.* Dresden: T. Steinkopff.

Bergh, S. van den. 1979. In Longair and Warner, p. 151.

Blaauw, A. 1965. In Blaauw and Schmidt, Chapter 20.

Blaauw, A., and Schmidt, M., eds. 1965. *Galactic Structure.* Chicago, IL: Univ. of Chicago Press.

Bok, B. J., and Basinski, J. 1964. *Mem. Mt. Stromlo Obs.*, **4**, 1.

Bok, B. J., and MacRae, D. A., 1941. *Ann. N.Y. Acad. Sci.*, **42**, 219.

Capaccioli, M., ed. 1984. *Astronomy with Schmidt-Type Telescopes*. Dordrecht: Reidel.

Carney, B. W. 1979. *Astrophys. J.*, **133**, 211.

Conlon, E. S., Brown, P. J. F., Dufton, P. L., and Keenan, F. P. 1988. *Astron. Astrophys.*, **200**, 168.

Elson, R., and Walterbos, R. 1988. *Astrophys. J.*, **333**, 594.

Elvius, T. 1965. In Blaauw and Schmidt, Chapter 3.

Eriksson, P-I. W. 1978. *Uppsala Astron. Obs. Report*, no. 11.

Feast, M. 1987. In Gilmore and Carswell, p. 1.

Fenkart, R. P. 1989a. *Astron. Astrophys. Supp.*, **78**, 217.

———. 1989b. *Astron. Astrophys. Supp.*, **79**, 51.

———. 1989c. *Astron. Astrophys. Supp.*, **80**, 89.

———. 1989d. *Astron. Astrophys. Supp.*, **81**, 187.

Frogel, J. A. 1988. *Ann. Rev. Astron. Astrophys.*, **26**, 51.

Garwood, R., and Jones, T. J. 1987. *Publ. Astr. Soc. Pacific*, **99**, 453.

Gilmore, G. 1981. *Mon. Not. Roy. Astron. Soc.*, **195**, 183.

———. 1984a. In Capaccioli, p. 77.

———. 1984b. *Mon. Not. Roy. Astron. Soc.*, **207**, 223.

Gilmore, G., and Carswell, B. 1987. *The Galaxy*. Dordrecht: Reidel.

Gilmore, G., and Hewett, P. C. 1983. *Nature*, **306**, 669.

Gilmore, G., and Reid, I. N. 1983. *Mon. Not. Roy. Astron. Soc.*, **202**, 1025.

Gilmore, G., and Wyse, R. F. G. 1987. In Gilmore and Carswell, p. 247.

Gliese, W. 1971. *Veroffentl. Astron. Rechen-Inst. Heidelberg*, no. 24.

Habing, H. J. 1987. In Gilmore and Carswell, p. 173.

Harmon, R. T., and Gilmore, G. 1988. *Mon. Not. Roy. Astron. Soc.*, **235**, 1025.

Hartwick, F. D. A. 1987. In Gilmore and Carswell, p. 281.

Hayakawa, S., Matsumoto, T., Murakami, H., Uyama, K., Thomas, J., and Yamagami, T. 1981. *Astron. Astrophys.*, **100**, 116.

Herschel, J. 1847. *Results of the Astronomical Observations at the Cape of Good Hope.* London.

———. 1849. *Outlines of Astronomy*. London.

Hill, G., Hilditch, R. W., and Barnes, J. V. 1979. *Mon. Not. Roy. Astron. Soc.*, **186**, 813.

Hiromoto, N., Maihara, T., Mizutani, K., Takami, H., Shibai, H., and Okuda, H. 1984. *Astron. Astrophys.*, **139**, 309.

Kapteyn, J. C., and van Rhijn, P. J. 1920. *Astrophys. J.*, **52**, 23.

Keenan, F. P., Lennon, D. J., Brown, P. J. F., and Dufton, P. L. 1986. *Astrophys. J.*, **307**, 694.

Kinman, T. D., Wirtanen, C. A., and Janes, K. A. 1966. *Astrophys. J. Supp.*, **13**, 379.

Koo, D. C., and Kron, R. G. 1982. *Astron. Astrophys.*, **105**, 107.

Koo, D. C., Kron, R. G., and Cudworth, K. 1986. *Publ. Astron. Soc. Pacific*, **98**, 285.

Kuijken, K., and Gilmore, G. 1989. *Mon. Not. Roy. Astron. Soc.*, **279**, 605.

Lance, C. M. 1988. *Astrophys. J.*, **334**, 927.

Longair, M. S., and Warner, J. W., eds. 1979. *Scientific Research with the Space Telescope*. NASA CP-2111.

Malmquist, K. G. 1936. *Stockholm Obs. Medd.*, no. 26.

Mihalas, D., and Binney, J. 1981. *Galactic Astronomy*. San Francisco, CA: W. H. Freeman and Co.

O'Connell, D. J. K., ed. 1958. *Stellar Populations*. Amsterdam: North Holland Press.

Oort, J. H., 1958. In O'Connell, p. 419.

Oort, J. H., and Plaut, L. 1975. *Astron. Astrophys.*, **41**, 71.

Plaut, L. 1965. In Blaauw and Schmidt, Chapter 13.

Pritchet, C. 1983. *Astron. J.*, **88**, 1476.

Reid, I. N., and Gilmore, G. 1982. *Mon. Not. Roy. Astron. Soc.*, **201**, 73.

Schmidt, M. 1963. *Astrophys. J.*, **137**, 758.

Schombert, J. M., and Bothun, G. 1987. *Astron. J.*, **93**, 60.

Shaw, M., and Gilmore G. 1989. *Mon. Not. Roy. Astron. Soc.*, **237**, 903.

Solomon, P., and Rivolo, R. 1987. In Gilmore and Carswell, p. 105.

Strugnell, P., Reid, I. N., and Murray, C. A. 1986. *Mon. Not. Roy. Astron. Soc.*, **220**, 413.

Trumpler, R. J., and Weaver, H. F. 1953. *Statistical Astronomy*. Berkeley: University of Calif. Press.

Vaucouleurs, G. de, and Pence, W. D. 1978. *Astron. J.*, **83**, 1163.

Wesselink, T., Le Poole, R. S., and Lub, J. 1987. In Azzopardi and Matteucci, p. 185.

Wyse, R. F. G., and Gilmore, G. 1989. *Comments on Astrophys.*, **13**, 135.

Yoshii, Y. 1982. *Publ. Astron. Soc. Japan*, **34**, 365.

Yoshii, Y., Ishida, K., and Stobie, R. S. 1987. *Astron. J.*, **93**, 323.

Yoshii, Y., and Rodgers, A. W. 1989. *Astron. J.*, **98**, 853.

Young, P. J., 1976. *Astron. J.*, **81**, 807.

Zinn, R. 1985. *Astrophys. J.*, **293**, 424.

CHAPTER 3

THE GALACTIC CENTER—
AND OUTWARD TO THE HALO

Ivan R. King

The region around the Galactic center is a complete mélange of components and populations. Nowhere is it more necessary to heed the precept that gravitation mixes an outer component into an inner component. At the Galactic center itself, there may be a massive but spatially tiny black hole (in any case there is a quite disturbed region), surrounded by a disk of gas and dust, possibly with some young stars, that blends outward into the gas-and-dust layer of the Milky Way as a whole. This in turn is surrounded by the old, metal-rich population of the central bulge, highest in density at the very center. And penetrating these components are the stars of the halo, also peaking in density at the Galactic center, but making only an inconsequential contribution to the total density there.

3.1 THE INNERMOST REGION OF THE GALACTIC CENTER[1]

The innermost hundred parsecs of the Galaxy are a confused region, and the interpretation of many of its observed characteristics is controversial.

The fundamental problem in interpreting our Galactic center, and in comparing it with the centers of other galaxies, arises from the limited and completely different way in which we observe the center of our own Galaxy. Other galaxies whose centers we observe are inclined to our line of sight, and we see easily into their centers; but in our own Galaxy we observe edge-on, and at the V band there are some 30 magnitudes of absorption between us and the Galactic center. As a result we can see nothing whatever at wavelengths short of $1\,\mu$m, and even at $2\,\mu$m the absorption is patchy enough to confuse the interpretation of the observations. At longer wavelengths, by contrast, detailed observations exist for our Galactic center; but data for other galaxies are not comparable, either in sensitivity or in resolving power.

---

[1]This section differs from the corresponding Saas-Fee lecture as actually given, because it is strongly influenced by papers presented at a symposium (Morris 1989) that was published after the time of the Saas-Fee course. It would be irresponsible not to include that material in this written account.

Because of this wavelength restriction, our information about the stellar component of the Galactic center is inaccurate and incomplete; the observations that exist refer chiefly to the interstellar material and only a little bit to the stars. The one common factor, however, is that all material, whatever its nature, is subject to gravitation, so that observations of the Galactic-center region do tell us something about the distribution of mass there.

As in so many observations of interstellar material, one of the basic problems of interpretation in the Galactic-center region is that we do not know relative positions along the line of sight. A great deal of effort has gone into figuring out— or guessing and arguing about—what is in front of what. Given this confused and confusing situation, I will not here inundate the reader with detailed facts, but will try, rather, to describe the general picture and refer to review articles for details.[2]

Before examining any of the phenomena, we should have the length scales in mind. If the Galactic center is 8 kpc away (not an "official" value, but probably a nearly correct one), a degree is 140 pc. Similarly, 1 arcmin corresponds to 2.3 pc, and 1 arcsec to 0.04 pc; conversely, a parsec is $26''$. In many good radio and infrared maps, the resolving power is about a second of arc, and very-long-baseline interferometry reckons in milliarcseconds.

The tiny central region of our Galaxy is clearly beset by some violent activity, but it is not at all clear what the nature of the activity is. Similarly, there has probably been recent star formation there, at the very center of a central bulge whose dominant population is certainly an old one. The Galactic center is strange because it contains the central part of the interstellar material of the disk, which coexists with the center of the otherwise quiescent bulge and creates its own peculiar fireworks there. Somehow—and the vagueness of this word again merits emphasis—the gas and dust at the very center have erupted, in the recent past and perhaps in a continuing way, into a high-energy activity and a process of star formation that we would not otherwise expect at the central point of the old bulge of an Sb or Sbc galaxy. It is not clear to what extent our Galactic center is like an active galactic nucleus. "Not much" must be the basic part of the answer, but there must also be a "somewhat."

An account of this central region should begin with the location of the center itself. This was identified decades ago with the radio source Sgr A, our first indication of high-energy events in the Galactic center. Further observation of this region has led the radio astronomers to the designation "the Sgr A complex." And complex it is, indeed. Modern resolving power divides Sgr A into three components. One of these, called "Sgr A East," is a nonthermal source, and has been commonly interpreted as a recent supernova remnant. The second part, Sgr A West, has a thermal spectrum that is characteristic of an H II region. It is very close to the center of the stellar distribution that is outlined by the $2\,\mu$m infrared distribution of light. At the center of Sgr A West is a milliarcsecond *non*thermal

---

[2]There is a great deal of detailed information about this situation in the review by Genzel and Townes (1987), in the symposium edited by Backer (1987), and particularly in the symposium edited by Morris (1989).

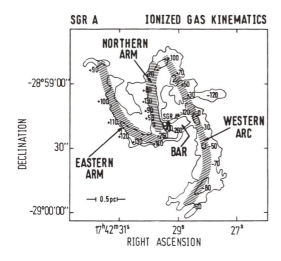

FIGURE 3.1
The tiny central spiral, with measured radial velocities indicated. (From Genzel and Townes 1987.)

source, which is called Sgr A* (pronounced "sadge-A-star"). Its spectrum suggests that what we observe may be scattered radiation around a source that is itself unresolved (Lo *et al.* 1981) and may be much smaller than the 10^{15}-cm diameter that is observed. Sgr A* has been presumed to be associated with the violent events that are characteristic of the center, so that it marks accurately the location of the Galactic center itself. The connection is supported by its measured radio proper motion (Backer and Sramek 1987), which is consistent with its being stationary with respect to the Galactic center. Its chief rival as a possible center-marker is a tight grouping of infrared sources called IRS 16, which lies only a few arcsec away from Sgr A*. The confusion in interpreting data on the Galactic-center region is increased by the fact that some studies of radial distributions have assumed a center at IRS 16 rather than at Sgr A*.[3]

Immediately surrounding the Galactic center the gas is ionized, by a source of excitation that is not well understood. The two principal candidates are (1) recently formed OB stars in a "starburst" region and (2) a high-energy source in the Galactic center itself. (For detailed arguments, see Genzel and Townes 1987.)

In this region a strange pattern is observed, both at far-infrared and at radio wavelengths. Extending out to a radius of about 1.7 pc, it looks for all the world like spiral arms (Fig. 3.1). Yet these cannot be conventional spiral arms. The scale of spiral arms is kiloparsecs, but here we have a distance scale of a few parsecs. More important, spiral structure is a phenomenon of the Galactic plane, whereas here we are dealing with a structure that we could not see if it were not inclined at a fairly large angle to that plane.

---

[3]The I.A.U. working group that defined the modern system of Galactic coordinates (Blaauw *et al.* 1959) took Sgr A to mark the Galactic center, but set the Galactic plane parallel to the mean Hɪ plane of the Galaxy. Relative to this plane, Sgr A* came out with a latitude of $-0°05$. This shift has been interpreted (Ratnatunga *et al.* 1989) as indicating a 7-pc height for the Sun with respect to the central plane of the Galaxy, although other estimates (see Chapter 15) have placed the Sun rather higher above the plane.

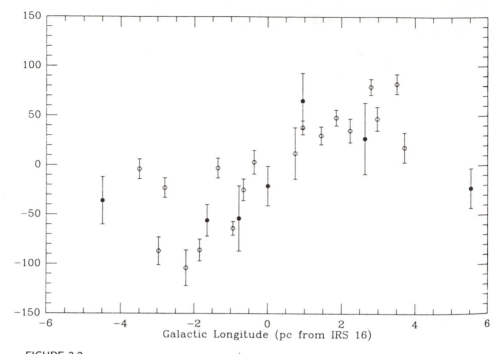

FIGURE 3.2

The mean velocities of stars near the Galactic center, referred to the local standard of rest. Filled circles are from individual stars, binned by longitude. Open circles are from the integrated light of a circle of diameter 0.8 pc; each individual point is shown. (From Sellgren 1989.) Note that this compilation uses IRS 16 as the center.

The motions of the central gas answer some questions but evoke many others. The spiral arcs have been observed in some detail, using a line of Ne ii at $12\,\mu$m; they show coherent large-scale motions (Fig. 3.1), but the pattern does not look like a simple rotation.

The velocities in the vicinity of the Galactic center have been interpreted in various conflicting ways. Both inflow and outflow have been suggested. More interestingly, it has been argued by many interpreters that the high velocities are due to the presence of a large mass in the Galactic center itself. The suggestion is frequently made that a black hole with a mass of at least several times 10^6 solar masses resides there and is responsible for the various kinds of activity.

It can be argued that the gas near the center is not necessarily in equilibrium with the gravitational field, that its motions are influenced by the magnetic fields that are known to be present, etc. No such objections can be made concerning the stars whose velocities have now been measured in the immediate vicinity of the center. (For a review, see Sellgren 1989.) These observations make use of the $2.3\,\mu$m absorption band of CO, either for individual stars or for the integrated starlight of a small area. For both of these the mean rotational velocity and the

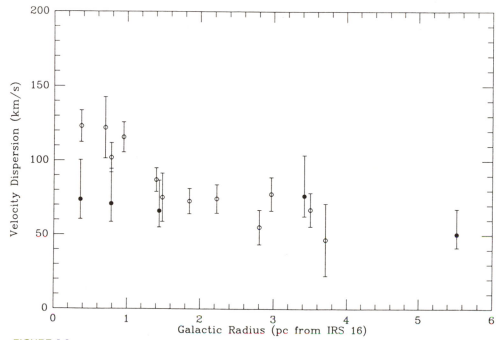

FIGURE 3.3
As for Fig. 3.2, but showing velocity dispersions.

velocity dispersion can be measured; the integrated spectra require a treatment similar to that used for optical spectra of elliptical galaxies (see Chapter 10).

The rotational velocities and the velocity dispersions within a few parsecs of the Galactic center are shown in Figs. 3.2 and 3.3, respectively. Except for a tendency for the integrated-light velocity dispersions to rise in the center, the two types of measurement are in good agreement.

Farther from the center, the so-called OH/IR stars give a velocity field that is presumably that of the Galactic bulge itself. These are red giants or supergiants that have OH masers in their circumstellar shells; they are identified as stars by the point-source infrared radiation from their shells, and their 18-cm radio radiation gives good velocities. Velocities of these objects are shown in Fig. 3.4.

From suitable dynamical analyses, we can use the radial velocities to calculate the function $\mathcal{M}(r)$, the mass contained within radius r. A large value of this function at a very small radius would be an indication of a black hole. In fact, $\mathcal{M}(r)$ appears to be about $2\times10^6 \, \mathcal{M}_\odot$ at a radius of about half a parsec, but there are insufficient observations closer to the center, and the results are quite sensitive to whether the center is assumed to be at Sgr A* or at IRS 16. Interestingly, when the gas motions are assumed to be in gravitational equilibrium and analyzed for $\mathcal{M}(r)$, they give nearly the same curve as the stars do; so perhaps the motions, although chaotic, *are* in equilibrium with the gravitational field.

Opponents of the black-hole interpretation assert that the central concentration of mass belongs to a "stellar cluster." Such a high density of stars may

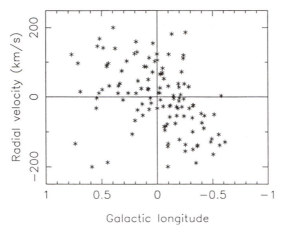

FIGURE 3.4

Radial velocities of OH/IR stars, indicating both random velocities and rotation. (From Lindqvist *et al.* 1989.)

yet be observed, but its existence is controversial. The brighter resolved stars in 2.2 μm ("K-band") maps do show a central peak whose core radius might be as small as 0.05 pc, but the mass in most stellar systems is largely in stars of lower mass, which cannot be observed directly here. Rieke and Lebofsky (1987) tried to derive a brightness profile by confining their attention to the bland areas between resolved stars; they concluded that the core radius of that distribution is actually 0.8 pc.

Studies in the K band have been helped immeasurably by the advent of infrared-sensitive imaging arrays, which allow mapping at a rate that is orders of magnitude faster than that of a single detector. On such a map of the Galactic-center region, it is striking how patchy the absorption is at K (Fig. 3.5).

Ionized gas extends for about 2 pc around the Galactic center. Farther out, the material turns neutral, and contains warm dust and much molecular material. There is a radio-emitting region 2 parsecs from the center that marks the inner edge of a rotating molecular disk extending out to 8 parsecs (Fig. 3.6).

Somewhere in this general region are also sources that emit gamma rays of at least two sorts. The positron–electron annihilation line at 511 kev (reviewed, along with the continuum radiation, by Lingenfelter and Ramaty 1989) varies on a time scale that shows that it comes from a region (or regions) smaller than a parsec; unfortunately, the positional accuracy of these observations is poor. At lower energies a map has been made (Cook *et al.* 1989), using the technique called coded aperture, in which a specially contrived mask allows accurate measurement of the direction from which the radiation came. It shows a single brightness peak, about 0°7 away from Sgr A*.

At higher energies the gamma-ray continuum is harder to observe, because the spectrum is a steeply dropping one. It has been detected, however, and is

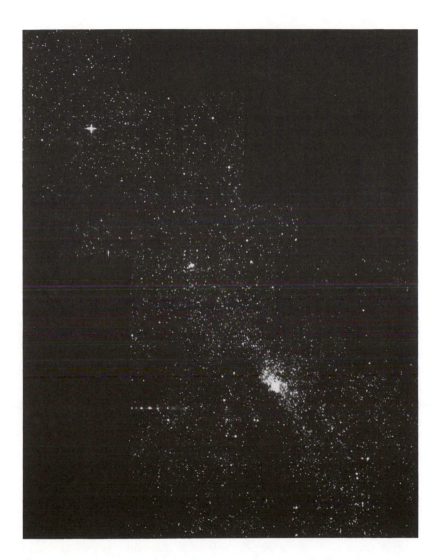

FIGURE 3.5
A mosaic of $2.2\,\mu$m images of the Galactic plane around the center, which is in the brightest region. The boundaries of the picture measure 0.6×0.75 degrees. The horizontal spike at the left was caused by scattered light. (From Gatley *et al.* 1989.)

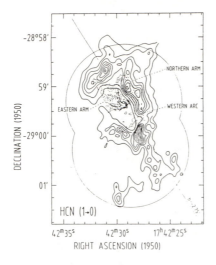

FIGURE 3.6

Radio and molecular emission from the molecular disk. One set of density contours refers to HCN at 3 mm, the other to the 6-cm radio continuum. The latter also shows the central "spiral." (From Townes 1989.)

highly variable. There is also a line at 1.8 Mev, due to decay of ^{26}Al, which has a half-life of 1.1×10^6 yr. (For a discussion see Diehl *et al.* 1989.)

The source of the gamma rays is not well understood, except that they probably do not arise from a single cause. The ^{26}Al is an indication of recent nucleosynthesis, either in supernovae or in evolved stars that dredge up processed material. Many high-energy processes are capable of producing positrons; among them is the interaction of energetic photons near a black hole.

One of the strangest features of all in the Galactic-center region is a set of parallel wisps (Fig. 3.7), observed in radio at 6 and 20 cm, that extend about 20 parsecs from the center, and then appear to make a right-angle bend (see Yusef-Zadeh 1989). Their smooth, regular structure suggests intuitively that a strong magnetic field plays a role here. Strong polarizations are in fact observed, and also Faraday rotation (see a summary in Townes 1989), and their magnitudes suggest fields that may reach as high as 10^{-2} Gauss. Needless to say, the Morris (1989) volume is not lacking in papers offering theoretical interpretations of the fields.

Farther out is the rapidly rotating "nuclear disk" of Hɪ, which extends out to about 700 parsecs. Within it, the rotation speed climbs from zero to the level of the flat rotation curve near the Sun.

In the emission spectrum of Sgr A, molecular observations show absorptions at negative velocities. These must come from material between us and the Galactic center, moving outward. One of these, the "3-kpc expanding arm," is also observed at 21 cm (Chapter 4).

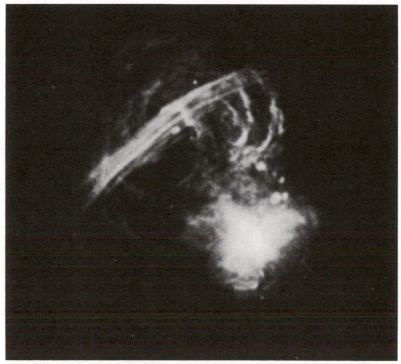

FIGURE 3.7
A 20-cm emission map of the central 60×60 pc of the Galaxy. (Data from observations by F. Yusef–Zadeh and M. R. Morris.)

3.2 THE CENTRAL BULGE OF THE MILKY WAY

As already indicated, the stellar density can be traced at $2.2\,\mu$m, even though the absorption at that wavelength is quite patchy. The K-band light comes from K and M giants, and perhaps supergiants. Near the center their density decreases outward approximately as $R^{-1.8}$; farther out the slope steepens, but the patchiness of the absorption makes the profile rather uncertain. Thus far, infrared mapping has been done only within a degree of the Galactic center. Nevertheless, we presume that like other galaxies of type Sb or Sbc, the Milky Way has a central bulge of old stars. Ultra-wide-angle infrared photographs, as in Fig. 3.8, show it peeping around the edges of the dust layer; and optical studies through the "windows" in that chaotic layer sample its population in regions that are fairly close to the Galactic center.

The best known of these windows is the one discovered and first exploited by Baade; it has borne his name ever since. At latitude $-3°9$ and zero longitude, its line of sight passes 550 parsecs below the Galactic center. Baade confirmed that he was seeing the bulge by discovering a large number of RR Lyrae stars, whose density peaked at the distance of the Galactic center. Perversely, this otherwise-brilliant discovery strengthened Baade's incorrect supposition that the dominant

FIGURE 3.8
A wide-angle near-infrared photograph of the southern Milky Way, taken with a fish-eye camera whose field of view stretches nearly from horizon to horizon. (From Schlosser *et al.* 1975.)

population in the central bulges of spiral galaxies was globular-cluster-like. Later studies have shown, however, that both in period (Blanco 1984) and ΔS values (Gratton *et al.* 1986) the RR Lyraes in Baade's window indicate a high metal abundance (even though some stars of low metal abundance are included).

Recently Baade's window, and other relatively clear fields at higher latitudes along the minor axis of the bulge, have received considerably more attention.[4] Among the most interesting studies are two Ph.D. theses by associates of Whitford, whose name cannot be omitted from a review of this area. Rich (1988) estimated abundances for 88 K giants in Baade's window, chosen as such from an R, I color–magnitude diagram; he asserts, from Galaxy modeling, that nearly all of these are actually in the bulge. He found a wide range of abundances, with

---

[4]For full details, see the excellent review by Frogel (1988), from which I have profited greatly in preparing this section.

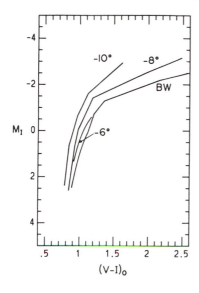

FIGURE 3.9

Giant branches of Terndrup's color–magnitude arrays at Galactic latitudes from $-4°$ (Baade's window) to $-10°$. (From Terndrup 1988)

[Fe/H] from -1 to $+1$. The mean abundance is twice solar, and 20 percent of his stars exceed in abundance the most metal-rich K giants in the solar neighborhood.

Terndrup (1988) took a different approach, using CCD color–magnitude diagrams in B, V and V, I. This approach, although less direct, is more economical of telescope time; so he was able to study not only Baade's window, but also fields at latitudes $-6°$, $-8°$, and $-10°$. His samples contained appreciable contamination from foreground disk stars, which he removed statistically using the Bahcall and Soneira (1980) model of the disk and halo of the Milky Way. Terndrup was able to delineate the red-giant branch in each of his fields. He found a steady progression of mean color with latitude (Fig. 3.9), in a way that would correspond to a drop in [Fe/H] by 0.6. Although there are some problems in the relation between giant-branch color and metallicity, the trend is clear. Terndrup does note, however, that its size depends sensitively on the reddenings adopted.

The giant branch in each of Terndrup's fields has a color spread that he interprets as being consistent with the metallicity spread that was found by Rich in Baade's window.

Terndrup's two lowest-latitude fields are too crowded for him to have observed the main-sequence turnoff reliably, but in the two fields at higher latitude he did delineate the turnoff. Fitting of isochrones gives an age in the general range 9 to 14 Gyr; the spread is due largely to uncertainties in the distance and in the metallicities.

Because of the interstellar extinction, which in the V band reaches 25 to 30 magnitudes at the Galactic center, the inner parts of the bulge can be studied only at longer wavelengths. Observing at J, H, and K (1.2, 1.6, and 2.2 microns, respectively), Frogel et al. (1989) have studied individual M giants down to

$b = -3°$. They find a systematic change of properties with latitude, in a sense that is consistent with a metal abundance that increases inward.

Infrared imaging and spectroscopy will eventually extend these studies closer to the center. But unfortunately the interpretation of the general spatial distribution is open to some question. For example, Blanco (1988) finds a very steep density gradient in M giants from latitude 3° to 13°; but he points out that the number of such stars in a population is strongly metal-dependent, since only at high metallicity do the evolutionary tracks penetrate deeply into the M-giant region. Thus a large part of his density gradient merely reflects the metallicity gradient and is probably not a property of the stellar distribution as a whole.

There are similar problems in the interpretation of the other infrared surveys, all of which find a steep density gradient, with scale heights of only 100 to 300 parsecs. (For details see the review by Frogel 1988.)

In the far infrared, the radiation that is observed comes from stars with circumstellar dust shells. Thus objects from the IRAS point-source catalog that radiate strongly at 12 microns are mainly asymptotic-giant-branch stars (Habing 1988). Their distribution in the Galactic bulge has been studied by Harmon and Gilmore (1988), who again find a steep density gradient. Using the IRAS variability index, they interpret their objects as long-period variables. They conclude that the majority of the periods are greater than 400 days, although such a conclusion, based on the statistics of three IRAS observations of each object, is necessarily uncertain. These long-period variables can shed some light on ages in the bulge, since such stars come from progenitors whose ages cannot be as great as those of the globular clusters. Harmon and Gilmore suggest ages less than 10 Gyr, but their conclusion is weakened by uncertainty about how a high metallicity will affect the masses that we should attribute to the progenitors of these AGB stars.

All in all, interpretation of the populations in the Galactic bulge and their density gradients is confusing. On the one hand, this region can be described as having a mixture of populations of different metal abundance, with the scale height increasing as the metal abundance decreases. On the other hand, we might speak of a single population in which the metal abundance decreases outward—but note the complication discussed a few paragraphs farther on.

One dynamical fact is clear, however. In order to be more concentrated to the center, the higher-metal-abundance stars must, at a given distance from the center, have a lower velocity dispersion than those of lower metallicity. This is, indeed, found by Rich (1985). Dividing a sample of 32 stars into a strong-line and a weak-line group, he reports a velocity dispersion for the former group that is only 60 percent of the velocity dispersion of the latter. Rich cites as a contradiction Rodgers' (1977) velocity dispersion of RR Lyrae stars, which is lower than that of the strong-line stars; but these RR Lyraes are much farther from the Galactic center and are therefore not directly comparable.

The Transition to the Halo

Since the densities of the various components of the bulge drop off rapidly with radius, eventually the metal-poor population of the halo must become dominant. There is already a hint of this in Terndrup's work; the color of the giant branch in his $-10°$ field suggests that the mean [Fe/H] there is negative by a few tenths. The dominance of the halo at higher latitudes is made clear by the colors measured by de Vaucouleurs and Pence (1978) in a photoelectric study of surface brightnesses along the great circle $\ell = 0°$. For latitudes between $10°$ and $20°$ they quote a mean color $(B - V)_0 = 0.65$. This is certainly a globular-cluster color rather than a galaxy-bulge color (which would be around 1.0); so at these latitudes the halo is already dominant.

We thus find, in just over a kiloparsec, a complete transition of populations. Although the bulge and halo both have approximately round shapes, and gravitation forces them to co-exist, this population profile makes it clear how inappropriate it would be to lump these components together under the single misleading title "the spheroid." But it would also be misleading to describe the population of the Galactic bulge in terms of any other population that we have encountered. It contains some stars of a higher metal abundance than we find anywhere else in the Milky Way, and closer to the center they may be predominant. We might be tempted to describe the central bulge as having a metal abundance that changes constantly with distance from the Galactic center. This is indeed factually correct, but it makes no dynamical sense to note only the abundance gradient. What we should note, more correctly, is that the stars of higher metal abundance have a smaller scale height *and* a lower velocity dispersion. Populations of higher metallicity are nested successively inside each other, probably in a continuous rather than a discrete fashion.

3.3 THE STELLAR HALO OF THE MILKY WAY

The most extended visible component of the Milky Way is the stellar halo. It is also the most tenuous; within the disk and the bulge the halo stars are heavily outnumbered.

Prominent in the halo are the globular clusters. Although they contain only about 1 percent of the total population of the halo, they occupy a disproportionate place in studies of the halo, for three reasons: (1) they are easy to find; (2) their distances are easily measured; and (3) each cluster (with a few exceptions) is a chemically homogeneous sample.

Field stars of the halo can be studied locally. They can be picked out easily by their proper motions, but this method of discovery unfortunately biases the sample. With more difficulty they can be picked out by spectroscopic or photometric searches, and these samples avoid motion bias. Local halo stars have the advantage that they can be studied down to absolute magnitudes fainter than can be reached in globular clusters; but they have the disadvantage that there is no way of placing an individual field star correctly within the wide range of population parameters exhibited by the globular clusters.

One type of star, however, allows us to study the field throughout the halo. RR Lyrae variables can be discovered and studied out to distances of many tens of kiloparsecs. The large amount of labor required for such studies has limited their number, however.

Before examining the structure of the halo, we need to note one distinction. The globular clusters divide, spatially, kinematically, and in abundances, into two classes: the disk globulars and the halo globulars; and we shall frequently need to make this distinction.

The Spatial Distribution of the Stellar Halo

The halo objects whose distribution is most easily studied are the globular clusters. The first step, of course, is to measure their distances. Although other criteria have been used, the outstanding distance criterion is the apparent magnitude of the horizontal branch, along with a knowledge of its absolute magnitude. The latter rests in turn on the absolute magnitudes of the RR Lyrae stars. These are moderately well known, from statistical parallaxes (Strugnell et $al.$ 1986; Hawley et $al.$ 1986) and from the Baade–Wesselink method (Jones et $al.$ 1988; Cacciari et $al.$ 1989, and references therein), both applied to field RR Lyraes. For a long time determinations of the absolute magnitudes of RR Lyrae stars have given values in the neighborhood of $M_V = 0.7$, with an uncertainty of about $0^m\!.1$; but for an even longer time there has been disagreement about whether this absolute magnitude is a function of metal abundance.

The absolute magnitudes of RR Lyrae stars will probably become somewhat better known if the Hipparcos satellite ever functions adequately. Several RR Lyraes are close enough that at least the mean of their Hipparcos parallaxes would have a significant accuracy.

Noting the reservation about RR Lyrae magnitudes, we can discuss the spatial distribution of globular clusters. It has been studied in detail by Zinn (1985). His paper makes a clear distinction between the disk and halo systems of globular clusters. He shows the bimodality of the general distribution of abundance values, and its association with a sharp kinematic difference between the two groups (Fig. 3.10). He makes his division between the two groups at [Fe/H] $= -0.8$. Equally clear is the difference in spatial distribution, shown here in Fig. 3.11. From all these differences it is clear that the disk globulars are either a separate population component or else associated in some way with the thick disk.

In a recent study devoted to the disk globulars alone, Armandroff (1989) argues strongly that the disk globular clusters are a part of the thick-disk population.

The distribution of globular clusters with distance from the Galactic center is shown in Fig. 3.12. (Note that this graph appears in Zinn's paper prior to his separating out the disk globulars; at $R < 7$ kpc about a third of the sample are disk globulars. Their presence does little damage, however, because most of

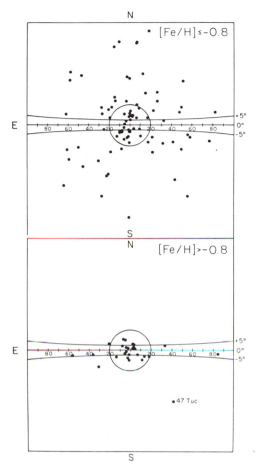

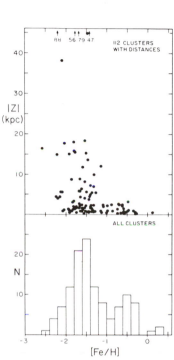

FIGURE 3.10

Distribution of metal abundances of globular clusters (lower panel). The upper panel shows their distribution in height above the Galactic plane. (From Zinn 1985.)

FIGURE 3.11

Comparison of the distributions on the sky of weak-lined (above) and strong-lined (below) globular clusters. (From Zinn 1985.)

the densities to which they contribute have little reliability, on account of the incompleteness of discovery of globular clusters in the crowded regions near the Galactic center.) The middle region follows a −3.5 power of distance, and the outermost points lie on that line too. The outer densities are, however, in a sense distorted by binning. Zinn notes that there are no globular clusters at all between 33 and 60 kpc from the center. This fact leads him to speculate, as others have before, that the drop shown by his solid circles indicates the edge of the halo and that the outermost globular clusters belong to a separate outer system, along with the dwarf spheroidal galaxies.

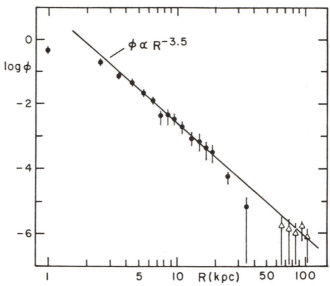

FIGURE 3.12

The number of globular clusters per kpc³, plotted against distance from the Galactic center. (From Zinn 1985.)

Saha (1985), who made a deep search for RR Lyrae stars, found them out to 41 kpc from the Galactic center. His rather small sample does seem, however, to be dropping off rapidly in density beyond 20 kpc.

We may also ask about the flattening of the halo. The halo globular clusters shown in the upper part of Fig. 3.11 show no appreciable flattening, as earlier studies had already found. Also, the system of halo globulars has a very low rotational velocity. Its numerical value is shaky because of arbitrary assumptions about the shape of the rotation law and uncertainty about the rotational velocity of the solar neighborhood, but the rotational velocity is definitely much lower than the velocity dispersion of these objects. This again suggests a round shape.

By contrast, the field stars of the halo do not appear to have such a round distribution. It was stated in Chapter 2 that the distribution of faint stars implies a substantial flattening, and it will be argued in Chapter 11 that the kinematics of local halo stars make a similar demand.

The shape of the system of RR Lyrae stars is not clear. Hartwick (1987) fitted a parametric density model to a set of RR Lyrae surveys, and concluded that the inner part of the density distribution was flattened, but his outermost equal-density contours are round. It is hard to say, however, how much of this behavior reflects the model chosen rather than the nature of the RR Lyraes themselves.

But in any case, it is distressing that the flattening of the system of globular clusters seems to differ from that of the halo field stars. If this turned out really to be so, we would then have to regard them as different population components, and the beautiful HR-diagram information that we get from the globular clusters would not be fully applicable to the field stars.

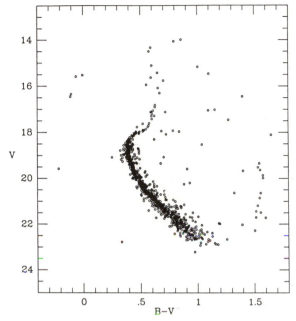

FIGURE 3.13
The turnoff and lower main sequence of the globular cluster M92, from CCD photometry by Stetson and Harris (1988).

A final question is the number of halo stars; all densities referred to thus far have been relative. The best absolute numbers that can be found come from studies of local subdwarfs. Such studies require extreme care to combat the selection effects that arise from discovering these stars by means of their large proper motions. They tend to give results of about 0.1 to 0.2 percent of the local disk density. (For a discussion of such studies, see Hartwick 1987.) When this normalization is applied to the halo distribution, it leads to a total halo mass of a few times 10^9 solar masses, about a hundredth of the mass of the disk and a hundred times the aggregate mass of the globular clusters.

Color–Magnitude Diagrams of Globular Clusters

The two vehicles that we have for studying the populations in globular clusters are their chemical abundances, deduced either spectroscopically or photometrically, and their color–magnitude diagrams. Abundances are tabulated by Zinn (1985), and a compilation of CMDs was given by Peterson (1986). The latter are almost impossible to keep up with, however. The advent of CCDs at major observatories has revolutionized faint photometry (see Fig. 3.13); it is now possible, with a reasonable amount of observing time, to secure material from which accurate photometry can be done below twentieth magnitude, with main sequences often detected as far as $V = 24$.

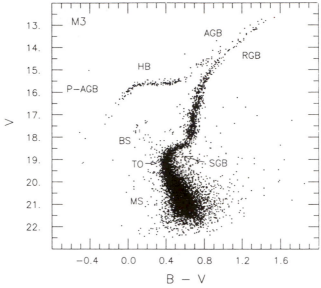

FIGURE 3.14
A photographic HR diagram of the globular cluster M3. Note the large observational scatter at the faint end. The labels indicate the main sequence, its turnoff, the blue stragglers, the subgiant branch, the red-giant branch, the asymptotic giant branch, the horizontal branch, and the post-AGB region. (From Renzini and Fusi-Pecci 1988.)

The important regions of a generalized CMD are shown in Fig. 3.14. Globular clusters differ from each other in the locations and prominence of the various sequences. The most striking differences correlate well (but not totally; see below) with metal abundance. A close correlation with metallicity is shown by the height of the red-giant branch. The conventional measure of this is ΔV, the magnitude difference between the horizontal branch and the giant branch measured at $(B-V)_0 = 1.4$. This quantity increases monotonically with decreasing metal abundance.

Another red-giant-branch criterion is its color at the absolute magnitude of the horizontal branch, $(B-V)_{0,g}$. Its interpretation is less clear, but it also seems to correlate with metal abundance.

The color of the main-sequence turnoff also depends on metal abundance (Fig. 3.15). This is a simple consequence of differences in line blanketing.

The horizontal branch does not behave in such a simple way, however. Although low-metal-abundance clusters generally tend to have a blue horizontal branch, and high-metal-abundance ones to have merely a red stub, some clusters violate this rule blatantly. In the middle abundance range, some clusters have very blue horizontal branches; and some low-abundance clusters have only the red stub that is otherwise characteristic of high metal abundance. These aberrations are referred to as the "second-parameter effect." It is still a mystery what

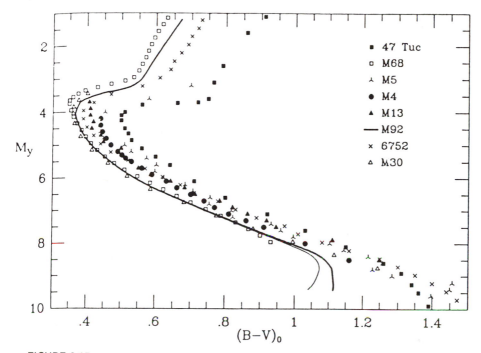

FIGURE 3.15

Comparison of the main-sequence turnoffs of several globular clusters.   (From Stetson and Harris 1988.)

the second parameter is. Various possibilities have been suggested: age, helium abundance, other abundance anomalies, stellar rotation, etc. Interestingly, there seems to be a tendency for second-parameter clusters to lie far from the Galactic center.

One cluster characteristic has a simple explanation: the number of RR Lyrae stars. Since they occupy the region of the horizontal branch that intersects the instability strip, their number is determined simply by how strongly the horizontal branch goes through that region. Every horizontal-branch star that lies within the instability strip is an RR Lyrae star (Roberts and Sandage 1955).

And there is one characteristic that has remained elusive. White dwarfs have not yet been observed in globular clusters. Claims of detection have been made, but they are not convincing, and the study of white dwarfs in globular clusters may have to await the color–magnitude diagrams that will be produced by the Hubble Space Telescope.

In addition to direct population studies, the CMDs of globular clusters have another important application: finding out their ages. This process requires fitting the observed main-sequence turnoff of a cluster to isochrones calculated from

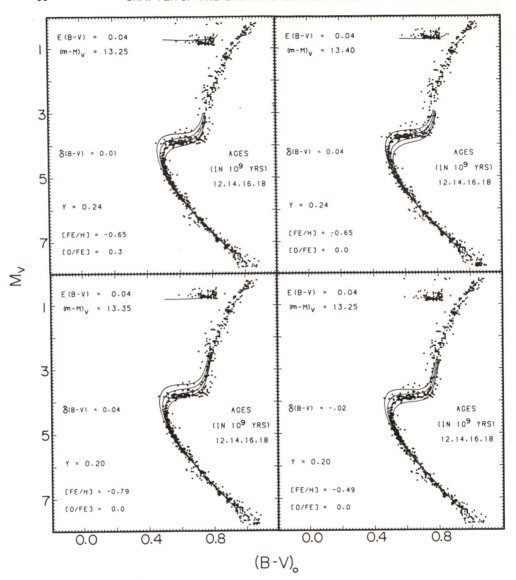

FIGURE 3.16
Fitting of isochrones to CCD photometry of the metal-rich globular cluster 47 Tucanae, for various assumed compositions and distances. (From Hesser *et al.* 1987.)

theoretical stellar models (Fig. 3.16). Recent values for their ages have tended to fall in the range from 13 to 16 Gyr. Unfortunately, they are all uncertain by 2 or 3 Gyr, for several reasons. Some troubles are on the observational side: distance modulus, reddening, and calibration of the photometric zero point and the color system. And some are on the theoretical side: uncertainties in helium abundance, oxygen abundance, mixing-length theory, etc. Finally, a further uncertainty is introduced in the transformation of the isochrones from the theoretician's M_{bol}, $\log T_e$ coordinates to the M_V, $B - V$ coordinates of the observer. This is another task of the theoretician, for it requires putting a theoretical stellar atmosphere on each star and calculating its flux through the observer's filters.

For the isochrones themselves, see reviews by King *et al.* (1988) and by VandenBerg (1988).

REFERENCES

Armandroff, T. E. 1989. *Astron. J.*, **97**, 375.

Backer, D. C., ed. 1987. *The Galactic Center*. AIP Conference Proceedings no. 155. New York: American Institute of Physics.

Backer, D. C., and Sramek, R. A. 1987. In Backer, p. 163.

Bahcall, J. N., and Soneira, R. M. 1980. *Astrophys. J. Supp.*, **44**, 73.

Blaauw, A., Gum, C. S., Pawsey, J. L., and Westerhout, G. 1959. *Mon. Not. Roy. Astron. Soc.*, **119**, 422.

Blanco, B. M. 1984. *Astron. J.*, **89**, 1836.

———. 1988. *Astron. J.*, **95**, 1400.

Cacciari, C., Clementini, G., and Buser, R. 1989. *Astron. Astrophys.*, **209**, 154.

Cook, W. R., Palmer, D. M., Prince, T. A., Schindler, S. M., Starr, C. H., and Stone, E. C. 1989. In Morris, p. 581.

Diehl, R., Ballmoos, P. von, Schönfelder, V., and Morfill, G. E. 1989. In Morris, p. 617.

Frogel, J. A. 1988. *Ann. Rev. Astron. Astrophys.*, **26**, 51.

Frogel, J. A., Terndrup, D., Blanco, V. M., and Whitford, A. E. 1990, *Astrophys. J.*, in press.

Gatley, I., Joyce, R., Fowler, A., DePoy, D., and Probst, R. 1989, In Morris, p. 361.

Genzel, R., and Townes, C. H. 1987. *Ann. Rev. Astron. Astrophys.*, **25**, 377.

Gilmore, G., and Carswell, B., eds. 1987. *The Galaxy*. Dordrecht: Reidel.

Gratton, R. G., Tornambè, A., and Ortolani, S. 1986. *Astron. Astrophys.*, **169**, 111.

Habing, H. J. 1988. *Astron. Astrophys.*, **200**, 40.

Harmon, R., and Gilmore, G. 1988. *Mon. Not. Roy. Astron. Soc.*, **235**, 1025.

Hartwick, F. D. A. 1987. In Gilmore and Carswell, p. 281.

Hawley, S. L., Jefferys, W. H., Barnes, T. G. III, and Lai, W. 1986. *Astrophys. J.*, **302**, 626.

Hesser, J. E., Harris, W. E., VandenBerg, D. A., Allwright, J. W. B., Shott, P., and Stetson, P. B. 1987. *Publ. Astron. Soc. Pacific*, **99**, 739.

Jones, R. V., Carney, B. W., and Latham, D. W. 1988. *Astrophys. J.*, **332**, 206.

King, C. R., Demarque, P., and Green, E. M. 1988. In Philip, p. 211.

Lindqvist, M., Winnberg, A., Habing, H. J., Matthews, H. E., and Olnon, F. M. 1989. In Morris, p. 503.

Lingenfelter, R. E., and Ramaty, R. 1989. In Morris, p. 587.

Lo, K. Y., Cohen, M. H., Readhead, A. S. C., and Backer, D. C. 1981. *Astrophys. J.*, **249**, 504.

Morris, M., ed. 1989. *The Center of the Galaxy*. I.A.U. Symposium no. 136. Dordrecht: Kluwer.

Peterson, C. J. 1986. *Publ. Astron. Soc. Pacific*, **98**, 1258.

Philip, A. G. D., ed. 1988. *Calibration of Stellar Ages.* Schenectady: L. Davis.

Ratnatunga, K. U., Bahcall, J. N., and Casertano, S. 1989. *Astrophys. J.*, **339**, 106.

Renzini, A., and Fusi-Pecci, F. 1988. *Ann. Rev. Astron. Astrophys.*, **26**, 199.

Rich, R. M. 1985. *Mem. Soc. Astron. It.*, **56**, 23.

———. 1988. *Astron. J.*, **95**, 828.

Rieke, G. H., and Lebofsky, M. J. 1987. In Backer, p. 91.

Roberts, M. S., and Sandage, A. R. 1955. *Astron. J.*, **60**, 185.

Rodgers, A. W. 1977. *Astrophys. J.*, **212**, 117.

Saha, A. 1985. *Astrophys. J.*, **289**, 310.

Schlosser, W., Schmidt-Kaler, T., and Hünecke, W. 1975. *Atlas der Milchstrasse.* Bochum: University of Bochum Observatory.

Sellgren, K. 1989. In Morris, p. 477.

Stetson, P. B., and Harris, W. E. 1988. *Astron. J.*, **96**, 909.

Strugnell, P., Reid, N., and Murray, C. A. 1986. *Mon. Not. Roy. Astron. Soc.*, **220**, 413.

Terndrup, D. M. 1988. *Astron. J.*, **96**, 884.

Townes, C. H. 1989. In Morris, p. 1.

VandenBerg, D. A. 1988. In Philip, p. 117.

Vaucouleurs, G. de, and Pence, W. D. 1978. *Astron. J.*, **83**, 1163.

Yusef-Zadeh, F. 1989. In Morris, p. 503.

Zinn, R. 1985. *Astrophys. J.*, **293**, 424.

CHAPTER 4

THE DISTRIBUTION OF
INTERSTELLAR MATERIAL

Ivan R. King

Although the interstellar material has only a few percent as much mass as the stars, it is quite important because of its role in star formation, and also because we use its motions to trace out the mass distribution of the Milky Way. The latter will be covered in Chapter 6; here we discuss the distribution of the interstellar material.

4.1 THE NATURE OF INTERSTELLAR MATERIAL

The interstellar material is basically gas, with a small admixture of dust. The dust plays a very important role, however. Its sub-micron-sized particles scatter visible light so efficiently that interstellar absorption, as we so often loosely call it, influences nearly all optical observations. It also plays a direct role in the formation of the giant molecular clouds (GMCs), in which most star formation appears to take place. (Also, incidentally, it very probably serves as a heating mechanism for diffuse gas.)

The interstellar gas behaves in quite different ways in different places. Most of it is neutral (H I), some of this cool (100 K) and some considerably warmer (several thousand K); but around sufficiently hot stars the hydrogen is ionized (H II regions, at nearly 10^4 K). Also, supernovae have heated some large regions to 10^6 K. Finally, in the GMCs the gas is very cool (20 K), and the hydrogen is predominantly molecular. Paradoxically, the very cool regions and much of the ionized hydrogen coincide in their distribution, in general and sometimes in detail, because the cool GMCs are the regions in which star formation is initiated, so that they often contain the H II regions that result from star formation.

4.2 INTERSTELLAR DUST

One difficulty that must be surmounted in all optical measurements is interstellar absorption. Its amount can usually be deduced only from the amount of reddening, measured either from comparison of observed color with spectral type or from multicolor photometry. With either method, we must know the ratio of absorption to reddening. The latter should depend on the nature of the dust grains, but that seems to be remarkably uniform, except for a very few anomalous places, like the Orion nebula.

The ratio of absorption to reddening is not easy to determine. It has been attempted in two ways. One makes use of the fact that the absorption approaches zero at very long wavelengths. The first step here is to choose two stars of identical spectral type but different apparent colors. The two are then compared at many different wavelengths. The comparison at the longest wavelength tells their true difference in magnitude; at shorter wavelengths they will differ by differing amounts, comparison of which shows by how much the absorption differs at those wavelengths. (In using this method, we must be very careful to avoid stars that have circumstellar infrared excesses, which can seriously falsify the result.) The second method involves observing upper-main-sequence stars in a cluster or association, and comparing their simultaneous displacements in color and in magnitude.

When reddening measures are lacking, as they often are for high-latitude fields, it is possible to make use of the close coupling of the gas and dust. Burstein and Heiles (1978) have calibrated the total column density of H I, measured at 21 cm, against the amount of absorption. Their method is often the best available for finding out how much reddening there is between us and globular clusters or galaxies, these being objects that we can be confident do lie beyond all the interstellar material that contributes to the column density.

4.3 PROBING THE INTERSTELLAR GAS

The interstellar gas radiates in many different ways, often characteristic of regions of different physical properties, and of differing uses and degrees of usefulness in mapping out the interstellar material of our Galaxy. H II regions, for example, tell us little about the general distribution of gas, but serve as excellent locators of the regions of star formation. As such, they are good spiral-arm tracers. Optically visible H II regions have the advantage over all other interstellar radiators that their distances can be calculated unequivocally, from observations of the imbedded stars that excite them. The absolute magnitudes of these stars can be estimated spectroscopically, and the interstellar absorption can be calculated from the observed reddening, so that the distance is easily calculated. This method has the disadvantage, however, at least for traditional optical observations, that interstellar absorption limits observations to H II regions that are relatively nearby on a Galactic scale. It is to be hoped that the increasing availability of infrared imaging arrays will remove this difficulty.

IR arrays are small, however, and are likely to remain so for some time. Their survey possibilities are thus quite limited, and it is hard to imagine wide-angle searches in the higher-series hydrogen lines that would be analogous to the surveys that have been made photographically with interference filters at Hα.

Other discovery methods promise to make infrared surveys possible, however. The radio lines of the CO molecule can be used to identify giant molecular clouds (GMCs) all over the Galaxy. Since these tend to be regions of star formation, a good technique will be to look for hydrogen emission in the infrared, searching only the regions where the CO emission says that it is likely to be found. In the inner parts of the Milky Way, moreover, H II regions are plentiful enough that they can be found directly by their radio recombination radiation (see below). We shall return to these possibilities later, in connection with tracing the spiral structure.

The workhorse of interstellar probing is, of course, the 21-centimeter line of hydrogen. It pays to review a few elementary facts that make it clear why this line is so useful. We see it as spontaneous emission when a hydrogen atom drops from the upper to the lower of the two hyperfine states within its ground energy level. An important characteristic of this line is its very low transition probability, which corresponds to a lifetime of 10^7 years in the upper state. This guarantees a correspondingly low absorption coefficient, so that the hydrogen is transparent almost everywhere to its own radiation. (Contrast the Lyman continuum, which wraps the Sun almost completely in a local cocoon.) Only in a few directions, where there is a very large column depth and very little velocity spread, does the hydrogen fail to be optically thin.

(A little elementary intuition makes the nearly forbidden nature of the 21-cm line easy to understand. The two states in question differ by the direction of the electron spin. Thus a 21-cm transition requires the improbable atomic acrobatics of flipping an electron head-for-foot.)

To go on with these elementary considerations, 90 percent of interstellar atoms are hydrogen; and in regions that are not ionized, the hydrogen is essentially all in its ground energy level. The upper state is easily populated by collisions; its Boltzmann factor is only infinitesimally less than unity; and its statistical weight insures that at any given moment 75 percent of the hydrogen atoms are poised to emit a 21-cm photon, subject only to the low transition probability. Finally, differential Galactic rotation allows us, in most directions, to separate out the hydrogen at different distances. No wonder we know so much more about the disk of the Milky Way than we did before the 21-cm line was first detected in 1950! (For a list of the extensive surveys that have been made at 21 centimeters, see Burton 1988.)

Radio radiation of hydrogen is also detected from H II regions. This radiation comes from very-high-level recombination transitions whose small energy jumps put them in the radio part of the spectrum. These lines cannot be observed in the laboratory; only the low density of interstellar space allows hydrogen atoms to hold an electron that is so far out, without constant interference from neighboring atoms. Fortunately, we can calculate everything that we need to know about the properties of these transitions, since hydrogen, as a two-body system, is the unique atom that has a full quantum-mechanical solution. Recombination lines of the heavier elements can be calculated too, because the distant nucleus

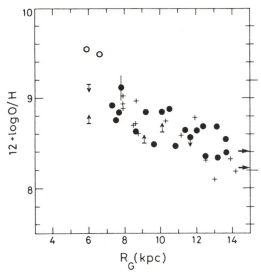

FIGURE 4.1

Heavy-element abundance as a function of distance from the Galactic center. Dots are
from H II regions whose temperatures are calculated from radio-recombination-line
measurements; pluses are previous temperatures arrived at by purely optical means.
(Shaver *et al.* 1983, Royal Astronomical Society.)

surrounded by $Z - 1$ electrons (where Z is the nuclear charge) looks almost
exactly like a heavier hydrogen nucleus.

The radio recombination lines are named by the number of the lower energy
level and a Greek letter indicating the amount of the jump; thus H109α, the
first of these lines to be detected, is the transition from the 110th to the 109th
energy level. For some years the radio recombination lines gave the best surveys of
distant H II regions, and they will probably continue to do so. Currently, however,
CO observations (discussed below) are receiving more attention, in the mapping
of the GMCs that tend to surround the H II regions.

Recombination lines have also contributed greatly to the study of chemical
abundance gradients in the Galaxy. The abundances come from optical observations of emission lines in H II regions, but the line strengths from which they
are derived are very sensitive to the value of the electron temperature, which
is usually difficult to calculate from optical observations. The electron temperature can best be derived from observations of recombination lines (Shaver *et al.*
1983). Abundance gradients established in this way agree with, and have greatly
extended, the results obtained by purely optical means. (See Fig. 4.1.) For a
more detailed discussion of radio recombination lines, see Gordon (1988).

The advent of millimeter-wave astronomy has added an important new probe
to Galactic-structure studies: the CO molecule. It is easy to observe, and it is
abundant in the GMCs. The lines that are observed are rotational transitions
at 1.3 and 2.6 mm. An important observing problem, however, is saturation.
The intensity of the line is proportional to the column density only if the CO is

optically thin, but these lines are permitted lines and tend to become optically thick. Fortunately, for these rotational transitions at radio wavelengths the isotope shifts are considerable, and we can switch to observing the less abundant ^{13}CO, or the even less abundant $^{12}C^{18}O$. Thus GMCs can be studied up to a large total depth in CO.

This, at least, has been the conventional wisdom about measuring amounts of CO. It has recently been argued, however (Solomon *et al.* 1987), that the amount of CO can be estimated simply by measuring the total CO luminosity of a cloud. The advocates of this technique argue that the CO is in dense cloudlets that are opaque to their own radiation but transparent to that of other cloudlets of different velocity; so the total CO luminosity is a measure of the number of cloudlets and therefore of the amount of CO.

Another advantage of CO is that the short wavelengths of its rotational lines allow a good beam size for mapping to correspond to a small-aperture telescope. The whole Galactic plane has been mapped with a 1-meter radio telescope. (For a tabulation of surveys, again see Burton 1988.) A large radio telescope, conversely, has too small a beam size for Galactic surveys, but is appropriate for mapping within GMCs or within other galaxies.

The interpretation of CO line strengths in terms of mass densities is far from trivial, however. First we must know the ratio of CO to H_2, which makes up most of the mass of a molecular cloud. This is too difficult to calculate; the chemistry of a cold mixture of gas and dust grains is just too poorly understood. (For a detailed discussion, see Turner and Ziurys 1988.) But amounts of CO can be empirically correlated with interstellar extinction, and this has been correlated in turn with the total $H_I + H_2$, by means of far-UV measurements made with the Copernicus satellite. (In a dense cloud the hydrogen has, of course, all turned into H_2.) There are also other methods, based on the virial equilibrium of a cloud or on the amount of mass needed to produce the observed gamma rays by cosmic-ray impact. (For a summary of results, see Scoville and Sanders 1987.) The ratios of the isotopic varieties of CO also have to be calibrated empirically.

4.4   THE LARGE-SCALE RADIAL DISTRIBUTION OF THE INTERSTELLAR MATERIAL

The different components of the interstellar medium have very different radial distributions in the Galaxy. If the mean densities are averaged in annuli, so that we are ignoring local irregularities or spiral structure, then the H_I and the H_{II} components exhibit quite different profiles. The density of H_I is low in the center, peaks near the solar circle, and slowly tails off outward. The H_{II} regions and GMCs, on the other hand, peak much farther in, and become relatively rare

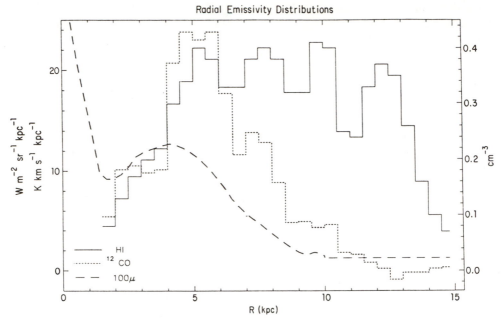

FIGURE 4.2
Radial distributions of H I, CO, and dust. (From Burton 1988.)

much beyond the Sun. Fig. 4.2 compares the GMC distribution with that of H I; the distribution of H II regions would look quite similar to that of the GMCs. The figure also shows the overall distribution of dust, as deduced from the $100\,\mu$m IRAS survey.

4.5   DELINEATING THE SPIRAL STRUCTURE

Although it had been known for decades that the Milky Way must be a spiral galaxy, it was not until 1952 that Morgan and his colleagues (Morgan *et al.* 1952) succeeded in plotting out the first fragmentary sketches of the spiral arms. This they did by measuring the distances of H II regions, by measuring the distances of the stars that excite them. Since then, other young, luminous objects have been used: long-period cepheids, young open clusters, OB stars, etc. It is essential that spiral-arm tracers be young. We must catch them while they are still at the locations where star formation has taken place quite recently (and before the regions of star formation have had a chance to move either). Since 1 km sec^{-1} is equivalent to a parsec in a million years, and even young stars have velocities of around 10 km sec^{-1}, a few times 10^7 years is the maximum age for spiral-arm tracers. (This is why shorter-period cepheids will not do; their progenitor stars were too old.)

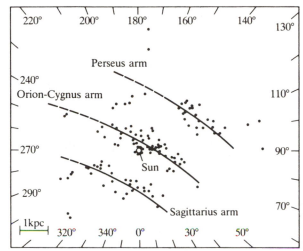

FIGURE 4.3
Local bits of spiral arms, as inferred from H II regions and young clusters.
(From Mihalas and Binney 1981, p. 248.)

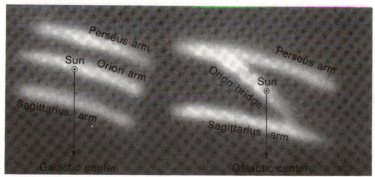

FIGURE 4.4
Alternative pictures of the local spiral structure. (From King 1976.)

Unfortunately, observations of all these types of objects are hampered by interstellar absorption, and optical maps of spiral structure are still confined to a couple of kiloparsecs around the Sun. They do not extend far enough to delineate arms into a clear spiral structure. Fig. 4.3 shows one such attempt, but I have always thought that we might perceive the structure somewhat differently if the lines had not been put there to guide the eye.

The astronomers who first discerned such structures named them, following the astronomical tradition of inept terminology, after the constellations that contained prominent features in them; and these names have stuck. But the interpretation has not. A frequently expressed opinion is that the arm nearest the Sun is not a real arm; it has the alternative name "Orion spur." (See Fig. 4.4.)

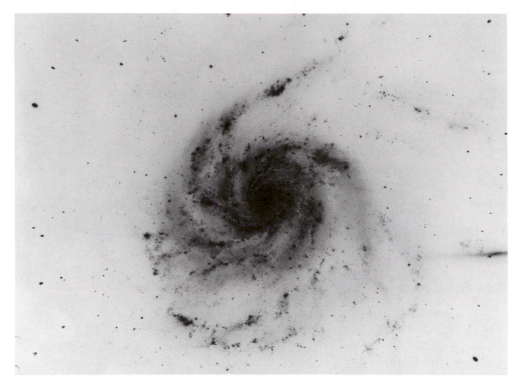

FIGURE 4.5
M101 is a typical giant spiral galaxy, with spirality everywhere but only a suggestion of a two-armed pattern.

In fact, the spiral structure around the Sun might not be simple at all. We should keep in mind that neat, continuous, two-armed spirals are the exception rather than the rule. We might well live in a galaxy that has a messy spiral structure (Fig. 4.5).

The remainder of mapping has been done with radio astronomy, using mainly the 21-cm line. The key to such mapping is to be able to find the distance of the hydrogen that emits at a particular Doppler shift along a particular line of sight. The velocities arise, of course, from the differential rotation of the Milky Way, so that interpreting them depends on knowing the rotation curve of the Galaxy. How it has been constructed will be discussed in Chapter 6; here we will merely note that in most directions the 21-cm profile, converted into a Doppler shift, can in principle give the distribution of hydrogen along the line of sight. Since the distances are found from a velocity-distance relationship along each line of sight, they are often referred to as kinematic distances.

In practice, however, the mapping is not nearly so easy. A serious problem is "velocity crowding." Where the velocity depends little on distance, velocity is a poor discriminant. For example, near longitudes 0° and 180°, the differential velocities go to zero, giving two pie-shaped wedges that are blank in the

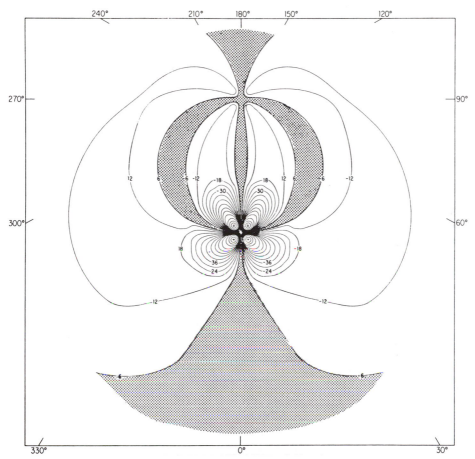

FIGURE 4.6
Map of velocity crowding. The numbers are indications of the precision with which kinematic distances can be calculated. In the shaded area kinematic distance cannot be calculated at all. (From Burton 1988.)

hydrogen map. Discrimination is also poor where the velocity along a line of sight goes through an extremum, which happens everywhere on the circle that is centered halfway from the Sun to the Galactic center. This creates another area of confusion in the 21-cm map. Fig. 4.6 illustrates the sensitivity of velocity to distance in the different parts of the Galaxy. But worst of all are non-circular velocities, which distort all distances. All in all, 21-cm maps of spiral structure are a tantalizing hodgepodge, and few have been drawn in recent years.

How, then, are we ever to delineate the spiral structure of our Galaxy? I believe, along with Wielen (1974), who emphasizes the point strongly in his review of theories of spiral structure, that the job will have to be done by laboriously discovering H II regions all over the Galaxy and measuring their individual distances. Interstellar absorption will drive this work into the infrared, but such

efforts are now becoming possible. The answer will probably be to discover the
H II regions by means of the radio recombination lines and then to observe them
at infrared wavelengths. Unfortunately, OB stars are not very bright in the infrared, and even at those wavelengths absorption will still be high, but successful
observations should be possible for the more luminous stars, in directions of for-
tuitous clarity (the so-called "windows.")

4.6   THE THICKNESS OF THE GAS LAYER

Measuring the distribution of gas in a direction perpendicular to the Galactic
plane falls into three separate observational problems: the inner disk, the solar
neighborhood, and the outer disk. As usual, the crucial problem is to know the
distance of the gas that is being observed.

For the solar neighborhood we use ultraviolet space observations of Lyman
alpha absorption, concentrating on stars at high Galactic latitude and at different
distances from the Galactic plane. A comparison of the strength of Lyα with the
total 21-cm intensity in that direction tells what fraction of the hydrogen is at
lower z than the star. Lockman et al. (1986) assembled a set of such data, and
found a patchy distribution in which half of the hydrogen could be reasonably
well approximated by a Gaussian distribution with $\sigma_z = 135$ pc and the other
half by an exponential with a scale height of 500 pc.

Observations of the inner disk make use of the tangent-point method. As
will be explained at a little greater length in Chapter 6, along a given line of
sight the greatest velocity of the hydrogen relative to us will be at the point
where the line of sight passes closest to the Galactic center (making it thus
tangent to a circle around the center). If we concentrate on this edge of the
Doppler-stretched 21-cm profile, then we know that we are looking at gas near
the "tangent point," and we can directly measure its vertical distribution. Such a
study (Lockman 1984) found that about half of the atomic hydrogen in the inner
part of the Milky Way lies within about 150 parsecs of the plane. In fact, this
could actually comprise more than half the hydrogen, because the optical depth
caused by velocity crowding at the tangent point could cause us to underestimate
the amount. Beyond 3 kpc from the center, there is an extended component, like
that in the solar neighborhood, with a scale height of 500 pc. But within 3 kpc
of the center this extended component is missing, and the gas near the plane is
also somewhat more concentrated than it is farther out.

For the outer parts of the Galaxy, we must rely on kinematical distances.
(The distances used in tangent-point studies are not kinematical distances, even
though they are found by means of velocities. They will be shown in Chapter 6 to
depend only on geometric ratios and not on the rotation curve; on the contrary,
they are used to deduce the rotation curve.) Kulkarni et al. (1982) studied the
outer hydrogen and found that exterior to the Sun the scale height of the gas
increases linearly, up to rather more than a kiloparsec at the outer limit of their
observations, which is at about three times the Sun's distance from the center.

They did not study the detailed height distribution, however, except for noting briefly that it seems to be somewhere between Gaussian and exponential.

4.7 HIGH-LATITUDE GAS

Although most of our Galaxy's interstellar material lies close to the Galactic plane, there are interesting manifestations of material away from the plane, which we can detect by looking at high latitudes. The most mysterious of these are the high-velocity clouds, whose nature has defied understanding for more than 25 years. (For a review of the facts, and speculations about them, see van Woerden *et al.* 1985.) The high-velocity clouds are detected by means of high-sensitivity observations of their 21-cm emission; they are scattered over the sky. The observations, of course, give us the amount of hydrogen along the line of sight and the velocity of the cloud, but there is no way of knowing its distance. In this respect kinematics do us no good, because it is clear that the high-velocity clouds do not partake of the relatively smooth rotation of the disk. The best hope will probably be to look for absorption lines in the spectra of emitting objects at various distances, in the direction of a cloud, to see which objects are behind the cloud and which in front. Unfortunately, attempts to do this have not yet succeeded. Without a distance we know very little about the properties of a cloud, most of which depend on some power or other of the distance. The most detailed maps of high-velocity clouds yet published are those by Hulsbosch (1985) and by Hulsbosch and Wakker (1988).

The distribution and kinematics of the high-velocity clouds show some tantalizing regularities. Although some velocities are positive, negative velocities predominate, and all the highest velocities (from 220 to more than 400 km sec^{-1}) are negative. There seem to be more high-velocity clouds in the first two quadrants of Galactic longitude than in the other half of the sky. Also, the velocities in the first two quadrants tend to be more negative. Unfortunately, these are only tendencies; individual clouds show a great scatter in their behavior.

The longitude asymmetry of the velocities suggests that the high-velocity clouds may be moving more slowly with respect to the Galactic center than are the stars of the solar neighborhood, but the asymmetry seems to be centered on a line about 30° away from the center-anticenter direction. The general predominance of negative velocities suggests infall, but the existence of large positive velocities shows that not all of the motion is infall.

A variety of suggestions has been made about the origin of the high-velocity clouds. Perhaps the least improbable of them (Bregman 1980) envisions hot supernova ejecta bubbling up out of the Galactic plane and cooling as they fall back in, a picture usually referred to as the Galactic fountain model. But understanding of these clouds is hobbled badly by lack of knowledge of their distances, as well as by knowing too little about the Galactic corona (see below) with which they probably interact.

Another enigmatic entity, probably not related to the high-velocity clouds, is the Magellanic Stream. This narrow band of HI, which extends almost 90° away from the Magellanic Clouds, is probably not a part of the Milky Way, although our Galaxy may well be related to its origin. As with the high-velocity clouds, the distances of the various parts of the stream from us are quite uncertain.

There is an indication that the Magellanic Stream is trailing behind the Clouds as they follow their orbit around (or past) the Milky Way. The edge of the Clouds that is opposite the Stream is abrupt, as if the Clouds were encountering the ram pressure of a tenuous medium at their present Galactocentric distance of about 50 kpc.

Many attempts have been made to interpret the Magellanic Stream, usually by some sort of dynamical modeling. Most models see the Stream as torn out of the Magellanic Clouds by an encounter with the Milky Way, some 2×10^8 years ago. But beyond this there is little agreement, largely because of the unknown distances, but also to some extent because we know too little about the outer extent of our dark halo. At a recent symposium, two review articles took opposing points of view that (1) the Magellanic Clouds are in a bound orbit around the Milky Way (Fujimoto and Murai 1984), and that (2) they must be making a hyperbolic passage, because in a bound orbit they would have had their gas swept out by now (Mathewson and Ford 1984).

The idea of ram pressure on the Magellanic Clouds implies that the Milky Way has gas as far out as that. Since this gas needs to be supported somehow, its existence would imply high velocities and perhaps high temperatures. High-latitude gas does indeed exist, although again its location along the line of sight is completely unclear. The existence of hot "coronal" gas closer to the Galactic plane was postulated by Spitzer (1956), who needed its pressure in order to contain cool clouds whose internal velocities would otherwise lead them to fly apart. It was not actually observed until many years later, when spacecraft were able to detect UV absorption lines in the spectra of the Magellanic Clouds and of a few stars far from the Galactic plane. To call this gas a corona implies that it is at a high temperature, an idea that would appear to be supported by the fact that the lines observed come from highly ionized species. It is also possible, however, that the ionization is produced by high-energy photons coming from hot stars in the disk, or else by cosmic rays.

To make it worse, we have no idea where the lines are produced along the line of sight. The gas could be just outside the disk, or it could be far away. The answer may well have to wait, like so many astronomical problems, for the long-delayed Hubble Space Telescope, which will do UV spectroscopy of large numbers of stars much fainter than IUE can reach.

This material, too, has velocities very unlike those of the material in the disk. Like the high-velocity clouds, its velocities generally lag behind Galactic rotation. Here the sample of directions is much smaller, though. For a review of the properties of the Galactic corona see de Boer (1985) and Savage (1987 and 1986).

REFERENCES

Bergh, S. van den, and Boer, K. S. de., eds. 1984. *Structure and Evolution of the Magellanic Clouds.* I.A.U. Symposium no. 108. Dordrecht: Reidel.

Boer, K. S. de. 1985. In van Woerden *et al.*, p. 415.

Bregman, J. N. 1980. *Astrophys. J.*, **236**, 577.

Bregman, J. N., and Lockman, F. J., eds. 1986. *Gaseous Halos of Galaxies.* Green Bank, WV: National Radio Astronomy Observatory. (This is, unfortunately, in few libraries.)

Burstein, D., and Heiles, C. E. 1978. *Astrophys. J.*, **225**, 40.

Burton, W. B. 1988. In Verschuur and Kellerman, p. 295.

Fujimoto, M., and Murai, T. 1984. In van den Bergh and de Boer, p. 115.

Gordon, M. A. 1988. In Verschuur and Kellerman, p. 37.

Hollenbach, D. V., and Thronson, H. A., eds. 1987. *Interstellar Processes.* Dordrecht: Reidel.

Hulsbosch, A. N. M. 1985. In van Woerden *et al.*, p. 409.

Hulsbosch, A. N. M., and Wakker, B. P. 1988. *Astron. Astrophys. Supp.*, **75**, 191.

King, I. R. 1976. *The Universe Unfolding.* San Francisco: W. H. Freeman and Co.

Kulkarni, S. R., Blitz, L., and Heiles, C. 1982. *Astrophys. J. (Letters)*, **259**, L63.

Lockman, F. J. 1984. *Astrophys. J.*, **283**, 90.

Lockman, F. J., Hobbs, L. M., and Shull, J. M. 1986. *Astrophys. J.*, **301**, 380.

Mathewson, D. S., and Ford, V. L. 1984. In van den Bergh and de Boer, p. 125.

Mihalas, D., and Binney, J. 1981. *Galactic Astronomy.* San Francisco: W. H. Freeman and Co.

Morgan, W. W., Sharpless, S., and Osterbrock, D. E. 1952. *Astron. J.*, **57**, 3.

Savage, B. D. 1986. In Bregman and Lockman, p. 17.

———. 1987. In Hollenbach and Thronson, p. 123.

Scoville, N. Z., and Sanders, D. B. 1987. In Hollenbach and Thronson, p. 21.

Shaver, P. A., McGee, R. X., Newton, L. M., Dans, A. C., and Pottasch, S. R. 1983. *Mon. Not. Roy. Astron. Soc.*, **204**, 53.

Solomon, P. M., Rivolo, A. R., Barrett, J., and Yahil, A. 1987. *Astrophys. J.*, **319**, 730.

Spitzer, L. 1956. *Astrophys. J.*, **124**, 20.

Turner, B. E., and Ziurys, L. M. 1988. In Verschuur and Kellerman, p. 200.

Verschuur, G. L., and Kellerman, K. I., eds. 1988. *Galactic and Extragalactic Radio Astronomy.* New York: Springer, 2nd ed.

Wielen, R. 1974. *Publ. Astron. Soc. Pacific*, **86**, 341.

Woerden, H. van, Allen, R. J., and Burton, W. B., eds. 1985. *The Milky Way Galaxy.* I.A.U. Symposium no. 106. Dordrecht: Reidel.

Woerden, H. van, Schwarz, U. J., and Hulsbosch, A. N. M. 1985. In van Woerden *et al.*, p. 387.

CHAPTER 5

PHOTOMETRIC COMPONENTS
IN DISK GALAXIES

Pieter C. van der Kruit

Various properties of galaxies can be used in comparing our Galaxy to external ones, and many of these will be used in this book to see where the Galaxy we live in fits into the variety we observe among such objects in the Universe. This in itself is justification enough to make such comparative studies. However, there is a second, possibly more important reason: by investigating properties of external galaxies, we may be better able to interpret observations in our own stellar system, where our internal position is sometimes a major disadvantage. In my chapters in this book, I will start out by describing what observations can be done in galaxies and how these are performed. Then I will review the findings of such studies, see what relevant observations are available for our Galaxy, and then compare these two sets. I will begin with the distribution of stars, as revealed by studies of surface photometry. Since our Galaxy is a spiral galaxy, I will in general not describe elliptical galaxies. Distances are based on a Hubble constant of 75 km sec^{-1} Mpc^{-1}. A related review in the Saas-Fee series is the one by Kormendy (1982).

5.1   METHODS OF SURFACE PHOTOMETRY

The classical technique is photographic surface photometry. The basic principle is simple: we take a photographic exposure and scan the relevant part of the plate to provide a digital record. It is no surprise that we have been able to use this technique on a large scale only since the 1970's, when fast and accurate scanning machines were being built, and large computers became available to control the scanning equipment and perform the off-line data reduction. Although this kind of work was done in earlier decades, it could not be done on a large scale. With a few notable exceptions, most of these old studies provided a luminosity profile along a few axes, usually the minor and major axis. Another reason for the advances in the 1970's was the availability of high-contrast, fine-grained emulsions, such as the IIIa series of Kodak, that made it possible to work down to fainter levels and generally improve the accuracies.

The digital matrix that makes up the recording of the photographic densities over the galaxy and its surroundings often comprises 512^2 or 1024^2 picture elements ("pixels"), or sometimes even more. The photographic density simply expresses the amount of light that emerges from the plate when a light source is made to shine through it. It is on a logarithmic scale, in which 0.0 stands for 100 percent, 1.0 for 10 percent, etc. Usually densities can be measured reliably up to values of 3 or 4. The procedure for arriving at these is to calibrate the scanning machine with a set of filters of known density. The reduction consists of two steps.

First, the level of background density must be measured. This is usually done by fitting a two-dimensional low-order polynomial to a region surrounding the galaxy, where we iteratively correct for the field stars.

Second, a conversion from photographic density to intensity must be performed. For this purpose the "characteristic curve" must be known; this curve expresses the relation between the logarithm of the exposure (incident intensity times exposure time) and density. During or shortly after exposure in the telescope, therefore, a set of spots of known exposure ratio, or a wedge along which the exposure varies in a known manner, is imprinted on a separate part of the plate. It is vital that this be done with the same wavelength dependence as prescribed by the telescope–filter combination and with the same exposure time. The latter is necessary because of low-intensity reciprocity failure, which says that at faint levels, such as from the night sky, a particular exposure results in a different density than at bright levels. The linear part of the characteristic curve between under- and over-exposure is often described by the slope γ, which is about 3 for Kodak IIIa-J or IIIa-F plates.

With this curve it is possible to convert the density of galaxy + sky and of sky alone into an intensity, and hence into a ratio of galaxy surface brightness to that of the sky during the observations. To find the absolute values of surface brightness (the zero point), we compare the integrated magnitude of (a part of) the galaxy to that arrived at by photoelectric measures. The main uncertainty usually comes from the background; every plate is usually prepared by "baking" for a few hours in a nitrogen or nitrogen/hydrogen environment at something like 65°C, and this and other inhomogeneities result in nonuniform emulsion response. For practical purposes this limits photometry from a single plate to about 5 magnitudes fainter than sky, although fainter work can be done from repeated exposures. Because of the limited dynamic range of a plate, we need a combination of short and long exposures in order to measure the entire surface brightness distribution of a single galaxy.

It is an incorrect belief that luminosity scales are in practice often seriously wrong. The reality is that characteristic curves can be calculated rather easily to high accuracy. Furthermore, as was (as far as I know) first pointed out by Ivan King, at faint levels magnitude scales are independent of the value of γ, which can be seen as follows. The characteristic curve in its linear part can be written as

$$D = \gamma \log E, \tag{5.1}$$

where D is density and E exposure. If surface brightness μ is expressed in magnitudes per square arcsecond, it follows that

$$D_{sky} = -\left(\frac{\gamma}{2.5}\right) \mu_{sky} + \text{constant}. \tag{5.2}$$

If the density of sky + object is $D_{sky} + \Delta D$, and if $\Delta D \ll D_{sky}$, then

$$\log(\Delta D) = \log\left(\frac{\gamma}{\ln 10}\right) + \frac{(\mu_{sky} - \mu)}{2.5}. \tag{5.3}$$

Now, for two faint regions in the outer parts of the galaxy, the difference in surface brightness in mag arcsec^{-2} is

$$\mu_1 - \mu_2 = 2.5 \log\left(\frac{\Delta D_2}{\Delta D_1}\right) \tag{5.4}$$

and independent of γ. The basic reason for this is, of course, that magnitudes are ratios of intensity. Clearly the background subtraction remains vital, but the intensity scale is correct regardless of the density to intensity conversion.

It is worth noting that if the value of γ *is* known, then Eq. (5.3) can be put in the form

$$I - I_{sky} = I_{sky}\left(\frac{\ln 10}{\gamma}\right) \Delta D; \tag{5.5}$$

from this the actual intensity at each faint point can be found, in terms of the surface brightness of the sky.

Recently we have been able to do digital recording at the telescope with electronic detectors, particularly CCDs (charge-coupled devices). The data reduction here has some special features not encountered with photographic plates (see also Okamura 1988). The detectors suffer from sensitivity variations due to electronic and optical effects during the exposure, and noise during the transfer of the data from the chip to disk storage (readout noise). The general noise level is measured from a so-called bias frame that has collected data for the same period as the observations, but with the camera shutter closed. This is subtracted from the raw data, but does, of course, leave some noise. Then there are cosmetic effects, such as in particular "fringing" (interference in the chip) and cosmic-ray events, and columns can be affected by a bright star in the field. These are removed by linear interpolations across the relevant pixels. Then there is the serious problem of sensitivity variations from pixel to pixel, which is calibrated by dividing the observed frame by a flat-field exposure on a uniformly illuminated screen in the dome or the twilight sky. Flat-fielding and readout noise are the most important limitations. Calibration of the photometric scale is usually done by making exposures during the night on fields with standard stars.

At present the most serious shortcoming of CCDs is that the field of view is small because the commercially available chips are small. To some extent this can be overcome by placing a few CCDs next to each other, such as in the "four-shooter" at the Palomar 5-m telescope. Alternatively, we could take overlapping exposures and combine these later. Current industrial developments may make larger chips accessible soon. The great advantages of electronic detectors are the good quantum efficiency, the large dynamic range, and the fact that digital data are provided right at the telescope. The linearity is often quoted also as an advantage. However, the response of photographic plates can be calibrated in a straightforward and reliable way, whereas—as shown above—at faint levels a magnitude scale is provided independent of the slope of the characteristic curve. The necessity to digitize the plates, the limited dynamic range, and the unavoidable nonuniformities remain the most serious limitations; and for study of individual objects of limited angular extent, electronic detection will be the technique to use. The high quantum efficiency does make the use of CCDs on small telescopes possible; with them a reasonable field of view of 10 or more arcmin can be obtained.

Published photographic and CCD observations now are in excellent agreement. Begeman (1987) has compared radial profiles from photographic photometry by Wevers et al. (1986) and CCD work by Kent (1987) for three spiral galaxies (see Fig. 5.1). Not only are the general shapes in excellent agreement, but also features in the profiles reproduce between the two techniques. The largest individual difference is only 0.2 mag. Kent's profiles reach into the center, where the photographic plates were overexposed, but (mainly because of angular size) the CCD data do not go as far out in galactocentric radius. The sky background for these data is about 21 mag arcsec^{-2}; so it can be seen that the agreement continues to a level of 5 mag below sky or 1 percent of the sky surface brightness.

5.2   BULGE AND DISK LUMINOSITY DISTRIBUTIONS

Because of the brighter surface brightness, the first attempts to measure light profiles in external galaxies were performed on bulges of spiral galaxies and on ellipticals. The oldest reference I know about is Reynolds (1913), who presented a radial light profile along the major axis of the bulge ("globular nucleus") of M31. It was based on photographic exposures of up to 100 minutes on a 28-inch reflector. It is interesting that Reynolds in his paper argued on the basis of his photographs that M31 is unlikely to be a "very distant galaxy of stars," and that "the light of the nebula is derived in some measure from a central star which is too much involved in nebular matter to be visible as such." He immediately tried to fit the radial profile with an empirical function, and found a satisfactory one to be $(x + 1)^2 y = $ constant, with x being radial distance and y the "light ratio" (surface brightness on a linear scale without known zero-point). Reynolds' measurements and his fit are shown in Fig. 5.2. His largest radial extent is 6.0 mm or 6.9 arcmin. From modern surface photometry (Walterbos and Kennicutt 1987) we know that at this radius the surface brightness is about 21 B-mag arcsec^{-2}. The fitted function was later also used by Hubble in the form $I(R) = I_0(R+a)^{-2}$.

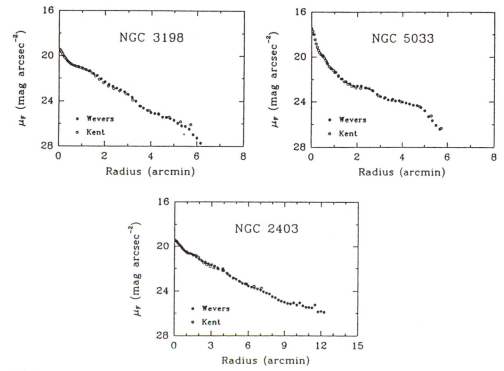

FIGURE 5.1
The radial surface brightness profiles of three spiral galaxies measured by photographic photometry (Wevers *et al.* 1986) and from CCD photometry (Kent 1987). The vertical scale is in the F-band of Wevers *et al.*, and the data from Kent have been scaled from the r-band to the F-band. The largest difference is 0.2 mag. (From Begeman 1987.)

(Here and in the following I will denote surface brightness on a linear scale, such as $L_\odot$ pc^{-2} with the symbol I, and use μ when the unit is mag arcsec^{-2}.)

The most successful empirical formula was formulated by de Vaucouleurs (1948) and is usually referred to as the "$R^{1/4}$ law." It appears to be able to represent actual profiles in bulges and ellipticals over a very wide range in surface brightness. Its equation is

$$\log\left[\frac{I(R)}{I_{\mathrm e}}\right] = -3.3307\left[\left(\frac{R}{R_{\mathrm e}}\right)^{1/4} - 1\right]. \qquad (5.6)$$

Here $R_{\mathrm e}$ is the effective radius that encloses half the total light, and $I_{\mathrm e}$ is the surface brightness at that radius. The central surface brightness then becomes $\mu(0) = \mu_{\mathrm e} + 8.3268$. The corresponding total luminosity is

$$L = 7.215\,\pi I_{\mathrm e}R_{\mathrm e}^2(b/a), \qquad (5.7)$$

where b/a is the apparent axis ratio.

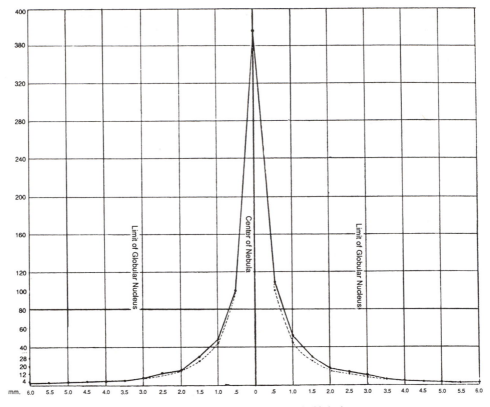

Light Curve of the Andromeda Nebula

FIGURE 5.2
The earliest observed distribution of surface brightness in the bulge of M31. The vertical
scale is linear in relative units, and the horizontal scale is in mm on the plate, where
1 mm corresponds to 69 arcsec. The dashed lines are fits with Reynolds' quadratic form
(see text). (From Reynolds 1913, Royal Astronomical Society.)

The wide applicability is surprising from a physical point of view. The corresponding volume luminosity density cannot be given analytically. Young (1976)
has given tables of various properties, including the space luminosity density.
Also he gives approximations at small and large radii, where the latter are most
useful for practical applications. With total luminosity L in $L_\odot$, R_e in pc, the
density at radius R (also in pc) is, for the spherical case in $L_\odot$ pc^{-3},

$$L(R) = 52.19 \left(\frac{L}{R_e^3}\right) \left(\frac{R}{R_e}\right)^{-7/8} \exp\left[-7.67 \left(\frac{R}{R_e}\right)^{1/4}\right]. \qquad (5.8)$$

For a flattened system with axis ratio (b/a), R should be replaced by α, where
$\alpha^2 = R^2(b/a)^2 + z^2$.

A close, alternative surface-brightness law that does have a manageable representation of the space distribution has been given by Jaffe (1983) in order to allow comparisons with dynamical models. It has the volume density distribution

$$L(R) = \frac{1}{4\pi} \left(\frac{R_{0.5}}{R} \right)^2 \left[1 + \left(\frac{R}{R_{0.5}} \right) \right]^{-2}, \tag{5.9}$$

where $R_{0.5} = 1.311 R_e$ encloses half the total emitted light in space, and the total integrated luminosity has been normalized to unity. The projected surface brightness can be calculated analytically, as can such properties as gravitational potential. Binney (1982) has devised a dynamical model that is in a sense "isothermal" and also corresponds closely to observed light profiles that can be fitted to the $R^{1/4}$ law. In his model the distribution function of stars is expressed as a function of the binding energy E, and is given by the Boltzmann formula $N(E) = N_0 \exp(-\beta E)$, where $\beta = 2.08 R_e / GM$, with M the total mass.

For ellipticals, and with less success also for bulges, there has been some use of the King model (King 1966), in which the starting point has been a distribution function of the particles that is isothermal, and in which all particles have the same mass. This distribution function is a lowered Maxwellian. Wilson (1975) extended these models to include rotation. More recently, Jarvis and Freeman (1985a,b) have constructed similar dynamical models for bulges in which rotation has been included, and the effects of the disk potential have also been taken into account. Their distribution function is

$$f(E, J) = \alpha \left[\exp(-\beta E) - \exp(-\beta E_0) \right] \exp(\gamma J), \tag{5.10}$$

where E ($< E_0$) is the energy per unit mass, and J the component of angular momentum per unit mass parallel to the axis of symmetry. Models with $\gamma = 0$ are King models. Assuming constant mass-to-light ratios M/L, Jarvis and Freeman were able to fit such models to the surface-brightness distribution and stellar kinematics of actual bulges of a few early-type systems, where the light distribution is dominated by that of the bulge. Data show that bulges are consistent with their being isotropic oblate spheroids in which flattening is mostly due to rotation.

The description of the radial surface-brightness distribution in disks followed that of bulges and ellipticals by about three decades. This is an understandable consequence of the much fainter surface brightness of the disks. There is one simple law for the radial distribution of surface brightness, namely, that of the exponential disk:

$$I(R) = I_0 \exp(-R/h). \tag{5.11}$$

The parameter h is usually referred to as the disk scale length. The integrated magnitude is

$$L = 2\pi h^2 I_0. \tag{5.12}$$

The exponential law was first described in an unpublished Harvard thesis based on observations of M33. A short report is available in Patterson (1940),

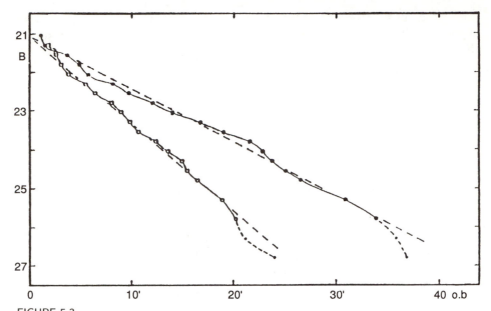

FIGURE 5.3
The surface brightness distribution of the disk of the Local Group Sc galaxy M33. The
two curves show the distributions measured along the minor and major axes separately.
The dashed lines indicate exponentials. (From de Vaucouleurs 1959b.)

and a plot of the data appears in the extensive description of light distributions
in galaxies by de Vaucouleurs (1959a, his Figure 10). The first detailed work
was performed by de Vaucouleurs in a series of papers in the late 1950's that
involved the suitable members of the Local Group: LMC, M31, and M33. This
fundamental work established once and for all that exponentials are the rule,
and occur in Sb, Sc, and Irregular galaxies. De Vaucouleurs' (1959b) exponential
light profile of the disk of M33 is reproduced from the original paper in Fig. 5.3.
Note that the data reach a level of about 27 B-mag arcsec^{-2}, which is not much
brighter than is attainable today in such studies. The two curves show the major
and minor axis profiles separately.

Many studies have in the meantime confirmed the universal applicability of
the exponential as a fitting function. It should, however, be noted that in all
galaxies for which an exponential has been fitted to the disk surface-brightness
profile, there are important, nonlocal deviations from the fit at a level of a few
tenths of a magnitude, and often the fit can be performed over an interval of a
few magnitudes (nearly equivalent to e-foldings) in the first place (*e.g.*, van der
Kruit 1987a; see also Figure 1 therein). The reason for the last point is that often
the bulge dominates the inner parts, and then there is little fall-off in surface
brightness left before it reaches the noise introduced by the sky background.
This means that the exponential is convenient to use because of its simplicity
as well as because of the limited range over which surface brightness profiles
can be measured. These facts mean that we should be hesitant to attach too

much physical meaning to the exponential shape. I will discuss the distribution
of the central surface brightness and the scale length later in this book (see
also van der Kruit 1987a); however, here I note that Freeman (1970) found that
large spiral galaxies appear to display a remarkably small range in extrapolated
central (face-on) surface brightness, namely, 21.65 ± 0.30 B-mag arcsec^{-2}.

Disks do have a finite thickness. The three-dimensional structure therefore requires a description of the vertical distribution. Van der Kruit and Searle (1981a)
proposed to use the formula for the (locally) self-gravitating, isothermal sheet
for this, so that the full description becomes

$$L(R, z) = L(0, 0) \exp(-R/h) \operatorname{sech}^2(z/z_0). \tag{5.13}$$

The face-on surface brightness follows from

$$I(R) = 2z_0 L(R, 0), \tag{5.14}$$

and the vertical scale parameter z_0 relates to the surface density $I(R)(\mathcal{M}/L)$
and the velocity dispersion $\langle W^2 \rangle^{1/2}$ as

$$\langle W^2 \rangle = \pi G I(R) z_0 (\mathcal{M}/L). \tag{5.15}$$

The reason for choosing the isothermal sheet was the observation in the
solar neighborhood (*e.g.*, Wielen 1977) that stars that are older than a few
Gyr all have roughly the same velocity dispersion, so that the mix of disk stars
should be dominated—at least moderately far from the plane—by essentially
a single stellar population with a single velocity dispersion. This component is
the "old-disk population" in terms of the population scheme put forward at the
famous Vatican symposium on stellar populations in 1957. Furthermore, earlier
observations (*e.g.*, van der Kruit 1979) had indicated that disks have roughly
exponential z-profiles with the scale parameter independent of galactocentric
distance. The latter agrees with the isothermal disk, because the sech^2 has the
property that it approaches an exponential with scale height $z_0/2$ for large z:

$$\operatorname{sech}^2(z/z_0) = 4 \exp(-2z/z_0) \quad \text{for} \quad z/z_0 \gg 1. \tag{5.16}$$

Actual fits of the proposed three-dimensional distribution to photometry of
edge-on galaxies with little or no bulge confirmed the applicability and in particular the remarkable property that z_0 is constant with R (see Fig. 5.4). However,
this description refers only to the old-disk population, and ignores effects from
young Population I stars and dust.

It was also found earlier that disks have rather sharp outer edges or trun-
cations (van der Kruit 1979; see also Fig. 5.4). From a sample of seven edge-on
spirals, van der Kruit and Searle (1982a) found that this radius $R_{\max}$ occurs at
4.2 ± 0.6 radial scale lengths h. This was not known from luminosity profiles of
inclined or face-on systems, and I proposed (1988) that it resulted from relatively small deviations from circular symmetry in the stellar disks. This can, for
example, be seen in the isophote map of the face-on spiral NGC 628 (Fig. 5.5,

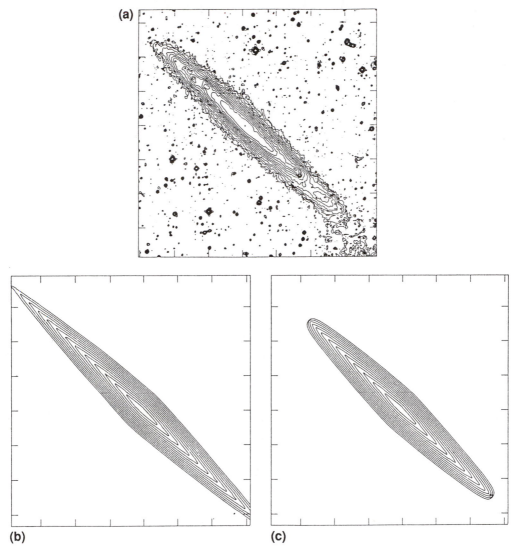

FIGURE 5.4

Optical isophotes of the edge-on, pure disk galaxy NGC 4244. (Top) The observed contour map after subtraction of the responses of some foreground stars that were superimposed on the image of the galaxy. The contours are spaced by 0.5 mag, and the faintest contour is at about 27.5 B-mag arcsec^{-2}. (Middle) Isophote map of an edge-on model disk with an exponential luminosity distribution in the radial direction and that of an isothermal-sheet approximation in the vertical direction. Although the inner isophotes are fitted very well (except in the plane, because of dust and young stars), the model extends to larger galactocentric distances than the observations. (Bottom) The same model as in the middle map here, but with a sharp edge. The observations are now represented very well by the model. (From van der Kruit and Searle 1981a.)

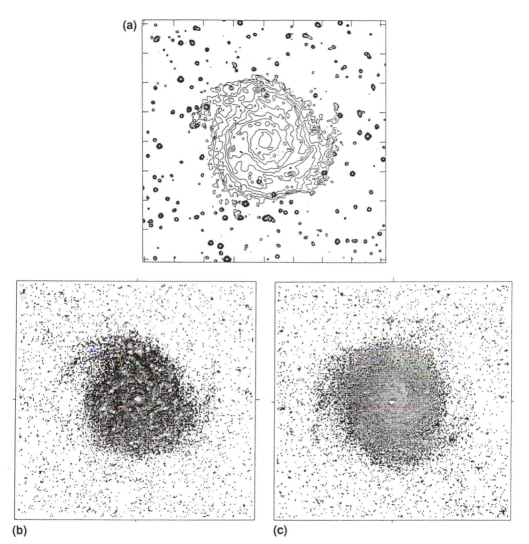

FIGURE 5.5

(Top) Isophote map of the face-on spiral NGC 628 after removal of some foreground stars. The contours are spaced by 0.5 mag, and the faintest contour is at 26.5 B-mag arcsec^{-2}. (From Shostak and van der Kruit 1984.) (Middle) Gray-scale representation of the color distribution in NGC 628 in pixels of 5 arcsec. The dark pixels have $(U - B)$ about 0.65 and the empty ones -0.85. Areas in which the surface brightness was too faint or where one of the plates was overexposed (in the nucleus and on bright field stars) were also left white. Note that the spiral arms show up very well because of the blue colors of the H$_{\mathrm{II}}$ regions. (Bottom) The same as in the middle picture, but now the range is $(B - V)$ from 1.5 (dark) to 0.3 (white). The effect of the H$_{\mathrm{II}}$ regions is now much less pronounced, and there appears a fairly uniform color across the disk.

top), where the outer three isophotes are much more closely spaced than the inner ones. On the other hand, these isophotes do deviate from pure circles. These deviations correspond to variations in R_{max} of order 10 percent with azimuthal direction. As a result, radial profiles obtained by azimuthal averaging of the light distribution fail to show the sharp declines. Beyond the "truncation radius," the radial e-folding distance of the light distribution drops to values generally less than 1 kpc.

In Fig. 5.5 an example of surface photometry of a face-on spiral with very little bulge is given, using the ScI galaxy NGC 628. Also, color distributions are shown. The spiral structure outlined by the H$_{II}$ regions shows up very well, especially in $(U - B)$. From such data the radial profiles can be calculated by azimuthal averaging of the intensities, as is shown in Fig. 5.6. Note the absence of a sharp edge, although (as was pointed out above) these profiles can be seen from the contour spacings in the maps, but have been smoothed out by the procedure. There is no significant color variation in $(U - B)$ with radius, and a reddening is observed in $(B - V)$ with an amplitude of 0.25 mag. This is marginally significant. The question of color variations has been investigated in a sample of 21 galaxies by Wevers (1984; see also Wevers *et al.* 1986). The result is that a few radial color gradients are indeed seen, but there certainly is no systematic trend in these, and in most disks there is no evidence at all for radial color gradients. This observation is of great interest, because it seems to indicate that at all radii there is the same mix of stars of various ages, and therefore no radial gradient in the time-dependence of the star-formation rate.

5.3   COMPONENT SEPARATION IN DISK GALAXIES

The luminosity laws described above make it possible to decompose a radial surface-brightness profile of a spiral galaxy into a disk component and a bulge component. The usual procedure is to select first that region of the profile that appears dominated by the exponential disk, and to estimate initial parameters for μ_0 and h. Subtraction of this initial disk model then gives rise to a first guess of the parameters μ_e and R_e for the bulge. Least-squares or χ^2 minimization techniques are then used to find a final solution to the observed profile by variation of the four parameters. Of course, this method relies heavily on the *a priori* assumption that these fitting functions are exact descriptions of the distributions within the two components.

For most galaxies the procedure works quite satisfactorily. Schombert and Bothun (1987) have recently studied it in detail both on artificial light profiles (with realistic noise introduced) and on those of actual galaxies. They find that for the bulge component, errors in effective radius and surface brightness occur in a way that conserves the total luminosity, and these errors are, of course, worse in systems with small bulges. In galaxies strongly dominated by the bulge, this procedure tends to underestimate the disk central surface brightness by 0.2

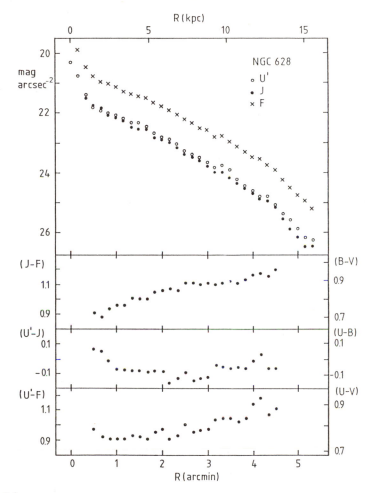

FIGURE 5.6

Radial profiles of surface brightness and color in the disk of NGC 628. At the top are three profiles in U' (about standard U), J (from IIIa-J plates and about B), and F (from IIIa-F plates and somewhat redder than V). Note the exponential declines and the absence of a sharp truncation. At small R the increase due to the small bulge can be seen. The bottom three profiles are the three color distributions. The colors have been translated into $(U - B)$ and $(B - V)$ at the right. (From Shostak and van der Kruit 1984.)

to 0.3 mag, and to overestimate the scale length by 10 to 30 percent. Fig. 5.7 illustrates these effects.

Fig. 5.8 gives a few examples of component decomposition in actual CCD luminosity profiles, also from Schombert and Bothun. Notice first that the bulges dominate usually only in the very inner region, even in the relatively strong bulge of NGC 2565. Another feature of the $R^{1/4}$ law is that at large radii it will eventually always dominate the disk contribution. Further it can be seen that disks seldom conform exactly to the exponential description in a detailed fashion.

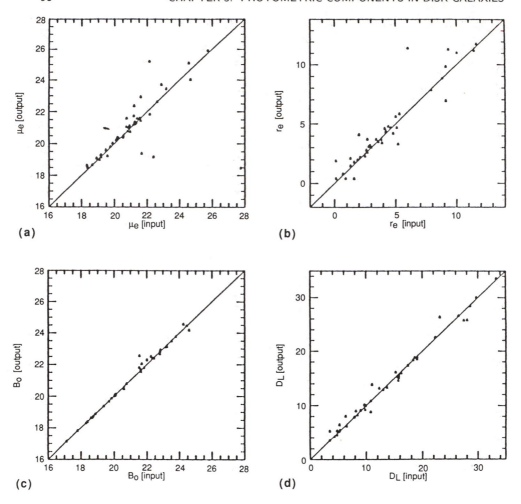

FIGURE 5.7
Results of component decomposition in artificial-galaxy surface-brightness profiles. The horizontal axes show the input parameters, and the vertical ones those derived by the fitting procedure. Parts (a) and (b) refer to the $R^{1/4}$ bulge; they compare μ_e and r_e (in our notation R_e). Parts (c) and (d) refer to the exponential disk; they compare B_0 (in our notation μ_0) and the scale length D_L (in our notation h). (From Schombert and Bothun 1987.)

FIGURE 5.8 (Facing page)
Results of component separation in a few galaxies. The $R^{1/4}$ bulge, the exponential disk, and the sum of the two model components are all indicated. (From Schombert and Bothun 1987.)

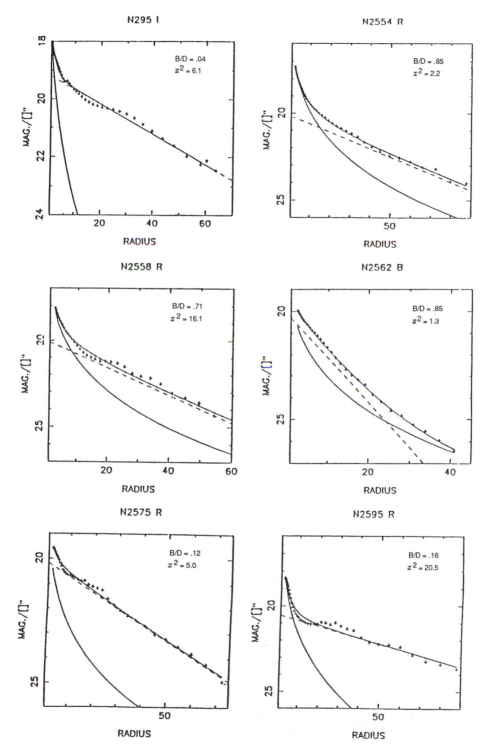

Most of this departure from a straight line in the plots is a result of structure in the disk, such as spiral arms or bars, but quite often it also is an intrinsic feature that is uncorrelated with such components. This stresses the limited value of the fitting function and the point already made that we should not attach too much physical significance to the exponent.

The reason for attempting component separation in the first place is that studies of our Galaxy have indicated that it—and therefore other spiral galaxies as well—can be described as consisting of two separate populations: halo and disk. In a first approximation these can be characterized by discretely different spatial distributions, kinematics, and distributions of stellar ages and of heavy-element abundances. The reasonable success of decomposition of radial luminosity profiles confirms that this basic picture also applies to spiral galaxies in general. However, these studies have not yet asked whether the two populations have different degrees of flattening, and for that purpose we need to look at edge-on galaxies, where a sizable part of the total solid angle is covered by the bulge alone and *vice versa*.

As examples I will take two edge-on galaxies studied by myself and Searle (1981b, 1982b), namely, the disk-dominated galaxy NGC 891 and the bulge-dominated system NGC 7814. Because these studies also addressed color distributions, I will also discuss this aspect. Fig. 5.9 illustrates the procedure for NGC 891. At the left we have the isophote map of the galaxy after subtraction of the responses of a large number of foreground stellar images. The regions away from the dust lane in the plane and the central bulge have been used to model the old-disk population. The best-fitting model, which is that of an exponential, locally isothermal, truncated disk with constant thickness seen edge-on, is shown in the middle illustration. For a distance of 9.5 Mpc ($H_0 = 75$ km sec^{-1} Mpc^{-1}), the parameters are $L(0,0) = 2.4 \times 10^{-2} L_\odot$ pc^{-3} (in about the B-band), $h = 4.9$ kpc, $z_0 = 0.99$ kpc, and $R_{max} = 21$ kpc; and the total old disk luminosity is then $6.7 \times 10^9 \ L_\odot$. At the right this model has been subtracted from the observations, and the bulge remains. It turns out that the minor-axis profile can be fitted well with an $R^{1/4}$ law, with $R_e = 2.3$ kpc in the same band. The isophotes have a mean axis ratio of 0.6, but the outer ones are definitely flatter than the inner ones. The integrated luminosity then is $1.5 \times 10^9 \ L_\odot$. The old disk contains 82 percent of the light in old stars.

Fig. 5.10 illustrates the color data on NGC 891. At the left is the total light. Note that the dust lane is red (dark), but that in its center there are clear indications of the bluer young Population I. At the right we see the bulge color distribution after disk subtraction. There is an obvious bluing toward larger radii. The color of the disk model is $(U - B) = 0.2$ and $(B - V) = 0.8$, which is indeed consistent with that of an old disk population. The color range in the observed part of the bulge is $(U - B) = 0.7$ to -0.1 and $(B - V) = 1.0$ to 0.7. These colors have zero-point uncertainties of about 0.2 mag, and the color variation itself is uncertain by about 0.3 mag. The color gradient is most likely a result of a radial variation in metal abundance. This is the combined result of three possible effects: (1) For lower abundance the effective temperature on the giant

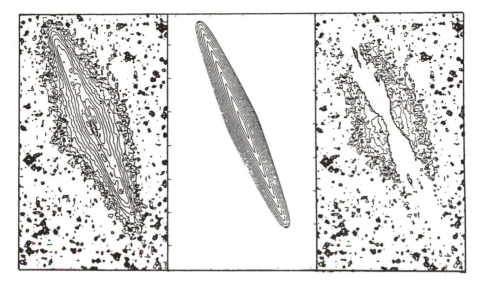

FIGURE 5.9

Isophote maps for NGC 891 in a band that is close to standard B. The faintest contour is at about 26.5 mag arcsec^{-2}, and the interval is 0.5 mag. At the left we have the total galaxy light after subtraction of foreground stars, in the middle the model for the old disk population, and at the right the light that remains after subtraction of the disk model from the observations. (After van der Kruit and Searle 1981b.)

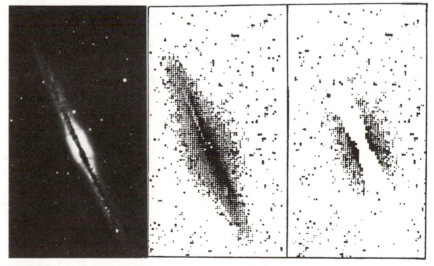

FIGURE 5.10

Gray-scale representation of the color distribution in NGC 891. The range of colors is from about $(U - V) = 0.6$ (light) to 1.4 (black). At the left is the total galaxy light, and at the right that of the bulge after disk subtraction. (After van der Kruit and Searle 1981b.)

branch increases. (2) For a lower metallicity there is among old populations a horizontal branch that reaches further toward blue colors (but beware of the so-called second-parameter effect in globular-cluster HR diagrams). (3) Line-blanketing, which is more severe at blue and UV wavelengths, is less effective in low-abundance stars. The range of colors above is comparable to integrated colors of Galactic globular clusters and therefore indicates a similar metallicity range, say, from [Fe/H] about −0.5 to −2 or so.

In Fig. 5.11 the procedure is illustrated for NGC 7814. Here the bulge dom-inates, and therefore a model is made for the bulge first. This model has an $R^{1/4}$ luminosity distribution and a uniform flattening of 0.57. In the B-band $R_e = 2.2$ kpc and $\mu_e = 22.1$ mag arcsec^{-2}; the total bulge luminosity, then, is 1.6×10^{10} $L_\odot$. The disk light remaining after subtraction of the bulge model from the observations is shown in the lower part. The center of the galaxy is missing, because the plates were overexposed there in this photographic work. The rough parameters for the disk are $L(0,0) = 6.6 \times 10^{-4}$ $L_\odot$ pc^{-3}, $h = 8.4$ kpc, $z_0 = 2.0$ kpc, and $R_{\max} = 18$–20 kpc, so that the total old disk luminosity is 1.2×10^9 $L_\odot$. The old disk now contains only 7 percent of the total light in old stars. That z_0 has a larger value than is usual among disks in later type spirals is likely to result from the dominance of the gravitational field of the rounder bulge, even in the plane of the disk.

The color distribution in the bulge shows again a progressive bluing with radius, but is remarkable in the sense that the isochromes have the same shape as the isophotes. The observed range is $(U - B) = 0.6$ to 0.3 and $(B - V) = 1.3$ to 0.5, with the same uncertainties as for NGC 891. The magnitude of the variation is again that observed among Galactic globular clusters. The disk colors are very uncertain, and are about $(U - B) = 0.6$, $(B - V) = 1.1$, again within the uncertainties consistent with an old disk population. As was true for NGC 891, the minor-axis profiles in three observed colors can be fitted rather well with $R^{1/4}$ laws, albeit with different slopes as a function of wavelength.

The general inference that can be drawn from this work is that the light distributions in spiral galaxies can apparently be described very well with only two basic components: a bulge and a disk. These two basic components have two discretely different flattenings. In the bulge we furthermore have found evidence for color gradients, which are interpreted as abundance gradients, and this is a feature that appears typical for bulges (see Wirth and Shaw 1983). Disks, on the other hand, are in general uniform in color with galactocentric distance. In this book, a considerable body of evidence will be presented that points to a third component, intermediate between disk and halo in flattening, metallicity, and

FIGURE 5.11 (Facing page)
Illustration of the component decomposition of NGC 7814, at the right as isophotes and at the left in intensity grey-scale. The faintest isophote is at about 26 mag arcsec^{-2}, and the interval is 0.5 mag. At the top we have again the total light, and in the middle is the model for the bulge. This model has an $R^{1/4}$ luminosity profile, and the isophotes have constant flattening. The bottom pictures show the disk remaining after subtraction of the bulge model from the observations. (After van der Kruit and Searle 1982b.)

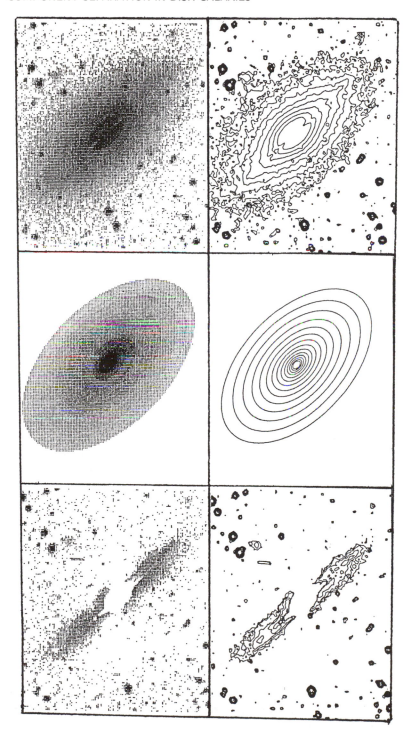

kinematics, sometimes referred to as the "thick disk." Detailed analysis of the photometric data of NGC 891 (van der Kruit 1984) shows that the bulge surface-brightness distribution can be equally well represented by a bulge with somewhat changing axis ratio, or by a combination of a bulge with constant axis ratio and a component rather similar to the "thick disk" proposed for our Galaxy. On the other hand, we find no evidence whatsoever of such a component in NGC 7814. My personal view is that this leaves unaffected the basic conclusion from Fig. 5.6, that after disk subtraction we have a stellar distribution that has a characteristic flattening that is noticeably less than in the disk. Furthermore, in our Galaxy the "thick disk" makes up only about 10 percent or so of the disk light, and if associated with disk formation it can equally well be seen as a subcomponent of the disk population, which, after all, goes down at the other side of the range of flattenings to a thickness of about 100 pc for Extreme Population I.

5.4   SURFACE PHOTOMETRY OF OUR GALAXY

The distribution of the stars in our Galaxy has traditionally been studied using star counts, starting with the work of William Herschel two centuries ago, and continued around the turn of this century by Kapteyn and others. It would be of obvious interest to supplement this work with surface photometry analogous to that in external systems. In particular this would allow a much more accurate value for the disk scale length than is possible with currently available star counts because at present all-sky counts are limited to stars brighter than about magnitude 11 (the program to provide a catalogue of guide stars for the Hubble Space Telescope will improve this situation), and faint counts are available in selected areas only, whereas the general distribution of stars in the intermediate-magnitude range is most sensitive to the disk scale length. These relevant stars do contribute significantly to the integrated surface brightness.

Surface photometry of the Galaxy is very difficult as a result of the low surface brightness, except maybe in the Galactic plane or the brighter parts of the bulge, and because it is not easy to obtain photometric stability over the very large angular scales involved. From the Earth, furthermore, we must worry about the zodiacal light, at least for higher Galactic latitudes, where because of absence of absorption the interesting information on Galactic structure can be obtained. For example, the surface brightness of the integrated starlight in the Galactic poles is in the B-band about 30 S_{10}, which means the equivalent of 30 stars of magnitude 10 per square degree. This is, in units more familiar to us, equal to about 24.8 B-mag arcsec^{-2}. For comparison, the brightest parts of the Galactic plane have surface brightnesses of about 800 S_{10} (about 21.2 mag arcsec^{-2}). On the other hand, the zodiacal light in the ecliptic poles provides about 80 S_{10} and about 210 S_{10} in the *Gegenschein* directly opposite the Sun on the sky. Obviously the (to us) interesting part of the integrated Galactic starlight is fainter than or comparable to the zodiacal light, and is therefore very difficult to study from the Earth.

Now, fortunately, the inverse problem occurs for researchers of the zodiacal light itself, for whom the Galactic background light is a nuisance. Therefore the Pioneer 10 (and 11) spacecraft that were sent to Jupiter and were equipped with a wide-field mapping capability in the optical were used to study this matter. Beyond the asteroid belt it turned out that there was no perceptible zodiacal light (except, of course, in directions near to that of the Sun), and the spacecraft was then used on its continuing journey to and beyond Jupiter to map the background starlight, in particular to assist in zodiacal-light studies from the Earth. The instrument used for this was a 1-inch telescope and an imaging photopolarimeter that in the configuration used provided a field of view of $2° \times 7°$. After many repeated scans the data resulted in two maps of the sky in the red and the blue with an angular resolution of $2°$. Many scans remain unreduced, and only a set from Pioneer 10 is available. After smoothing to an angular resolution of $8°$, the maps in Fig. 5.12 result, presented there as isophote maps in mag arcsec^{-2}.

Positional data by themselves can in principle not be used to derive length scales. It turns out that the Pioneer data can be used to constrain the ratio of radial scale length h to vertical scale parameter z_0 (van der Kruit 1986). Estimates for z_0 can be obtained from star-count studies toward the Galactic poles. Of course, the areas of high extinction $[E(B - V) > 0.09$ mag$]$ have been excluded, restricting the analysis essentially to latitudes above about $20°$. The exclusion of low latitudes is the reason that the ratio of scale length to the solar Galactocentric radius cannot be extracted from the data. It turns out also that the bulge is too faint to contribute sufficiently for a detailed study. The method of analyzing the Pioneer data (van der Kruit 1986) has been to use the Bahcall and Soneira (1984) computer code to derive surface brightnesses for various combinations of length scales.

The major result, then, is a definite value for the ratio h/z_0 of 8.5 ± 1.3. Using the value for the exponential old-disk scale height in the solar neighborhood from Gilmore and Reid (1983) and others of 350 ± 50 pc, which equals 0.5 z_0, results then in a radial scale length h of the disk of 5.5 ± 1.0 kpc. There are various other, independent arguments that can be used to derive values for h, and I have concluded from a detailed discussion of these (1987b) that the best value is 5.0 ± 0.5 kpc. Other results of the Pioneer 10 data are the central surface brightness of the Galactic disk (now the total disk, and not the old-disk population only), inferred from the surface brightness at the Galactic poles and a value for the scale length, of 22.1 ± 0.3 B-mag arcsec^{-2}, and the color index of the disk at the solar position of $(B - V) = 0.84 \pm 0.15$. The latter is interestingly red, and may indicate that the contribution from young Population I to the surface brightness is not the major one. It may be atypical for the disk in general, although Sb galaxies do have disks with similar colors. The uncertainty of 0.15 mag is large, but can in principle be improved using the unreduced data from Pioneer 10 and 11. The total luminosity of the disk is $(1.8 \pm 0.3) \times 10^{10}$ $L_\odot$. For the old disk the value of $L(0,0)$ is about 5×10^{-2} $L_\odot$ pc^{-3}, and the total luminosity is about 1×10^{10} $L_\odot$.

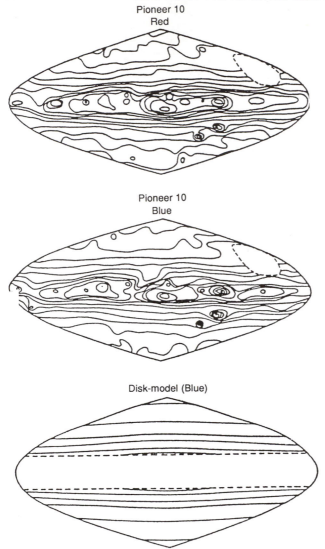

Pioneer 10
Red

Pioneer 10
Blue

Disk-model (Blue)

FIGURE 5.12
Isophote maps of the distribution of surface brightness from the integrated starlight from our Galaxy, as derived by Pioneer 10 on its way to and beyond Jupiter. Stars brighter than $V = 6.5$ have been removed. The direction toward the Galactic center is in the middle of the picture. The angular resolution is $8°$, and the contour interval is 0.25 mag. In the red band the faintest thick contour corresponds to 24 V-mag arcsec^{-2} and in the blue band to 24 B-mag arcsec^{-2}. The missing part is the general direction of the Sun as seen from Jupiter around Pioneer 10 encounter. The Galactic bulge, Carina spiral-arm region, and the Magellanic Clouds are prominent. (From van der Kruit 1986.)

The value for R_{max} of the disk of the Galaxy is more difficult to derive. Young stars are observed in the direction opposite the Galactic center to distances of at least 22 kpc (Chromey 1979), and a similar distance is found for HII regions. This gives an estimate for R_{max} of 20–25 kpc, in agreement with the observed ratios R_{max}/h of 4.2 ± 0.6 in external edge-on galaxies.

Surface photometry of the bulge has been attempted photoelectrically by de Vaucouleurs and Pence (1978), and has been successful in the brighter parts at low latitudes. They compared the projected surface brightness on the sky to that of an $R^{1/4}$ bulge, and came up with the values $R_e = 2.7$ kpc and $\mu_0 = 15.1$ B-mag arcsec^{-2}. From comparison with external galaxies, we can estimate a value for the flattening from the kinematics, using the relation between V_m/σ and the flattening, where V_m is the maximum rotation velocity of the bulge, and σ the central velocity dispersion (Kormendy and Illingworth 1982). Current best values are $V_m = 60 \pm 30$ km sec^{-1} and $\sigma = 110 \pm 10$ km sec^{-1} (e.g., Freeman 1987), and lead to $b/a = 0.7 \pm 0.15$. The total luminosity then is about 3×10^9 $L_\odot$. The old-disk population contains about 80 percent of the total light in old stars, comparable to the number for NGC 891.

In Chapter 12, I will discuss in more detail the statistical distribution of luminosity parameters, especially those of the disks. Here I want to make two remarks. The first is that the photometric parameters of our Galaxy are rather similar to those of NGC 891, which is classified as Sb. This similarity is discussed more extensively in van der Kruit (1984). The scale length of the disk of M31 is slightly larger, namely, 6.0±0.5 kpc (Walterbos and Kennicutt 1987). The second is that we can use these data to constrain the Hubble constant by comparing these values to those for the largest spirals in the Virgo cluster (van der Kruit 1986; photometry by Watanabe 1983). The six largest Sb galaxies in the Virgo cluster have a scale length of 52±5 arcsec, and the five largest Sc galaxies 50±5 arcsec. If our Galaxy is comparable to these systems, the distance to the Virgo cluster would be 20 ± 3 Mpc (from Sb galaxies) or 21 ± 3 Mpc (Sc galaxies), using $h = 5.0 \pm 0.5$ kpc. The scale length of M31 leads to 24 ± 3 Mpc. For a distance of 22 ± 4 Mpc with V_{rad}(Virgo) $= 1000 \pm 50$ km sec^{-1} and a Local Group infall of 330±40 km sec^{-1}, this results in a Hubble constant H_0 of 65±10 km sec^{-1} Mpc^{-1}. Comparing our Galaxy and M31 to galaxies with smaller scale lengths results, of course, in smaller values for H_0. We see already that these two galaxies in the Local Group probably rank among the largest spirals in the Local Supercluster. A Hubble constant of 100 km sec^{-1} Mpc^{-1} makes these probably the two largest in this volume, a very unlikely and unsatisfactory result.

It is an interesting fact that the surface brightness of the background integrated starlight is almost entirely (for more than 99 percent) provided by the stars brighter than magnitude 20 or so. In fact, before the Pioneer measurements this was the way in which it was estimated (e.g., Roach and Megill 1961), using the extensive work earlier this century on the Plan of Selected Areas (e.g., van Rhijn 1925). The difficulty was, however, that the old magnitude scales were unreliable. For historical purposes it is of interest to see just how reliable these are. This can be checked by calculating the star counts from a Bahcall and Soneira-type model of the stellar distribution in the Galaxy, using the parameters for h, z_0, and the local disk luminosity function as given above in agreement with the Pioneer surface brightness. In Fig. 5.13 various counts predicted from the model are compared with the ones given by van Rhijn (1925) as a function of latitude at a number of magnitudes. From this figure it is obvious that the scales go systematically wrong at fainter magnitudes, in the sense that the magnitudes estimated by van Rhijn are too bright. The difference amounts to 0.5–0.75 of a magnitude at the limits of the counts. It is true, however, that the counts are very consistent as a guide to relative variations that result from the stellar distribution in space.

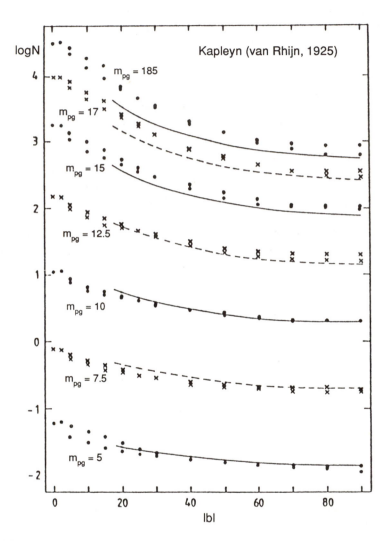

FIGURE 5.13
Comparison of the Kapteyn star counts in the Selected Areas reported in van Rhijn (1925) to the model of the Galaxy with the parameters derived in this section. The two values at each position come from positive and negative latitudes. (From van der Kruit 1986.)

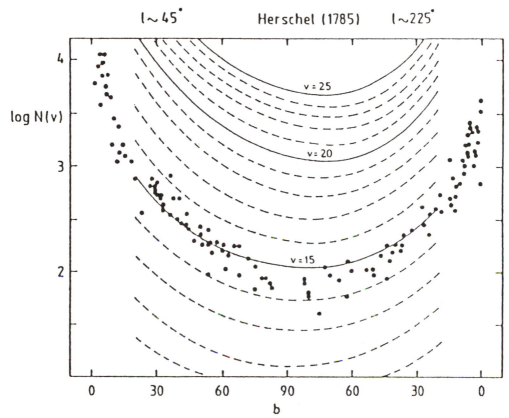

FIGURE 5.14
Comparison of the star counts made by Herschel (1785) with predictions from the Galaxy
model with the parameters derived in this section. This great circle crosses the Galactic
plane at longitudes about 45° and 225° and misses the poles by about 5°. Herschel
counted stars to a limiting magnitude V of about 15. (From van der Kruit 1986.)

    It is also of interest to compare the model for the Galaxy with the counts
that Herschel made two centuries ago. The necessary information can be found
in Herschel (1785), where he publishes his famous cross-cut of the "Sidereal
System." In his analysis he assumed that his telescope was able to see all the
way to the edges of the Sidereal System, and that therefore the number of stars
seen in a particular part of the sky could be used to calculate the distance of
this edge in that direction. The procedure for working back from this cross-cut is
described in more detail in van der Kruit (1986). The result is given in Fig. 5.14,
where we can see that Herschel observed consistently to a limiting magnitude V
of about 15. This is a remarkable achievement.

REFERENCES

Bahcall, J. N., and Soneira, R. M. 1984. *Astrophys. J. Supp.*, **55**, 67.

Begeman, K. 1987. Ph. D. thesis, Univ. of Groningen.

Binney, J. 1982. *Mon. Not. Roy. Astron. Soc.*, **200**, 951.

Chromey, F. R. 1979. *Astron. J.*, **84**, 534.

Freeman, K. C. 1970. *Astrophys. J.*, **160**, 811.

———. 1987. *Ann. Rev. Astron. Astrophys.*, **25**, 603.

Gilmore, G., and Carswell, B., eds. 1987. *The Galaxy.* Dordrecht: Reidel.

Gilmore, G., and Reid, N. 1983. *Mon. Not. Roy. Astron. Soc.*, **202**, 33.

Herschel, W. 1785. *Phil. Trans.*, **75**, 213.

Jaffe, W. 1983. *Mon. Not. Roy. Astron. Soc.*, **202**, 995.

Jarvis, B. J., and Freeman, K. C. 1985a. *Astrophys. J.*, **295**, 314.

———. 1985b. *Astrophys. J.*, **295**, 324.

Kent, S. 1987. *Astron. J.*, **93**, 816.

King, I. R. 1966. *Astron. J.*, **71**, 64.

Kormendy, J. 1982. In Martinet and Mayor 1982, p. 113.

Kormendy, J., and Illingworth, G. 1982. *Astrophys. J.*, **256**, 460.

Kruit, P. C. van der. 1979. *Astron. Astrophys. Supp.*, **38**, 15.

———. 1984. *Astron. Astrophys.*, **140**, 470.

———. 1986. *Astron. Astrophys.*, **157**, 230.

———. 1987a. *Astron. Astrophys.*, **173**, 59.

———. 1987b. In Gilmore and Carswell, p. 27.

———. 1988. *Astron. Astrophys.*, **192**, 117.

Kruit, P. C. van der, and Searle, L. 1981a. *Astron. Astrophys.*, **95**, 105.

———. 1981b. *Astron. Astrophys.*, **95**, 116.

———. 1982a. *Astron. Astrophys.*, **110**, 61.

———. 1982b. *Astron. Astrophys.*, **110**, 79.

Martinet, L., and Mayor, M., eds. 1982. *Morphology and Dynamics of Galaxies.* Geneva: Geneva Observatory.

Okamura, S. 1988. *Publ. Astron. Soc. Pacific*, **100**, 524.

Patterson, F. S. 1940. *Harvard Bull.*, no. 914, p. 9.

Reynolds, R. H. 1913. *Mon. Not. Roy. Astron. Soc.*, **74**, 132.

Rhijn, P. J. van. 1925. *Publ. Groningen Astron. Obs.*, no. 43.

Roach, F. E., and Megill, L. R. 1961. *Astrophys. J.*, **133**, 228.

Schombert, J. M., and Bothun, G. D. 1987. *Astron. J.*, **93**, 60.

Shostak, G. S., and Kruit, P. C. van der. 1984. *Astron. Astrophys.*, **132**, 20.

Vaucouleurs, G. de. 1948. *Ann. d'Astrophys.*, **11**, 247.

———. 1959a. *Handbuch der Physik*, **53**, 511.

———. 1959b. *Astrophys. J.*, **130**, 728.

Vaucouleurs, G. de, and Pence, W. D. 1978. *Astron. J.*, **83**, 1163.

Walterbos, R., and Kennicutt, R. C. 1987. *Astron. Astrophys.*, **198**, 61.

Watanabe, H. 1983. *Annals Tokyo Obs.*, 2nd Ser., **19**, 121.

Wevers, B. M. H. R. 1984. Ph.D. thesis, Univ. of Groningen.

Wevers, B. M. H. R., Kruit, P. C. van der, and Allen, R. J. 1986. *Astron. Astrophys. Supp.*, **66**, 505.

Wielen, R. 1977. *Astron. Astrophys.*, **60**, 263.

Wilson, C. P. 1975. *Astron. J.*, **80**, 175.

Wirth, A., and Shaw, R. 1983. *Astron. J.*, **88**, 171.

Young, P. 1976. *Astron. J.*, **81**, 807.

CHAPTER 6

THE DISTRIBUTION OF MASS

Ivan R. King

Our Galaxy reveals much of its structure through its gravitational field, to which the most important clue is the rotation curve: the rotational velocity as a function of distance from the center—or, more precisely, the velocity V_c of a hypothetical object that follows a circular orbit. This is directly connected with the potential function ψ by the relation

$$\frac{\partial \psi}{\partial R} = \frac{V_c^2}{R}.$$

(6.1)

In addition, as was explained in Chapter 4, the assignment of distances to radio-emitting interstellar gas depends on a knowledge of the rotation curve.

6.1   THE FORMULAS OF GALACTIC ROTATION

Suppose that everything is moving in circular orbits around the Galactic center, with a velocity curve $V(R)$. (We drop the subscript c of the preceding paragraph.) Because of the differential rotation of the Galaxy, an object at another point will in general have both radial and transverse components of motion with respect to the solar neighborhood. We will derive expressions for these.

Let subscript zero refer to values in the solar neighborhood. Fig. 6.1 shows an object at distance r from us, at longitude ℓ. It is clear from the figure that the radial velocity of the object with respect to us will be

$$v_{\rm R} = V \cos \alpha - V_0 \sin \ell.$$

Since $\sin(90° + \alpha) = \cos \alpha$, the law of sines gives

$$\frac{\cos \alpha}{R_0} = \frac{\sin \ell}{R}.$$

With this substitution the velocity equation becomes

$$v_{\rm R} = \left(\frac{V R_0}{R}\right) \sin \ell - V_0 \sin \ell,$$

103

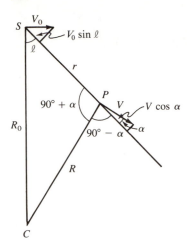

FIGURE 6.1
Galactic rotation velocities, and their projections along and perpendicular to the line of sight.

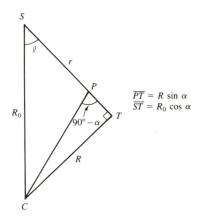

$\overline{PT} = R \sin \alpha$
$\overline{ST} = R_0 \cos \alpha$

FIGURE 6.2
The geometry that leads to Eq. (6.3).

or, in terms of angular velocities, $\omega = V/R$,

$$v_R = R_0(\omega - \omega_0)\sin \ell. \tag{6.2}$$

Similarly, for the transverse velocity (taken positive in the direction of increasing ℓ) Fig. 6.1 gives

$$v_T = V \sin \alpha - V_0 \cos \ell.$$

From Fig. 6.2 it is clear that

$$R \sin \alpha = R_0 \cos \ell - r, \tag{6.3}$$

so that

$$v_T = \left(\frac{V}{R}\right)(R_0 \cos \ell - r) - V_0 \cos \ell \tag{6.4}$$

$$= R_0(\omega - \omega_0)\cos \ell - \omega r.$$

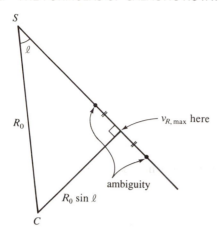

FIGURE 6.3

Showing $v_{R,\max}$ and points on either side of it that have the same Galactic-rotation speed.

Equations 6.2 and 6.4 are the basic equations of differential rotation. Eq. (6.2), in particular, is essential both for deriving the rotation curve and for locating the hydrogen that is observed in the 21-cm line. The latter procedure was mentioned in Chapter 4; now we will see how it is done. Along the line of sight at a particular longitude, different points will have different R's and therefore different ω's. For each point of a 21-cm profile, we can go from the v_R to the ω to the R from which radiation at a particular wavelength originates.

For directions within 90° of the Galactic center there is an ambiguity, however, because there are pairs of points along a line of sight that have the same value of R (situated symmetrically relative to the foot of the perpendicular from the Galactic center to the line of sight, as shown in Fig. 6.3). This ambiguity can be resolved by the extent of the hydrogen in latitude, which will be greater for the nearer of the two points.

For the near neighborhood of the Sun these two equations can be approximated so that they depend only on the local characteristics of Galactic rotation. We make two approximations. One is that $|R - R_0| \ll R_0$; this allows us to write

$$\omega - \omega_0 \approx (R - R_0) \left(\frac{d\omega}{dR}\right).$$

The other approximation is that r is small, so that $R_0 - R \approx r \cos \ell$. Substitution of these approximations in Eq. (6.2) gives, with some straightforward manipulation,

$$v_R = \left[\frac{V_0}{R_0} - \left(\frac{dV}{dR}\right)_0\right] r \sin \ell \cos \ell \qquad (6.5)$$
$$= A r \sin 2\ell,$$

where

$$A = \frac{1}{2}\left(\frac{V}{R} - \frac{dV}{dR}\right)_0. \qquad (6.6)$$

Similarly, substituting in Eq. (6.4) and dropping terms of higher order gives

$$v_T = \left[\frac{V_0}{R_0} - \left(\frac{dV}{dR} \right)_0 \right] r \cos^2 \ell - \frac{V_0}{R_0} r \tag{6.7}$$

$$= r(A \cos 2\ell + B),$$

where

$$B = -\frac{1}{2} \left(\frac{V}{R} + \frac{dV}{dR} \right)_0 . \tag{6.8}$$

A and B are the well-known Oort constants. As we shall see in a later section, their approximate numerical values are $A = 14$ km sec^{-1} kpc^{-1} and $B = -12$ km sec^{-1} kpc^{-1}. (Note that these rather strange operational units have the dimension of inverse time.)

Kinematically, A is a measure of the local shear rate and B of the local vorticity. Or, from another point of view, they together express the local angular velocity and the local slope of the rotation curve, since adding and subtracting them leads to the equations

$$\omega_0 = \frac{V_0}{R_0} = A - B \tag{6.9}$$

and

$$\left(\frac{dV}{dR} \right)_0 = -(A + B). \tag{6.10}$$

Note also that A has the alternative form

$$A = -\frac{1}{2} \left(R \frac{d\omega}{dR} \right)_0 . \tag{6.11}$$

The v_T equation has an alternative form, expressed in terms of proper motion. Since a transverse velocity in km sec^{-1} is related to a distance r in parsecs and a proper motion μ in arcsec yr^{-1} by

$$v_T = 4.74r\mu,$$

Eq. (6.7) can be expressed as a proper motion in Galactic longitude:

$$\mu_\ell = \frac{A \cos 2\ell + B}{4.74 \cdot 1000} \tag{6.12}$$

(where the extra factor of 1000 allows for the kiloparsec units in A and B, and has the effect of making Galactic-rotation proper motions numerically rather small). Thus, within the validity of the near-neighborhood approximations, the differential-rotation proper motions are independent of distance. Note also that here the double sine wave has a biasing term B added; the presence of this term makes it impossible to define a fundamental reference system by assuming that distant stars should look stationary.

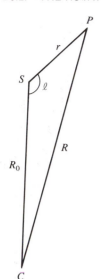

FIGURE 6.4
Geometric relationships for Galactic rotation outward of the Sun.

6.2   THE ROTATION CURVE

To derive the rotation curve of the Milky Way observationally, we need to observe objects that are moving in circular orbits around the center. In practice, any objects of low velocity dispersion will suffice. As we shall see in Chapter 7, the mean velocity of such objects is quite close to the circular velocity.

In practice the rotation curve is derived in two parts: interior to the solar circle, 21-cm observations are used; outside it, we observe H$_{II}$ regions or other optical objects.

In the inner part of the Milky Way, the kinematics themselves give us the rotation curve directly. Eq. (6.2) shows that, along any given line of sight, the observed radial velocity must be greatest at the point where ω differs most from ω_0. But in any well-behaved density distribution, ω must fall off monotonically with increasing R; hence the velocity extremum will correspond to the value of ω at the point where the line of sight comes closest to the Galactic center. From Fig. 6.3 it is clear that at this point $R = R_0 \sin \ell$; ω follows from substituting this maximum velocity into Eq. (6.2), and the pair (ω, R) then gives us a point on the rotation curve. If this observation is carried out at longitudes from 0° to 90° and, redundantly (but see below), from 270° to 360°, the inner rotation curve can be calculated. This method is often referred to as the tangent-point method, because at the point of extreme velocity the line of sight is tangent to a circle with $R = $ const.

Outside the solar circle, however, there are no tangent points. To measure a rotational velocity we must observe the radial velocity of an object whose distance is measured in an independent way. Then its R can be measured from the trigonometry indicated in Fig. 6.4, and its ω found from Eq. (6.2).

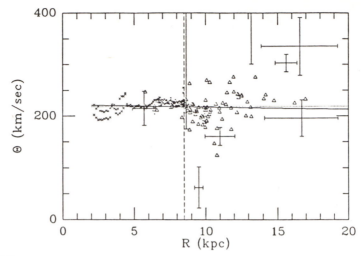

FIGURE 6.5
The rotation curve of the Milky Way. The two curves are different fits to the data. The vertical line marks the assumed position of the Sun. (From Fich *et al.* 1989.)

For this sort of measurement, the objects that have been most useful are H II regions. Once the stars that excite them have been identified, the distances of these stars can be measured. Most H II regions are associated with giant molecular clouds, and a good radial velocity can be obtained from a CO observation in the GMC. Other objects have been used also, such as Cepheids and carbon stars.

For a recent derivation of the rotation curve, see Fich *et al.* (1989). Like other galaxies, the Milky Way shows a rotation curve that is essentially flat; there is even some suggestion that the outermost part of the rotation curve might rise. (See Fig. 6.5.)

Calculation of the rotation curve is not without its problems, however. A serious one is the north-south asymmetry: between longitudes 30° and 50° (and the corresponding longitudes in the fourth quadrant), there is a discrepancy of nearly 10 km sec^{-1} between the two rotation curves. Two possible explanations have been suggested for this difference. One is that the solar neighborhood has a systematic outward motion of a few kilometers per second. The other is that the dark halo is triaxial, so that Galactic rotation takes place in a non-axisymmetric potential field. (Note that such a hypothesis would also allow naturally for a noncircular motion of the solar neighborhood.)

Another problem is the "3-kpc expanding arm." This shows up as the ridge in Fig. 6.6 that extends upward from $(\ell, v) = (345°, 120$ km sec$^{-1})$ and crosses $\ell = 0°$ at -50 km sec^{-1}. Its continuity suggests a spiral arm, yet where it crosses the Galactic center it is seen in absorption, against the background continuum of Sgr A, at -50 km sec^{-1}. This absorption velocity suggests that this arm is not only participating in the Galactic rotation, but also expanding outward from the Galactic center at 50 km sec^{-1}. It has been interpreted alternatively as a general outflow or as a strong set of noncircular motions, perhaps associated with

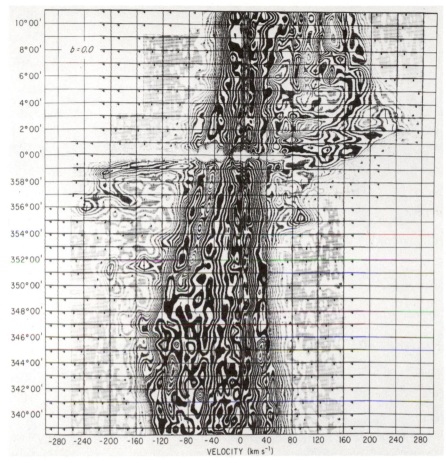

FIGURE 6.6
Longitude–velocity diagram for a region around the Galactic center, showing the "3-kpc arm" (see text for details). (Data from observations by R.P. Sinha.)

a central bar in the Milky Way. It is presumably because of this problem that the rotation curve of Fich *et al.* does not attempt to portray the region interior to 3 kpc.

6.3 THE ROTATION CONSTANTS

It is obvious that before we can use the basic rotation Eq. (6.2) we must know the values of R_0 and ω_0. These are not easy to evaluate directly; furthermore, they are related to other observable constants, with which they should be consistent. These interrelations are

$$\omega = \frac{V_0}{R_0} = A - B,$$

$$\left(\frac{dV}{dR}\right)_0 = -(A+B),$$

$$A \cdot R_0 = AR_0,$$

$$\frac{\sigma_\theta^2}{\sigma_R^2} = \frac{-B}{A-B}.$$

The third of these relations looks trivial, but is not so at all, because the quantity (AR_0) is subject to direct measurement. By treating Fig. 6.3 with the reasoning that was applied to Fig. 6.1, we can express the tangent-point velocity as

$$v_{R,\max} = V(R_{\min}) - V_0 \sin \ell,$$

where

$$R_{\min} = R_0 \sin \ell.$$

Then differentiation gives

$$\frac{dv_{R,\max}}{d(\sin \ell)} = \frac{dV}{dR} R_0 - V_0$$

$$= R_0 \frac{dV(R)}{dR} - V_0.$$

At the limit $R \to R_0$, with the use of the definition of A, this becomes

$$AR_0 = -\frac{1}{2}\left[\frac{dv_{R,\max}}{d(\sin \ell)}\right]_0. \qquad (6.13)$$

This equation allows a direct measurement of AR_0.

The problem of reconciling all the observations into a consistent, and (we hope) correct, set of rotation constants has been with us for a long time. In 1964 the I.A.U. adopted a set of values and recommended that they be used by everyone, so that maps made by different astronomers would be directly comparable. Those values were $R_0 = 10$ kpc, $V_0 = 250$ km sec^{-1}, $A = 15$ km sec^{-1} kpc^{-1}, $B = -10$ km sec^{-1} kpc^{-1}. They were used for many years, but it has become evident that the values of R_0 and V_0 are too high. In 1985 the I.A.U. re-examined the problem, in order to update the values of the constants. Kerr and Lynden-Bell, the leaders of the working group, wrote up the results (Kerr and Lynden-Bell 1986), with a long discussion of how they were arrived at. Their basic approach was to collect, for each quantity, all direct observational measurements to which they believed that they could attach any legitimacy at all. For each observed quantity they took an unweighted mean of the results. They then checked the interconnections and verified that they fit reasonably well. Their paper gives all the details of the methods and the results.

The I.A.U. recommendation is to use the values $R_0 = 8.5$ kpc and $V_0 = 220$ km sec^{-1}. For A and B it is less necessary to have agreement; the results of Kerr and Lynden-Bell were $+14$ and -12 km sec^{-1} kpc^{-1}, for A and B, respectively, but they note that the flatness of the rotation curve suggests that it might be preferable to use $+13$ and -13.

In this connection they make an important distinction: local versus global values of the rotation constants. We have seen again and again indications of departures from circular motion. Such may well exist locally, in which case they might distort the values of the Oort constants from what they would be if averaged around the whole solar circle.

6.4   THE DARK HALO

The flat rotation curve in the Milky Way (and in other galaxies) indicates a far more extended distribution of mass than the density distribution of the disk would imply. It is this sort of kinematic behavior that leads us to believe in the existence of massive, extended, dark halos in galaxies. To examine the density distribution that is required for a flat rotation curve, let us make the simple assumption of a spherical distribution of matter. (Even for a flattened distribution this is not too bad an approximation; when a sphere is flattened down into an infinitely thin disk, the attraction at its edge changes by a factor of less than 2.) We can write

$$\frac{G\mathcal{M}(R)}{R^2} = \frac{V^2}{R},$$

where $\mathcal{M}(R)$ is the mass contained within radius R, and V is the rotational velocity. For a flat curve,

$$\frac{G\mathcal{M}(R)}{R} = V^2 = \text{const.}$$

Thus $\mathcal{M}(R) \propto R$. The distribution for which this is true is easily seen to have a density that is proportional to $1/R^2$; this is characteristic of the outer envelope of an isothermal sphere.

The correspondence just derived should hold for the outer parts of the dark halo, where it is responsible for most of the gravitational force. Farther in, however, the disk makes an increasing contribution (in the Milky Way, becoming dominant for $R < 10$ kpc), and the dark-halo density must drop below the inward extrapolation of this R^{-2} law. Schmidt (1985) argues that there is no need to assume an R^{-2} law at all, because the disk and stellar-halo contributions are so significant; but his argument is directed specifically at the Milky Way, where the rotation-curve observations do not go nearly so far out as they do in some other galaxies (see Chapter 10). Schmidt also calls attention, as have many others, to the fact that much of the flat part of the rotation curve occurs in a region where the total force comes from the combined force of three independent components. How did the three components "conspire" to keep the curve flat?

The detailed nature of the dark halo is a mystery. Aside from its radial distribution, little is known about it. It is likely to have little or no flattening, because otherwise it would then be difficult to maintain the warps that are a common feature of so many galactic disks. (It will be noted in Chapter 9, however, that a small amount of flattening of the dark halo is required by one promising theoretical approach to the origin of warps.)

The extent of the dark halo is also uncertain, and even controversial. The rotation curve of the Milky Way has not been measured beyond 16 kpc, but something of the gravitational field can be deduced from the radial velocities of the outlying globular clusters and the dwarf spheroidal galaxies. Earlier estimates had led to a quite large total mass for the Milky Way, but Little and Tremaine (1987) have argued from these data that the dark halo does not extend out many more tens of kiloparsecs, and that the total mass of the Milky Way is only 2 or 3 times $10^{11} \mathcal{M}_\odot$.

The dark halo is also presumed to emit little or no light, since no light has ever been detected coming from the massive outer halos of other galaxies. (Detections of halo light do exist, but they have the R^{-3} to $R^{-3.5}$ dropoff that is characteristic of a stellar halo rather than a dark halo.)

Another reasonable assertion is that the dark halo contributes very little to the local mass density in the solar neighborhood (in the Bahcall–Schmidt–Soneira model that will be discussed below, only 6 percent). If this is so, then any local "missing mass" in the Oort limit (Chapter 8) could not be attributed to the dark halo. This exclusion would not be true, however, if the dark halo turned out somehow to be highly flattened.

What is the dark halo made of? Various suggestions have been made, all of them speculative. Red dwarf stars will not do; to have enough total mass, they would have to be so numerous that their presence would be obvious. Extreme brown dwarfs might do, but their existence would imply an unexpected turnup of the luminosity function far beyond its observed dropoff. Thus it might be "Jupiters." But it could equally well be rocks. Dust is excluded, however; it would totally obscure the light of distant stars. Perhaps it is black holes (the remnants of massive "Population III" stars that made the first metals). Still another possibility is that the dark halo is made up of the same elementary particles that many cosmologists believe give closure density to the Universe; indeed, some simulations of the formation of structure in the Universe have this material falling into galaxies as they form.

Although the dark halo of the Milky Way is poorly understood, it is a very important component, and much of our understanding of the Milky Way is going to depend on our understanding it better.

6.5 MASS MODELS OF THE MILKY WAY

As already indicated, an important application of the Galactic rotation curve lies in the information that it gives about the distribution of mass in the Galaxy. Clearly the rotation curve is not enough; Poisson's equation states that the density can be derived only from a knowledge of both the forces parallel to the Galactic plane and the forces perpendicular to the plane. Since much less is known about the latter, the customary approach to the overall potential field has been to assume that the Galaxy is made up of several components whose general shapes are known and to adjust the density distribution within each component to get the best fit to what we know about its actual distribution and the overall force field, particularly the rotation curve.

There were earlier attempts at modeling the Galaxy, but serious modeling can be said to have begun with the 1956 Schmidt model. It was created for an ulterior purpose; the Dutch astronomers needed the rotation curve of the Milky Way in order to find out the locations of all the hydrogen clouds in their 21-cm survey. They could find the inner rotation curve from the tangent-point method, but they had no direct way of getting the outer rotation curve. (The use of H II regions for this purpose is a much more recent development.) To fill out the rotation curve, their idea was (1) to fit a mass model to the inner part of the curve, and to everything else that was known about mass distributions in the Galaxy, and then (2) to assume that the potential of this model would give the outer rotation curve. Schmidt made a later revision (Schmidt 1965), but both these models preceded the discovery that the Milky Way has a flat outer rotation curve; so they are no longer valid. Before presenting a more modern model, we shall review the problems of calculating one.

6.6   THE POTENTIAL FIELDS OF GALAXY-LIKE CONFIGURATIONS

The gravitational attraction of an arbitrary mass configuration is difficult to calculate; in general it can be done only numerically. Even if we wish to simplify, not many shapes and density laws give rise to simple formulas; the sphere and the infinite plane are almost unique in their simplicity. Part of the game of calculating Galactic models is to find configurations whose gravitational attraction leads to manageable formulas.

One figure of obvious applicability to galaxies is the oblate spheroid. A general discussion of the potential of ellipsoidal configurations is given by Ramsey (1961, Chapter 7). For spheroids, a detailed discussion was given by Schmidt (1956), who considers the forces that arise from a configuration whose surfaces of equal density are concentric, similar spheroids, with an arbitrary radial density distribution.

Schmidt starts from the formulas for homogeneous spheroids, and represents a spheroid of arbitrary density profile by a nest of coexistent, homogeneous spheroids with semi-major axes a from 0 to a_r, and with densities chosen such that the densities of all spheroids containing the point $(R, z) = (a, 0)$ add up to the actual density at that point, which we shall call $\rho(a)$. Then integration over this set of spheroids gives

$$K_R = -4\pi G e^{-3}\sqrt{1 - e^2} \, R \int_0^\gamma \rho(a) \sin^2 \beta \, d\beta,$$
$$K_z = -4\pi G e^{-3}\sqrt{1 - e^2} \, z \int_0^\gamma \rho(a) \tan^2 \beta \, d\beta,$$

$$(6.14)$$

where γ is evaluated for an interior point from $\sin\gamma = e$ and for an exterior point from

$$\varpi^2 \sin^2\gamma + z^2 \tan^2\gamma = a_r^2 e^2. \tag{6.15}$$

In the integrations for K_R and K_z, a is evaluated from

$$R^2\sin\beta + z^2 \tan^2\beta = a^2 e^2. \tag{6.16}$$

Equations 6.14–6.16 lead to an especially simple formula for the circular velocity in the Galactic plane. At $z = 0$ we have $R^2 \sin^2\beta = a^2 e^2$, and the force law simplifies to

$$K_R = -4\pi G\sqrt{1-e^2}\, R^{-1} \int_0^R \frac{\rho(a)a^2 da}{\sqrt{R^2 - a^2 e^2}}, \tag{6.17}$$

or

$$V_c^2 = 4\pi G\sqrt{1-e^2} \int_0^R \frac{\rho(a)a^2 da}{\sqrt{R^2 - a^2 e^2}}. \tag{6.18}$$

Another configuration, applicable to galaxies, that has a tractable expression for its gravitational force is an exponential disk. If its surface density is $\sigma = \sigma_0 \exp(-R/a)$, then (see Binney and Tremaine 1987, Eq. 2-169)

$$K_R = -4\pi G\sigma_0 y^2 [I_0(y)K_0(y) - I_1(y)K_1(y)],$$

where $y = R/2a$, and I_i and K_i are modified Bessel functions of the first and second kinds.

A modern model of the Milky Way has been fitted by Bahcall, Schmidt, and Soneira (Bahcall *et al.* 1982, 1983; they state the model in the first paper and discuss it in the second). They use a spherical central mass, an exponential thin disk, a spherical stellar halo (which they call "the spheroid"), and a spherical dark halo ("the halo"). Their model gives a reasonable rotation curve, and they argue that the components agree with reality. It probably needs some fine tuning with a thick disk added, however.

REFERENCES

Bahcall, J. N., Schmidt, M., and Soneira, R. M. 1982. *Astrophys. J. (Letters)*, **258**, L23.

———. 1983. *Astrophys. J.*, **265**, 730.

Binney, J., and Tremaine, S. 1987. *Galactic Dynamics*. Princeton, NJ: Princeton Univ. Press.

Blaauw, A., and Schmidt, M., eds. 1965. *Galactic Structure*. Chicago, IL: Univ. of Chicago Press.

Fich, M., Blitz, L., and Stark, A. 1989. *Astrophys. J.*, **342**, 272.

Kerr, F. J., and Lynden-Bell, D. 1986. *Mon. Not. Roy. Astron. Soc.*, **221**, 1023.

Little, B., and Tremaine, S. 1987. *Astrophys. J.*, **320**, 493.

Ramsey, A. S. 1961. *Newtonian Attraction.* Cambridge, Eng.: Cambridge Univ. Press.

Schmidt, M. 1956. *Bull. Astron. Inst. Neth.*, **13**, 15 (no. 468).

———. 1965. In Blaauw and Schmidt, p. 513.

———. 1985. In van Woerden *et al.*, p. 75.

Woerden, H. van, Allen, R. J., and Burton, W. B., eds. 1985. *The Milky Way Galaxy.* I.A.U. Symposium no. 106. Dordrecht: Reidel.

CHAPTER 7

BASIC GALACTIC DYNAMICS

Ivan R. King

A stellar-dynamical approach has two things to offer to the study of the overall structure of the Milky Way. (1) In some situations we can, from the positions and the velocities of the stars, deduce the form of the gravitational field in which they move. Just as we can find the radial mass distribution in the Milky Way from its rotation curve, we can also deduce, from the statistics of the motions of stars up and down through the Galactic plane, the distribution of mass in this z-direction. (2) From the properties of stellar distribution functions, we can transform local knowledge of stellar velocity distributions into knowledge about densities and velocities at other points of the Milky Way.

In treating the dynamics of the Milky Way it is customary to make two simplifying assumptions. First, it is assumed that each star moves in a smooth potential field that results from the collective gravitation of all the other stars (plus any other material that may be present). Second, relaxation caused by stellar encounters is ignored. The first assumption is an idealization that is justified by the large number of stars present; the second is supported by calculations showing that the relaxation time for stellar encounters in the Milky Way is orders of magnitude longer than its age.

7.1   THE COLLISIONLESS BOLTZMANN EQUATION

The distribution functions of stellar dynamics are not completely free; they are subject to a continuity condition that has far-reaching consequences. This condition is easily expressed by considering the distribution of stars (or more strictly, of points that represent them) in a six-dimensional "phase space" that is defined by conjoining the three components of a star's position and the three components of its velocity.

In the phase space the points that represent the stars (hereafter we shall often refer to them simply as the stars) flow in a restricted way, because they must obey the laws of motion, which connect their movement in the position components with the values of their velocity components. There is, in fact, a continuity equation that the flow of stars must obey in the phase space. (The derivation, which will not be given here, goes exactly like that of the equation of continuity in hydrodynamics, except that we have six dimensions to deal with here.) If we adopt Cartesian coordinates x, y, z, u, v, w for the positions and velocities, then the equation takes the form

$$0 = \frac{\partial f}{\partial t} + u\frac{\partial f}{\partial x} + v\frac{\partial f}{\partial y} + w\frac{\partial f}{\partial z} - \frac{\partial \psi}{\partial x}\frac{\partial f}{\partial u} - \frac{\partial \psi}{\partial y}\frac{\partial f}{\partial v} - \frac{\partial \psi}{\partial z}\frac{\partial f}{\partial w}, \qquad (7.1)$$

where $\psi(x, y, z)$ is the potential function, whose negative gradient yields the gravitational acceleration of a particle.

Eq. (7.1) is called the collisionless Boltzmann equation, since it is precisely Boltzmann's equation of kinetic theory with the term omitted that takes into account collisions between particles. (It has also appeared in the astronomical literature under the names "Liouville equation" and "Vlasov equation." For a discussion of the terminology see Hénon 1982.)

This equation has an important direct physical interpretation, which becomes clear if we compare it with the equations of motion of a star,

$$\frac{dx}{dt} = u, \quad \frac{dy}{dt} = v, \quad \frac{dz}{dt} = w, \quad \frac{du}{dt} = -\frac{\partial \psi}{\partial x}, \quad \frac{dv}{dt} = -\frac{\partial \psi}{\partial y}, \quad \frac{dw}{dt} = -\frac{\partial \psi}{\partial z}. \quad (7.2)$$

The terms on the right of Eq. (7.1) contain all possible partial derivatives of f with respect to the independent variables, and each turns out to be multiplied by a quantity that is equal to the derivative of that variable with respect to t, along the path of motion of an individual point. Thus Eq. (7.1) states that along the path of any star through the phase space, the total derivative of f (what is often referred to in hydrodynamics as the "Lagrangian derivative") is zero. The equation is thus equivalent to what in statistical mechanics is called Liouville's theorem: as the point representing a star moves through the phase space, the density around it remains constant. In other words, the flow of stars through the six-dimensional phase space is incompressible. (This is *not* true of the flow through the three-dimensional position space!) It is from this property that a great deal of the value of this approach derives; in effect, the stars at any given position and velocity are carrying with them, in their phase density, information about the phase density at other velocities and at other spatial points. In a certain sense, it is because of this property that we are able to use the velocity distribution of local stars to probe the characteristics of remoter parts of our Galaxy.

The collisionless Boltzmann equation is fortunately of a type whose general solution is known; it is a linear first-order partial differential equation. (It is not even, in fact, the most general form of such an equation, in which a function of the variables would appear on the left-hand side too.) The general solution of this type of equation will be given here without a proof, which can be supplied by any of the numerous standard texts on partial differential equations.

Consider the equation

$$A_1\frac{\partial f}{\partial x_1} + A_2\frac{\partial f}{\partial x_2} + \cdots + A_n\frac{\partial f}{\partial x_n} = 0, \qquad (7.3)$$

where the A_i are functions of $x_1, x_2, \ldots, x_n$. To find the general solution, form the "subsidiary equations"

$$\frac{dx_1}{A_1} = \frac{dx_2}{A_2} = \cdots = \frac{dx_n}{A_n}. \qquad (7.4)$$

These are $n - 1$ independent ordinary differential equations. The solution algorithm says to find $n - 1$ independent integrals of these ordinary differential equations; let them be expressed in the form

$$I_i(x_1, x_2, \ldots, x_n) = \text{const.}, \quad i = 1, 2, \ldots, n - 1. \qquad (7.5)$$

Then the general solution of Eq. (7.3) is

$$f(x_1, x_2, \ldots, x_n) = F(I_1, I_2, \ldots, I_{n-1}), \qquad (7.6)$$

where F is an arbitrary function of its arguments. (This is, of course, in accord with the principle that where the general solutions of ordinary differential equations have arbitrary constants, the general solutions of partial differential equations instead have arbitrary functions.) The solution given by Eq. (7.6) may not at first appear to offer much information, since it contains a function that is completely arbitrary; but it is in fact a strong restriction, since it says that what might have appeared to be a function of n variables must in fact depend on only $n - 1$ variables. In the following section, furthermore, we shall see how this restriction can become even stronger.

The specific application of this solution algorithm to the collisionless Boltzmann equation goes as follows. The subsidiary equations are

$$\frac{dt}{1} = \frac{dx}{u} = \frac{dy}{v} = \frac{dz}{w} = \frac{du}{(-\partial\psi/\partial x)} = \frac{dv}{(-\partial\psi/\partial y)} = \frac{dw}{(-\partial\psi/\partial z)}. \qquad (7.7)$$

But these can be rearranged into the form

$$\frac{dx}{dt} = u, \quad \frac{dy}{dt} = v, \quad \frac{dz}{dt} = w, \quad \frac{du}{dt} = -\frac{d\psi}{dx}, \quad \frac{dv}{dt} = -\frac{d\psi}{dy}, \quad \frac{dw}{dt} = -\frac{d\psi}{dz}, \qquad (7.8)$$

which are nothing but the equations of motion of a star. Thus the theorem about partial differential equations says that the general solution of the collisionless Boltzmann equation requires that f be expressible as a function, F, of the integrals of the equations of motion of a star. But these integrals, taken all together, define the motion of the star through phase space. Thus f remains constant as the star moves. But this is just a repetition of Liouville's theorem. We have recovered the same result via this different route; but of course it had to be so, if our whole picture is to be consistent.

There are two points particularly to be noted here. First, let it be clear (and clearly remembered) that the integrals referred to are integrals of the motion of an individual test star, and have nothing to do with collective integrals, such as the conserved energy of the whole N-body system. Second, note clearly that the conserved density that is referred to is a density in the six-dimensional phase space, not in the position space.

7.2   JEANS' THEOREM AND THE CONCEPT OF ISOLATING INTEGRALS

In practice the general solution of the collisionless Boltzmann equation is of little use in the general, time-dependent situation. Where the solution becomes particularly powerful is in a time-independent situation, such as that of a stellar system that has existed long enough to have settled down into a statistically steady state. The stars then move about individually, but the overall distribution function remains constant, as can be expected in any stellar system in which the stars have circulated long enough for their orbits to have become well mixed.

In the time-independent situation Liouville's theorem has much stronger consequences. In the time-dependent situation the phase density f is indeed constant along its path, but at each point of the path it has this value *only at the moment when the star passes that point.* Before and after, the phase density at that point will in general be different. But with time-independence, f is constant, at all times, at every point on the path of a star in the phase space. Thus in a sense the stellar orbits map out the distribution function.

We will follow this idea through mathematically. The collisionless Boltzmann equation now lacks the time derivative, and takes the form

$$0 = u\frac{\partial f}{\partial x} + v\frac{\partial f}{\partial y} + w\frac{\partial f}{\partial z} - \frac{\partial \psi}{\partial x}\frac{\partial f}{\partial u} - \frac{\partial \psi}{\partial y}\frac{\partial f}{\partial v} - \frac{\partial \psi}{\partial z}\frac{\partial f}{\partial w}, \tag{7.9}$$

and the subsidiary equations are

$$\frac{dx}{u} = \frac{dy}{v} = \frac{dz}{w} = \frac{du}{-\partial\psi/\partial x} = \frac{dv}{-\partial\psi/\partial y} = \frac{dw}{-\partial\psi/\partial z}. \tag{7.10}$$

Recall that the subsidiary equations are the equations of motion of an individual star. Any elementary treatment of dynamics shows that one integral is

$$I_1 = \tfrac{1}{2}(u^2 + v^2 + w^2) + \psi(x, y, z) = \text{const.} \tag{7.11}$$

This is the familiar energy integral, and we shall often refer to it familiarly as E. If the potential is independent of time, this integral always exists, whatever the form of the potential may be. (Note that this integral, in the form in which we use it, is actually energy per unit mass. For ease of reference, however, we shall call it simply energy.)

Suppose now that the energy integral were the only integral that existed, that is, the only quantity conserved along the orbit of a star. If we consider the six-dimensional phase space from the point of view of analytic geometry, then $I_1 = \text{const}$ represents a five-dimensional hypersurface. Since I_1 is conserved, the motion of the star must be confined to that surface, since it could go nowhere else without having a different value of I_1.

Granted the confinement to this energy surface, let us consider now the other characteristics of the motion of the star. If there are no other restrictions on its motion, then it can be expected eventually to visit every point on the whole energy surface.

(Such travel presents some problems of mathematical detail. It is not in fact possible for a path of finite length to visit every point, but it is possible for a path to come arbitrarily close to each point on the surface. Such a path is called an *ergodic* path. A more proper statement would be that a test star with no other restrictions will traverse an ergodic path on its energy surface.)

But consider now the consequence of this effectively complete coverage of a surface by the orbit of a star. According to Liouville's theorem the value of the phase-space density f must then be constant everywhere on that surface. On any other energy surface, however, and everywhere on that surface, f can have some other value. No two energy surfaces can intersect, because no point in the phase space can correspond to two different values of the energy. Thus each energy surface has its own value of the phase-space density f. The simple mathematical way of summing up these statements is

$$f(x, y, z, u, v, w) = F(I_1). \tag{7.12}$$

This is an amazingly strong result, and we must examine how it could possibly be true. Our general solution said that F should have as its independent variables five integrals, not just one; where have the others gone? The answer to this question depends on the distinction between *isolating* and non-isolating integrals. I_1 can properly be called "isolating," because each value of the quantity I_1 isolates a surface characterized by that particular value, and the possible values of I_1 decompose the phase space into a disjoint set of energy surfaces. Whether other integrals do this or not depends on their individual nature.

The question of integrals of motion is a complicated and puzzling one. In some mathematical sense, there must always be six integrals of the equations of motion of a star, because the three second-order differential equations of motion of a star require six integrations to solve them. But, in fact, the integration of an orbit in an arbitrary potential can be carried out only numerically, and the numerical integration does not give rise to analytic relations like Eq. (7.11). When such analytic integrals can be found, they are isolating, but the other integrals that are implicit in a numerical integration of an orbit are usually non-isolating.

Thus the powerful restriction expressed by Eq. (7.12) really does hold— provided that I_1 is the only integral that exists. There are dynamical situations, however, in which it is easily shown that another isolating integral exists. For example, elementary mechanics easily shows that an axisymmetric potential always has an integral that expresses the conservation of the axial component of angular momentum. In our present coordinates, it is (again, per unit mass)

$$I_2 = xv - yu = \text{const.} \tag{7.13}$$

Again, from the point of view of analytic geometry this can be considered to be the equation of a five-dimensional hypersurface, and the motion must be confined to it. But the motion is also confined to a particular I_1 surface; so it must in fact take place on the four-dimensional hypersurface that is the intersection of the I_1 surface and the I_2 surface. Again, Liouville's theorem requires that f have the same value everywhere on this surface, and the mathematical expression of this fact is

$$f(x, y, z, u, v, w) = F(I_1, I_2). \tag{7.14}$$

The general principle exemplified by these two cases was stated a long time ago by Jeans: In a steady state, the phase-space distribution function is a function of the isolating integrals of the motion of a test star—and of only the isolating integrals. Jeans' theorem is one of the cornerstones of stellar dynamics.

The crux of the problem is clearly how many isolating integrals exist. Several specific forms of the potential have additional isolating integrals. As we will see, however, the problem is not this simple. In addition to the truly isolating integrals, others appear to exist for many or most orbits, in most dynamical problems. To pursue this train of investigation, though, we shall find it most convenient to treat the collisionless Boltzmann equation in a system of coordinates that is suitable for discussions of local Galactic dynamics, which is the context in which the fascinating question of "non-classic integrals" first arose.

7.3   THE COLLISIONLESS BOLTZMANN EQUATION IN GALACTIC COORDINATES

There is a coordinate system whose use is fairly standard in discussions of the local dynamics of the Milky Way. For the position coordinates we use a cylindrical polar system R, θ, and z. For the velocity coordinates we take Cartesian but noninertial components, each in the increasing direction of the corresponding position coordinate, and we use as symbols the corresponding capital letters U, V, W. These are related to the positional derivatives by the equations

$$U = \frac{dR}{dt}, \quad V = R\frac{d\theta}{dt}, \quad W = \frac{dz}{dt}. \tag{7.15}$$

Their noninertial character arises, of course, from the fact that the directions of U and V change as a star moves along its orbit.

Since the Galaxy rotates in a clockwise direction as seen from the north, we take clockwise rotations to be positive and use a left-handed system of coordinates, with the z-axis pointing toward the north Galactic pole. This causes no inconvenience in our physics, since none of our dynamical treatments will involve cross-products of vectors.

Converting the collisionless Boltzmann equation into these coordinates is straightforward but tedious, and we will simply give the result for an axisymmetric, steady-state stellar system, the condition that we will assume for the Milky Way. Its form is

$$0 = U\frac{\partial f}{\partial R} + W\frac{\partial f}{\partial z} - \left(\frac{\partial \psi}{\partial R} - \frac{V^2}{R}\right)\frac{\partial f}{\partial U} - \frac{UV}{R}\frac{\partial f}{\partial V} - \frac{\partial \psi}{\partial z}\frac{\partial f}{\partial W}. \tag{7.16}$$

(Note that whereas the axial symmetry makes all derivatives with respect to θ zero, there is still a derivative with respect to V.) Eq. (7.16) is the equation with which we will do most of our Galactic dynamics.

For its general solution, we write down the subsidiary equations,

$$\frac{dR}{U} = \frac{dz}{W} = \frac{dU}{(V^2/R - \partial\psi/\partial R)} = \frac{dV}{-UV/R} = \frac{dW}{-\partial\psi/\partial z}. \tag{7.17}$$

These are again equivalent to the equations of motion of a star. We could integrate combinations of these equations to derive the integrals, but we already know them to be the energy integral

$$E = \tfrac{1}{2}(U^2 + V^2 + W^2) + \psi(R, z), \tag{7.18}$$

and the angular-momentum integral

$$h = RV. \tag{7.19}$$

These are the only two isolating integrals that can be found for all orbits in a general steady-state axisymmetric potential.

7.4   THE PROBLEM OF THE THIRD INTEGRAL

Application of Jeans' theorem now says that

$$f(R, z, U, V, W) = F(E, h). \tag{7.20}$$

But this immediately makes a strong prediction: whereas the dependence on V can be quite arbitrary, because it appears independently in h, the roles of U and W must go identically in lock-step, since they enter only in the combination $(U^2 + W^2)$, in E. More specifically, this requires that the velocity dispersions of U and W be the same, which observationally they are not, by a factor of about 2.

Clearly something is wrong here. A first thought might be that the Galaxy is not yet in a steady equilibrium, but after some fifty revolutions this seems quite unlikely. The more likely avenue to pursue is that there is a third integral of the motion of a star—or, at the very least, a quantity that is conserved for all practical purposes for a time of the order of the present age of the Milky Way. It is just this view that Oort and Lindblad took in 1926–28, when they were laying the foundations of the theory of Galactic dynamics. For stars that do not stray far from the Galactic plane, the z-force, which depends on local mass density, is effectively decoupled from the R-force, which comes from the matter all the way from here to the Galactic center. Mathematically this statement of independence can be expressed in terms of the cross-derivative of the potential,

$$\frac{\partial^2 \psi}{\partial R \partial z} = 0, \tag{7.21}$$

or equivalently in terms of the potential itself:

$$\psi(R, z) = \psi_1(R) + \psi_2(z), \tag{7.22}$$

which is easily verified to be the general solution of the differential equation (7.21).

If this is true, however, then the subsidiary equations yield (from equating their second and last members) an additional integral for the z-motions:

$$I_3 = \tfrac{1}{2}W^2 + \psi_2(z), \tag{7.23}$$

which is a decoupled z-energy. With I_3 as a third argument of F, the dependence of the distribution function on W can now be quite different from its dependence on U.

Since most stars in the solar neighborhood remain in the low-z realm where Eq. (7.23) is a good approximation, this rationalization was complacently accepted for several decades. The structure began to crumble only when the increasing power of computers made it possible to compute orbits, in a reasonable approximation to the Galactic potential, for stars whose motions above and below the Galactic plane had amplitudes of a kiloparsec or more (Contopoulos 1958). These orbits clearly conserved a quantity that resembled the z-energy, but had a more complicated behavior and no obvious mathematical expression. A large set of Galactic orbits of stars was computed, and nearly all of them had a "third integral."

One way in which it became clear that these orbits were not ergodic was that they did not fill all the space within their limiting equipotential. Instead of filling the whole round lobe, orbits were confined to a "box," or sometimes to a "tube" (Fig. 7.1).[1]

A sharper test can be made, however. It is hard to examine a multidimensional phase space, but if two dimensions can be extracted from it, they are easy to look at. What we do is to plot a "surface of section." Consider the case of Galactic orbits. An orbit crosses the Galactic plane again and again; are these crossings as ergodic as they can be, or do they exhibit some regular restriction? (Strictly speaking, "ergodic" refers to filling an E surface, but we will find it convenient to use the word here for the mathematical idea of going arbitrarily close to every point in a specified manifold.)

To make the question clearer, let us state it as follows: when the star crosses the Galactic plane ($z = 0$) at some particular value of R, the magnitude of its velocity is determined by the value of the potential at that point. But the value of V is also determined by the angular-momentum integral, so that between the two, the value of $U^2 + W^2$ is also fixed. The crux of the problem then becomes: if the motion were ergodic, then the star would eventually come back through the plane at this point with all possible combinations of U and W that satisfy the restriction $U^2 + W^2 = \text{const}$. Does this happen? In fact, we need only examine

---

[1]These two terms are aptly descriptive of the orbits shown, but in recent years "box" and "tube" orbit have taken on a different pair of meanings.

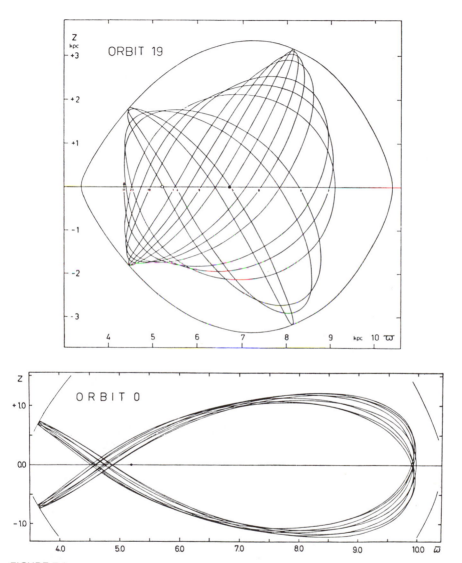

FIGURE 7.1
A box-shaped orbit and a tube-shaped orbit, as exhibited by Ollongren (1965). Each graph is in a meridional (*i.e.*, R, z) section, and all or part of the limiting surface is shown.

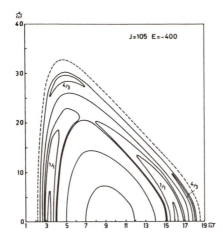

FIGURE 7.2
A surface of section, taken from a different study of galactic orbits (Mayer and Martinet 1973). The continuous curves correspond to box-shaped orbits, and the sets of islands to tube-shaped orbits. The dashed line is the limiting envelope.

the values of U, since for a fixed value of $U^2 + W^2$, W is determined by U. Thus we plot against the R coordinate of each plane-crossing the value of U with which the star crosses. (Because of symmetry, in fact, we need plot only the positive values of U.) The question to examine, then, is whether at a given R all values of U appear (up to the maximum permissible value), or whether there is some restriction. Plotting of such surfaces of section for numerically calculated Galactic orbits shows that most of them give points that do not scatter all over the surface of section, but instead lie on an "invariant curve," the name given to its track in the surface of section. Only rarely is the behavior "wild." Thus for nearly all Galactic orbits something additional is conserved, as a third integral, or whatever its nature may be.

Fig. 7.2 shows a typical surface of section, with the invariant curves of a number of orbits plotted.

It is further puzzling that some orbits do *not* show a third integral. How can there be an integral of the equations of motion, such that it exists for some initial conditions but does not hold for others? For this and other reasons, the third integral that we find in most Galactic orbits is called a "nonclassical integral." It is not a true integral in the classical sense of mechanics, but it is nevertheless conserved.

At first there was a suspicion that the flattened, and nearly spheroidal, potential of the Milky Way might have some special properties that were causing these complications. But all such suspicion vanished, in 1964, with a classical study by Hénon and Heiles, who chose a two-dimensional potential that was too simple to be accused of having any special properties (Fig. 7.3), and calculated a large number of orbits in it. Their orbits again showed invariant curves in the surfaces of section. Hénon and Heiles turned up a new phenomenon, however. For the larger values of energy, many of the orbits jumped about the surface of section, on their successive crossings of $x = 0$, in a way that did not appear ergodic at all (Fig. 7.4). The tendency toward the occurrence of such orbits was,

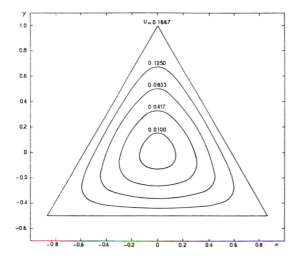

FIGURE 7.3
Equipotentials of the
Hénon–Heiles potential.

in fact, a function of energy; at the higher values of E, the orbits made a rather rapid transition from being all "regular" (*i.e.*, showing invariant curves) to being all "stochastic," as this less-restricted variety of motion is called (Fig. 7.5).

Since then, many other potentials have been studied. Not only is the presence of an extra integral characteristic of the Galactic problem; a nonclassical integral turns up in almost every dynamical problem that has been investigated. The naive inquiry after a Galactic "third integral" has grown into a completely new field of dynamics and mathematics, chaos theory, whose ramifications go far beyond what can be discussed in a simple discourse on stellar dynamics.

Let us sum up the present understanding of the situation, and its impact on stellar dynamics, as follows. In addition to the classical integrals, such as energy and angular momentum, most potentials do admit "nonclassical integrals." These tend to exist for most orbits, but at the higher energies (and where it is relevant, at the lower angular momenta) some percentage of the orbits show stochastic behavior. A really proper distribution function can be built up only by a numerical investigation of orbit types, followed by a classification of them. Fortunately, however, good approximate characterizations of the third integral in the Milky Way do exist, and a lot of understanding can be built up on the basis of them.

7.5 THE THIRD INTEGRAL IN THE MILKY WAY

Since a nonclassical integral can be studied only numerically, the question arises, is there some potential that resembles that of the actual Milky Way but that has an analytic third integral? If we had such a potential available, then we could study the properties of the third integral in a general way. Happily, a potential of this sort is well known. As we shall see in a later discussion, the assumption

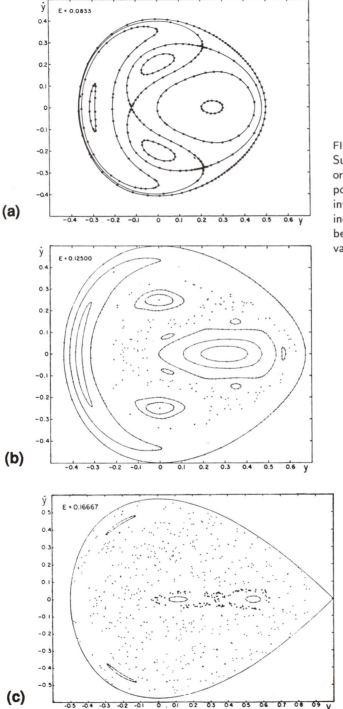

(a)

(b)

(c)

FIGURE 7.4
Surfaces of section for
orbits in the Hénon–Heiles
potential, showing
invariant curves and
increasingly stochastic
behavior at increasing
values of the energy.

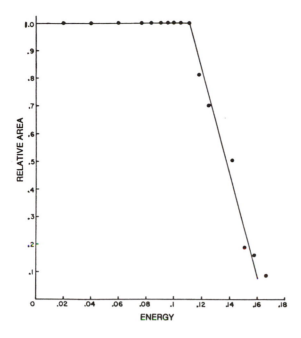

FIGURE 7.5
Relative area covered by regular orbits and stochastic ones, as a function of energy, in the Hénon and Heiles study.

that the velocity distribution can be expressed exactly as a quadratic form in the velocity components leads to a restriction on the nature of the potential, in the form of a differential equation that it must satisfy. The solutions of this differential equation have long been known; they are called Stäckel potentials (after the nineteenth-century mathematician who studied their properties). They take on a simple form only when expressed in terms of a set of coordinates called confocal coordinates, whose coordinate surfaces are a set of ellipsoids and hyperboloids of both kinds, all of which share the same foci. Near the Galactic plane these coordinate surfaces are nearly the cylinders R = const and the planes z = const, and the third integral approaches the Oort–Lindblad third integral. Away from the plane, however, the behavior of Stäckel potentials and their integrals is too complicated to study in this book. Suffice it to say that they offer a good approximation to the actual potential of the Milky Way and to its nonclassical third integral, and that methods exist for coping with the problems raised by the existence of the third integral. These problems are still under intensive study, however, and it may be some time before the practical application of Stäckel potentials can be easily described.

7.6 LOCAL GALACTIC DYNAMICS

We turn now to the specific discussion of the local dynamics of the Milky Way. The particular questions that we ask are conditioned by the observational facts and limitations: (1) We have considerable information about the velocity distributions of various types of stars in the solar neighborhood, but almost no direct information about how these properties differ at other points in space. (2) However, we do have reasonably good knowledge of the mean rotational velocity at various distances from the Galactic center. (3) Substantial data are available about density gradients in the z-direction, but there is almost no direct information about radial density gradients in the Galaxy.

One of the main tasks of Galactic dynamics is to supplement this meager observational information by using theory to extend it. Conceptually our basic leverage here, regardless of the actual method that we may choose to employ, is Liouville's theorem. Each star that passes through the solar neighborhood brings with it, for every place that it ever visits in the position space, information about one point in the phase-space density at that spatial location. If we can somehow put together such information, while assuming some reasonable regularities in phase-space densities, then we can deduce from our local observations some of the properties of remoter places, which are unobservable in any direct way. In the following sections we shall explore—and exploit—three approaches. In one we shall assume a plausible functional form for the velocity distribution, and see what characteristics it implies for the overall structure of the distribution function in the Milky Way. In the second approach we shall deal with moments of the velocity distribution—which are often all that we can calculate from observations with any reliability—and see how they can be related to more extended properties such as density gradients. The third approach, which will be postponed to Chapter 9, will be to look at individual stellar orbits and see what their statistics can tell us about the extended properties of our local area.

7.7   THE FUNCTIONAL APPROACH: THE OORT SUBSTITUTION

In his original discussion of the dynamics of a rotating galaxy, Oort simply followed through the consequences of Schwarzschild's classical assumption that the local velocity distribution of stars can be represented by a trivariate Gaussian. (Schwarzschild's original function was actually a prolate spheroid, but others soon recognized that there was no reason for two of the axes to be equal.)

Although we know now that this treatment predicts that $\langle U^2 \rangle$ should be independent of R, contrary to observational fact, Oort's approach is still of value for two reasons. (1) It leads to a solution that predicts many details of Galactic structure and kinematics, and it is possible that a realistic model could be built up as a superposition of such solutions. (2) It leads in a natural way to Stäckel potentials and the quadratic third integral that goes with them; these can be used, in the moment approach, to approximate terms that could not otherwise be evaluated.

Oort assumed a velocity distribution of the form

$$f = f_0 \exp(-Q), \tag{7.24}$$

where

$$f_0 = h\,k\,l\,\nu/\pi^{\frac{3}{2}}, \tag{7.25}$$

$$Q = h^2 U^2 + k^2 (V - V_0)^2 + l^2 W^2 + mU(V - V_0) + nUW + p(V - V_0)W, \tag{7.26}$$

and h, k, l, m, n, p, V_0, f_0 are functions of R and z. Thus we have assumed that the velocity distribution is a trivariate Gaussian everywhere, but that its size, shape, orientation, and mean rotational velocity can vary from place to place. Note that the second equation relates f_0 to a normalized Gaussian distribution, so that ν is the space density of stars (which we have elsewhere denoted by N).

In the solar neighborhood this is a fair approximation, and it is reasonable to assume that the velocity distribution takes a similar form at other locations in the Milky Way. The constants h, k, l, $\ldots$, however, will depend on position. Oort's approach was to substitute the form given by Equations 7.24–7.26 into the steady-state, axially symmetric collisionless Boltzmann equation,

$$0 = U\frac{\partial f}{\partial R} + W\frac{\partial f}{\partial z} + \left(\frac{V^2}{R} + K_R\right)\frac{\partial f}{\partial U} - \frac{UV}{R}\frac{\partial f}{\partial V} + K_z\frac{\partial f}{\partial W}, \tag{7.27}$$

and see what the consequences are for the spatial behavior of the constants.

There are two small points to note here. First, we are using Oort's notation for acceleration, K, instead of the negative gradient of the potential. Second, following Oort we have tacitly omitted from Q the quantities U_0 and W_0; they could have been included and, at the cost of some small additional complication, shown to be equal to zero.

When f is substituted into Eq. (7.27), nearly all the terms contain $-f$ times a derivative of Q, except for a few that have $\exp(-Q)$ times a spatial derivative of f_0. The result is therefore simpler if we divide through by $-f$. When the terms are then collected by powers of U, V, and W, the result is

$$0 = U^3\frac{\partial h^2}{\partial R} + U^2 V\left(\frac{\partial m}{\partial R} - \frac{m}{R}\right) + U^2 W\left(\frac{\partial h^2}{\partial z} + \frac{\partial n}{\partial R}\right)$$

$$+ UV^2\left[\frac{\partial k^2}{\partial R} + \frac{(2h^2 - 2k^2)}{R}\right] + UVW\left(\frac{\partial m}{\partial z} + \frac{\partial p}{\partial R} - \frac{p}{R}\right)$$

$$+ UW^2\left(\frac{\partial l^2}{\partial R} + \frac{\partial n}{\partial z}\right) + V^3\frac{m}{R} + V^2 W\left(\frac{\partial k^2}{\partial z} + \frac{n}{R}\right)$$

$$+ VW^2\frac{\partial p}{\partial z} + W^3\frac{\partial l^2}{\partial z} - U^2\frac{\partial(mV_0)}{\partial R} + 2UV\left[\frac{\partial(k^2 V_0)}{\partial R} - \frac{k^2 V_0}{R}\right]$$

$$- UW\left[\frac{\partial(mV_0)}{\partial z} + \frac{\partial(pV_0)}{\partial R}\right] - V^2\frac{mV_0}{R} - 2VW\frac{\partial(k^2 V_0)}{\partial z} - W^2\frac{\partial(pV_0)}{\partial z}$$

$$+ U\left[\frac{\partial(k^2 V_0^2)}{\partial R} + 2h^2 K_R + nK_z - \frac{1}{f_0}\frac{\partial f_0}{\partial R}\right] + V(mK_R + pK_z)$$

$$+ W\left[\frac{\partial(k^2 V_0^2)}{\partial z} + 2l^2 K_z + nK_R - \frac{1}{f_0}\frac{\partial f_0}{\partial z}\right] - mV_0 K_R - pV_0 K_z. \tag{7.28}$$

But now we can set the total coefficient of each of the individual powers of the variables equal to zero. The argument goes as follows. Consider, for example, any fixed values of V and W. For these values the equation is an algebraic equation in U. Being of only third order, it can have only three roots. But we know that it must hold for all values of U; hence it must be an identity rather than an equation, so that each of the powers of U must be identically zero. Similarly for the powers of the other variables, and by extension for all combined powers. The identity can thus be satisfied only by setting the total coefficient of each power of U, V, and W equal to zero.

This procedure leads to a large set of equations, one for each power that appears. Most of them involve spatial derivatives of the quantities h, k, l, m, n, p, V_0, and f_0. The equations can be solved in a systematic way, to derive the spatial properties that the coefficients of the velocity ellipsoid must satisfy.

The solution is long and tedious, and will not be reproduced here. Before giving the results, however, we note that each integration of a differential equation gives rise to a constant of integration. A number of these are shown to be equal to zero, but five of them survive and are designated below as c_1–c_5. These constants should be carefully distinguished in their nature from h, k, l, ... The latter are functions of position, whereas the c's are absolute constants.

The results of the solution are

$$h^2 = c_1 + \tfrac{1}{2}c_5 z^2, \tag{7.29}$$

$$k^2 = c_1 + c_2 R^2 + \tfrac{1}{2}c_5 z^2, \tag{7.30}$$

$$l^2 = c_4 + \tfrac{1}{2}c_5 R^2, \tag{7.31}$$

$$n = -c_5 R z, \tag{7.32}$$

$$m = p = 0, \tag{7.33}$$

$$V_0 = \frac{c_3 R}{c_1 + c_2 R^2 + \tfrac{1}{2}c_5 z^2}. \tag{7.34}$$

The first five equations give strong information about the velocity ellipsoid. First, the fact that m and p are zero shows that one axis of the velocity ellipsoid always points toward the axis of rotation of the Galaxy and another axis in the direction of rotation. Moreover, at $z = 0$ the first axis points along the center–anticenter line and the third axis perpendicular to the plane; but away from the plane these axes may tilt.

It is well known, however, that many types of stars have velocity ellipsoids whose long axis does not point toward the axis of rotation. The present theory does not allow such a "vertex deviation," which is presumably due to departures from a steady state, from axial symmetry, or both.

The first three equations also tell how the velocity dispersions vary with position. Since Equations 7.24 and 7.26 define a Gaussian distribution, it follows that $h^2 = 1/2\sigma_R^2$, etc. Thus Eq. (7.29) tells us that the U dispersion does not depend on R (or "is isothermal in R"), although it will change with z; similarly, Eq. (7.31) says that σ_z is isothermal in z but should change with R.

It is here that we have a contradiction with observational fact. It has long been known that $\langle Z^2 \rangle$ increases with increasing distance from the Galactic plane, and Lewis and Freeman (1989) have recently shown that $\langle U^2 \rangle$ decreases with increasing R. In studies of z-motions it has therefore been customary to employ a sum of Gaussians of the type discussed here (Oort 1932, 1960). Radial studies have not yet used this technique, but it is clear that any application of Gaussian velocity distributions will have to do so. The isothermality properties exist in any given component; but the relative proportions of the components vary spatially, and so, accordingly, do the velocity dispersions.

The behavior of σ_θ is different; it is connected with Galactic rotation. We investigate it by using Eq. (7.34) at $z = 0$, expressing the Oort A and B constants,

$$A = -\tfrac{1}{2} R \frac{d}{dR} \left(\frac{V_0}{R} \right),$$
(7.35)

$$B = A - V_0/R,$$
(7.36)

in terms of the c's and R and z, and noting that h and k are inversely related to the velocity dispersions. The result is

$$\frac{\sigma_\theta^2}{\sigma_R^2} = \frac{-B}{A - B}.$$
(7.37)

Thus the ratio of the U and V axes of the velocity ellipsoid is directly connected with the law of differential rotation of the Galaxy. We shall refer to Eq. (7.37), which will follow also from our other treatments of galactic dynamics, as "the star-streaming equation," as Kapteyn named this first velocity anisotropy ever discovered.

There is one way in which the quantities with which we have dealt above are not exactly the Oort constants. The latter are defined in terms of the circular velocity V_c, whereas here we have used the mean velocity V_0. For stellar types of low velocity dispersion, however, we shall see that the difference is small. For types that have a higher velocity dispersion, the distinction is an important one; it shows that their velocity ellipsoid gives us some information about their rotation curve.

The manipulations that resulted in Equations 7.29–7.34 used all the powers of the velocities in Eq. (7.28) except U and W. The latter two terms are different, in that they contain the spatial gradients of f_0, which corresponds closely to the star density. Thus from these equations we can derive information about density gradients. For the radial density gradient, the equation that comes from the coefficient of U can be manipulated, in ways that are again complicated and that use some of the intermediate results from the previous lengthy solution. Again, only the result will be quoted:

$$V_c^2 - V_0^2 = \frac{1}{2h^2} \left(-\frac{\partial \ln \nu}{\partial \ln R} - 1 + \frac{h^2}{k^2} - \frac{1}{2} \frac{c_5 R^2}{l^2} \right).$$
(7.38)

Here we have replaced K_R by its equivalent $-V_c^2/R$, where V_c is the circular velocity at the Sun's distance from the Galactic center.

In actual quantitative fact, the dominant term on the right is the density gradient, which is of course negative, so that both sides of the equation are positive. Thus the mean velocity of a stellar type, V_0, is less than the circular velocity V_c. For disk stars the two quantities do not differ by much, however; so we can write

$$V_c^2 - V_0^2 = (V_c + V_0)(V_c - V_0)$$

and approximate this by $2V_c(V_c - V_0)$. It then follows from Eq. (7.38) that the mean velocity of a group of stars lags behind the circular velocity by an amount that is proportional to its σ_R^2. This property of stellar motions is sometimes called the Strömberg quadratic relation, after the astronomer who discovered it empirically in the 1920's, before its reason was understood.

Because of the association of this lag with the general tendency for the distribution of V velocities to be different in the forward and backward directions, we shall refer to Eq. (7.38) as "the asymmetric-drift equation." It has important applications, which we will discuss after rederiving it in a sounder way by the velocity-moment technique. That equation will be directly applicable, whereas Eq. (7.38) requires, as already indicated, a more complicated superposition of solutions.

Another of our conclusions needs modification too. In Eq. (7.34) we have in fact derived the whole rotation curve of the Galaxy, from its center out to infinity. (This form is sometimes called a Chandrasekhar rotation curve, from its derivation by that astronomer as part of a more general discussion.) Such a form is clearly too restrictive, but the contradiction is easily understood by examining the extent to which our original assumption about the velocity distribution can be taken seriously. What is literally true is that if the *whole* velocity distribution were described by Equations 7.24 and 7.26, then the rotation curve would have to take exactly the form of Eq. (7.34). The fallacy is that, although Equations 7.24–7.26 represent the actual velocity distribution reasonably well for the lower velocities, the higher-velocity stars show all sorts of asymmetries and non-Gaussian characteristics. Here, then, is just where we can use Liouville's theorem to understand the paradox of the too strongly determined rotation curve. The stars farthest from the mean velocity satisfy the assumptions the least well. But they are the stars that go farthest from the solar neighborhood; so it is natural that our rotation curve is much less reliable far from the solar neighborhood. In fact, except for a general crude shape, we can use it reliably only to relate the Oort constants to the shape of the velocity ellipsoid, as we have already done.

Finally, the assumption of a quadratic velocity distribution has one further important consequence. (Our Gaussian assumption was stronger than merely assuming a quadratic form, but we shall see later that the quadratic assumption suffices for this conclusion.) If in Eq. (7.28) we equate the coefficients of U and

W separately to zero, differentiate the first with respect to z and the second with respect to R, and equate the cross derivatives that are in common, the result is

$$0 = 3c_5 \left(z\frac{\partial\psi}{\partial R} - R\frac{\partial\psi}{\partial z} \right) + c_5 Rz \left(\frac{\partial^2\psi}{\partial R^2} - \frac{\partial^2\psi}{\partial z^2} \right)$$
$$+ c_5(z^2 - R^2)\frac{\partial^2\psi}{\partial R\partial z}$$
$$+ 2(c_1 - c_4)\frac{\partial^2\psi}{\partial R\partial z}. \tag{7.39}$$

Unless $c_5 = c_1 - c_4 = 0$, which corresponds to the two-integral model that we already found to contradict the observed velocity ellipsoid, this equation sets a restriction on ψ. In his original treatment in 1928, Oort took $c_5 = 0$, $c_1 \neq c_4$, which means $\partial^2\psi/\partial R\,\partial z = 0$ and leads to the Oort–Lindblad third integral. But if we retain the generality of $c_5 \neq 0$, Eq. (7.39) is the equation that defines the Stäckel potentials previously referred to. We shall return later to the relation of the value of c_5 to the values of the other constants.

7.8 MOMENT EQUATIONS

We now introduce another of the ways of manipulating the collisionless Boltz-mann equation. In this approach we integrate over the velocities. Although the full truth about a stellar system is indeed contained in its full phase-space dis-tribution function, it is rare that we have such detailed information available, and in practice the properties for which we are searching often relate only to spatial variations of density, of mean velocity, or of velocity dispersions. In order to isolate information about the spatial properties of the system, it thus makes good sense to integrate the collisionless Boltzmann equation over the velocities, term by term. The resulting terms all contain moments of the velocity distribu-tion, usually including the zero-th moment, which is of course the spatial density. These moment equations have often been referred to under the heading of "equa-tions of stellar hydrodynamics," but we shall avoid that term here, because it belittles the versatility of moment equations. This versatility arises from the fact that we are free to multiply the collisionless Boltzmann equation through by any powers of the velocities before integrating, and each choice of powers leads to a different result.

In the derivation of moment equations the integrations that need to be per-formed can be carried out by following a simple set of rules. First, we note the general definition of a moment:

$$\langle U^j V^k W^l \rangle = \frac{\iiint\limits_\infty U^j V^k W^l \, f \, dU \, dV \, dW}{\iiint\limits_\infty f \, dU \, dV \, dW}. \tag{7.40}$$

Note that the denominator is just the number density N, so that:

(a) The integral of $U^j V^k W^l$ is $N\langle U^j V^k W^l \rangle$. (Special case: the integral of unity is N.)

We can apply two other general rules in performing the integrations:

(b) A differentiation operator with respect to *position* can be brought outside the integral signs.

(c) Pure functions of position can be brought outside the integral signs (but not outside position differentiations, of course).

It turns out that there are just three types of integral to cope with, according to the relation between velocity differentiations and the presence of those velocities.

Type 1: no velocity derivatives. Apply rules *b* and *c* as appropriate, then apply rule *a*. Example:

$$\iiint_\infty U \frac{\partial f}{\partial R}\, dU\, dV\, dW = \frac{\partial}{\partial R} \iiint_\infty U f\, dU\, dV\, dW$$

$$= \frac{\partial(N\langle U \rangle)}{\partial R}. \tag{7.41}$$

Type 2: differentiation with respect to velocity, but that velocity does not appear explicitly. Rule: the result is zero. Example:

$$\iiint_\infty \frac{V^2}{R} \frac{\partial f}{\partial U}\, dU\, dV\, dW = \frac{1}{R} \iint_\infty V^2\, dV\, dW \int_{-\infty}^{\infty} \frac{\partial f}{\partial U}\, dU$$

$$= \frac{1}{R} \iint_\infty V^2\, dV\, dW \int_{U=-\infty}^{\infty} df$$

$$= \frac{1}{R} \iint_\infty V^2\, dV\, dW \left[f \right]_{U=-\infty}^{\infty}$$

$$= 0, \tag{7.42}$$

since any well-behaved f is zero at $U = \pm\infty$.

Type 3: differentiation with respect to velocity, and that velocity is present explicitly. Rule: differentiate that power of the velocity, reverse the sign, and then apply Rule *a*. Example:

$$\iiint_\infty \frac{(UV^2)}{R} \frac{\partial f}{\partial V}\, dU\, dV\, dW = \frac{1}{R} \iint_\infty U\, dU\, dW \int_{-\infty}^{\infty} V^2 \frac{\partial f}{\partial V}\, dV$$

$$= \frac{1}{R} \iint_\infty U\, dU\, dW \int_{V=-\infty}^{\infty} V^2\, df$$

$$= \frac{1}{R} \iint_\infty U\, dU\, dW \left\{ \left[V^2 f \right]_{V=-\infty}^{\infty} - 2 \int_{-\infty}^{\infty} V f\, dV \right\}$$

$$= -2 \frac{N\langle UV \rangle}{R}. \tag{7.43}$$

(In the next-to-last line, the integrated part was zero, because f must go to zero faster than V^2, or else the kinetic energy would be infinite. For even higher powers we can fall back on the weaker, but still adequate, argument that f must be identically zero beyond some finite escape velocity.)

7.9   THE GALACTIC MOMENT EQUATIONS

The most useful forms of moment equations apply to problems of local Galactic dynamics, where to a good approximation we can confine ourselves to axial symmetry and a steady state. As we have seen, the collisionless Boltzmann equation is then

$$0 = U\frac{\partial f}{\partial R} + W\frac{\partial f}{\partial z} + \left(\frac{V^2}{R} - \frac{\partial \psi}{\partial R}\right)\frac{\partial f}{\partial U} - \frac{UV}{R}\frac{\partial f}{\partial V} - \frac{\partial \psi}{\partial z}\frac{\partial f}{\partial W}. \qquad (7.44)$$

The moment equations that are most useful are those that are found by integrating after multiplication by U and W, respectively. These equations are (as you can easily see by applying the rules quoted above)

$$0 = \frac{\partial(N\langle U^2\rangle)}{\partial R} + \frac{\partial(N\langle UW\rangle)}{\partial z} + \frac{N}{R}\left(\langle U^2\rangle - \langle V^2\rangle\right) + N\frac{\partial \psi}{\partial R}, \qquad (7.45)$$

$$0 = \frac{\partial(N\langle W^2\rangle)}{\partial z} + \frac{\partial(N\langle UW\rangle)}{\partial R} + \frac{N}{R}\langle UW\rangle + N\frac{\partial \psi}{\partial z}. \qquad (7.46)$$

These are often referred to as the stellar hydrodynamic equations, or sometimes, as in the following chapter, as the Jeans equations. These equations almost separate the motions parallel to the Galactic plane from the perpendicular motions; only the $\langle UW\rangle$ terms couple the two equations. In the early development of Galactic dynamics it was customary to neglect the coupling terms and treat the parallel and perpendicular motions as completely independent; here, however, we note that the $\langle UW\rangle$ terms can be treated by applying the theory of the quadratic third integral. (Eq. [7.46] will be discussed in the following chapter.)

7.10   THE ASYMMETRIC-DRIFT EQUATION

We now treat the "U-equation," 7.45, in order to derive the velocity-moment form of the asymmetric-drift equation. To do this, we make the substitutions

$$\frac{\partial \psi}{\partial R} = \frac{V_c^2}{R}, \qquad (7.47)$$

$$\langle V^2\rangle = V_0^2 + \langle V'^2\rangle, \qquad (7.48)$$

where V_c is the circular velocity, V_0 is the mean velocity, and V' is the residual velocity of a star with respect to this mean. (Note that Eq. [7.48] is a general property of all distribution functions.) After a little manipulation the result becomes

$$V_c^2 - V_0^2 = \langle U^2 \rangle \left[-\frac{\partial \ln(N\langle U^2 \rangle)}{\partial \ln R} - \left(1 - \frac{\langle V'^2 \rangle}{\langle U^2 \rangle} \right) - \frac{R}{N\langle U^2 \rangle} \frac{\partial(N\langle UW \rangle)}{\partial z} \right]. \quad (7.49)$$

This is the moment form of the asymmetric-drift equation. It corresponds closely to Eq. (7.38), which we derived from the Oort substitution; but there are some subtle differences. Here we have $\langle U^2 \rangle$ instead of $1/2h^2$, but the Gaussian form of the Oort substitution made them equal in that more restrictive case. The R derivative now includes $\langle U^2 \rangle$; when we made the Oort substitution, the R gradient of this quantity was found to be zero, but our present result applies to more general velocity distributions. The middle term is the same as before, on account of the close correspondences between $1/2h^2$ and $\langle U^2 \rangle$, etc., that we have noted. The last term in the brackets looks different from the one that resulted from the Oort substitution, but when we come to evaluate these correction terms, we will find that it is almost equivalent.

The asymmetric-drift equation is quite valuable, beyond merely showing that high-velocity-dispersion stars lag. Its quadratic character allows another important result to be extracted from it: we can locate the circular velocity, whose relation to the Sun's velocity is by no means obvious. Since the various species of stars that go to make up the Galactic disk have generally similar density gradients (as evidenced, indeed, by the very existence of the Strömberg quadratic relation), we should be able to plot V_0 against σ_R^2 for different types of stars and get a straight line. All these V_0's are necessarily referred to the Sun, which is the only available reference point. The extrapolation of the line to zero velocity dispersion should then give the circular velocity, referred to the Sun.

Even more important, perhaps, is the fact that the asymmetric-drift equation allows us to find density gradients from velocities, without directly measuring densities at remote points. On the right side of Eq. (7.49) the other terms can be evaluated. We have already noted the Lewis and Freeman (1989) observations evaluating $\partial \ln\langle U^2 \rangle / \partial \ln R$. The second term is, of course, known from local observation, and we shall see in a moment how to approximate the final term. Furthermore, these other terms turn out to be considerably smaller in magnitude than the gradient term, so that they need not be evaluated with high precision. Thus the logarithmic density gradient $\partial \ln N / \partial \ln R$ can be found from other observed quantities. This is a direct method of evaluating the scale length of the Galactic disk.

It is interesting to note that we can also recover the equivalent of the "star-streaming equation," Eq. (7.37), by a moment integration. Here we multiply each term by $UV - U\langle V \rangle$ before integrating. (The manipulation is long and tedious.) The result is the same as before, except for some small correction terms involving third moments (all of which would be zero for the previous quadratic velocity distribution). And again, the ratio $-B/(A - B)$ involves $\langle V \rangle$ rather than V_c.

7.11 ALLOWING FOR THE EFFECTS OF THE THIRD INTEGRAL

If we avoid the simplification of setting $c_5 = 0$, then Eq. (7.39) opens broader possibilities for the dynamics of the Milky Way. The equation is trying to tell us something about the form of the potential, which turns out to be of the Stäckel form previously referred to. If we want to have a triaxial Gaussian velocity distribution, without introducing the oversimplification of the Oort–Lindblad potential, then ψ must satisfy Eq. (7.39). Furthermore, we found earlier that the shape of the velocity ellipsoid implies the existence of another integral. Its particular form should also follow from the form of the potential. To investigate these questions, we can take a somewhat more general approach than that of the Oort substitution.

Remember also that the unrealistic isothermalities implied by the Oort substitution can be removed by using a superposition of solutions. The same remark applies here, again with the restriction that all solutions that are used must correspond to the same solution of Eq. (7.39).

It turns out that Oort's discussion can be repeated under slightly less restrictive assumptions, with almost the same results. We can drop the Gaussian part of his assumption, merely requiring that the velocity distribution be expressible as $F(Q)$, where Q is the quadratic form given by Eq. (7.26). It is then easily shown that Q itself must satisfy the collisionless Boltzmann equation. This means in turn that Q must, by Jeans' theorem, be expressible as a function of the integrals, including the new one that we are seeking. Furthermore, since Q is quadratic in the velocities, the new integral must also be quadratic. We can, in fact, use Jeans' theorem to find its analytic form. Since Q must be expressible as a function of the integrals of the equations of motion, we can simply express as much of it as possible in terms of

$$I_1 = \tfrac{1}{2}(U^2 + V^2 + W^2) + \psi(R, z),$$
$$I_2 = RV,$$

and identify the left-over terms as the sought-after third integral. The result is a complicated expression that will not be given here, but it is actually a quadratic third integral.

The result of substituting Q turns out to be almost the same as that of the Oort substitution. The only terms missing are $-U(\partial \ln f_0/\partial R) - W(\partial \ln f_0/\partial z)$; all the other coefficients of powers of the velocities (which we can again set equal to zero) come out exactly as they did in the Oort substitution. We thus recover all the same equations for the spatial changes of the velocity dispersions, as well as for the axial ratio of the velocity ellipsoid, plus the full Chandrasekhar rotation curve (which is again subject to the same criticisms of excessive extent). The only thing lacking from the Oort-substitution results is the asymmetric-drift equation, whose detailed form depends on the profile of the velocity distribution, not just on its quadratic shape.

The fitting of the Milky Way by a Stäckel potential can be done by using one of the mass models discussed in Chapter 6 to calculate the values, in the solar neighborhood, of the derivatives of the potential that appear in Eq. (7.39); the equation then gives a relation between c_5 and the other constants. By means of some rather tedious manipulations, and some simplifying assumptions, it can then be shown that the c_5 term in Eq. (7.38) can be approximated by

$$\tfrac{1}{2} c_5 \frac{R^2}{l^2} \cong \left(1 - \frac{h^2}{l^2}\right).\tag{7.50}$$

Similarly, the $\langle UW \rangle$ term in Eq. (7.49) can be approximated by

$$\frac{R}{N\langle U^2\rangle} \frac{\partial(N\langle UW\rangle)}{\partial z} \cong \left(1 - \frac{\langle W^2\rangle}{\langle U^2\rangle}\right).\tag{7.51}$$

Again, the reasoning used before, about the correspondence between h^2 and l^2 on the one hand and the second moments on the other hand, shows that the $\langle UW \rangle$ correction terms in both forms of the asymmetric-drift equation are equivalent.

Finally, the $\langle UW \rangle$ terms in the W hydrodynamic equation can be shown to be approximately

$$\frac{\partial(N\langle UW\rangle)}{\partial R} + \frac{N\langle UW\rangle}{R} \cong -9N\frac{z}{R^2}\langle W^2\rangle.\tag{7.52}$$

These are rather crude approximations, however. In practice, the velocity distribution must be represented as a sum of quadratic components. (But note, again, that they must all correspond to the *same* Stäckel potential.) As a result, the two preceding equations are only coarse approximations, usable only in the immediate vicinity of the Sun. Furthermore, the relation of this approximation to that used in Eq. (8.4), in the following chapter, is obscure, since they result from different starting assumptions.

It is interesting to note that Dejonghe and de Zeeuw (1988) have fitted a potential of the Stäckel type to the rotation curve of the Bahcall–Schmidt–Soneira model that was described briefly in Chapter 6. This is not necessarily a suitable potential for local studies of Galactic dynamics, however. It is a global fit, whereas there may be Stäckel potentials that give a better fit to local circumstances. Furthermore, the potential alone tells only a part of the story; the distribution function depends on how the orbits are populated. An incisive discussion of such problems is given by Statler (1989).

REFERENCES

Blaauw, A., and Schmidt, M., eds. 1965. *Galactic Structure*. Chicago, IL: Univ. of Chicago Press.

Contopoulos, G. 1958. *Stockholm Obs. Annals*, **20**, no. 5.

Dejonghe, H., and de Zeeuw, P. T. 1988. *Astrophys. J.*, **329**, 720.

Hénon, M. 1982. *Astron. Astrophys.*, **114**, 211.

Hénon, M., and Heiles, C. 1964. *Astron. J.*, **69**, 73.

Lewis, J. R., and Freeman, K. C. 1989. *Astron. J.*, **97**, 139.

Mayer, F., and Martinet, L. 1973. *Astron. Astrophys.*, **27**, 199.

Ollongren, A. 1965. In Blaauw and Schmidt, Chapter 21.

Oort, J. H. 1932. *Bull. Astron. Inst. Neth.*, **6**, 249 (no. 238).

———. 1960. *Bull. Astron. Inst. Neth.*, **15**, 45 (no. 494).

Philip, A. G. D., and Lu, P., eds. 1989. *The Gravitational Force Perpendicular to the Galactic Plane*. Schenectady, NY: L. Davis Press.

Statler, T. S. 1989. In Philip and Lu, p. 133.

CHAPTER 8

MOTIONS PERPENDICULAR
TO THE GALACTIC PLANE

Gerard Gilmore

The distribution of mass in the Galactic disk is characterized by two numbers, its local *volume* density ρ_0 and its total *surface* density $\Sigma(\infty)$. Both these dynamical quantities are derived from a measurement of the vertical Galactic force field, $K_z(z)$. They are fundamental parameters for many of the aspects of Galactic structure discussed in these lectures, such as chemical evolution (is there a significant population of white-dwarf remnants from early episodes of massive star formation?); the physics of star formation (how many brown dwarfs are there?); disk galaxy stability (how important dynamically is the self-gravity of the disk?); the properties of dark matter (does the Galaxy contain *dissipational* dark matter, which may be fundamentally different in nature from the dark matter assumed to provide flat rotation curves?); and non-Newtonian theories of gravitation (where does a description of galaxies with Newtonian gravity and no dark matter fail?).

Although $\Sigma(\infty)$ and ρ_0 are different measures of the distribution of mass in the Galactic disk near the Sun, they are related. Of the two, the most widely used and commonly derived is the local *volume* mass density, *i.e.*, the amount of mass per unit volume near the Sun, which for practical purposes is the same as the volume mass density at the Galactic plane. This quantity has units of $\mathcal{M}_\odot \, \mathrm{pc}^{-3}$, and its local value is often called the "Oort limit," in honor of the early attempt at its measurement by Oort (1932). The contribution of *identified* material to the dynamically determined Oort limit may be calculated by summing all local observed matter, an observationally difficult task, and one that leads to considerable uncertainties. The uncertainties arise in part from difficulties in detecting very low-luminosity stars, even very near the Sun, in part from uncertainties in the binary fraction among low-mass stars, in part from uncertainties in the stellar mass–luminosity relation, but mostly from uncertainties in calculating the volume density of the interstellar medium (ISM). This last uncertainty is exacerbated by the fact that the physically important quantity for dynamical purposes is the mean volume density of the patchily distributed ISM at the solar Galactocentric distance. The best available value for the local mass density in identified material is $\sim 0.1 \, \mathcal{M}_\odot \, \mathrm{pc}^{-3}$, with a very poorly defined uncertainty of perhaps as much as 25 percent. Comparison of this value with that calculated from dynamical analyses has provided the evidence for the existence of dark matter associated with the Galactic disk.

The second measure of the distribution of mass in the solar vicinity is the integral surface mass density. This quantity has units of $\mathcal{M}_\odot \, \mathrm{pc}^{-2}$, and is the total amount of disk mass in a column perpendicular to the Galactic plane. It is this quantity which is required for the interpretation of rotation curves and the large-scale distribution of mass in galaxies. Recent calculations of this surface mass density lead to values in the range $45 \, \mathcal{M}_\odot \, \mathrm{pc}^{-2}$ to $80 \, \mathcal{M}_\odot \, \mathrm{pc}^{-2}$. As an indication of the global dynamical significance of this mass density, the contribution of a disk potential generated by some known local mass density to the local circular velocity, assuming an exponential disk with the Sun 2.5 radial scale lengths from the Galactic center, is

$$V_{c,\mathrm{disk}} \sim 150 \left(\frac{\Sigma_{\mathrm{local}}}{60 \, \mathcal{M}_\odot \, \mathrm{pc}^{-2}} \right)^{\frac{1}{2}} \mathrm{km \, sec}^{-1}. \tag{8.1}$$

The local circular velocity is ~ 220 km sec^{-1}. The contribution of the potential due to a given enclosed mass $\mathcal{M}$ to the circular velocity is approximately $V_c^2 = G\mathcal{M}/r$, so that velocity contributions to the observed local circular velocity from the various mass components generating the Galactic potential add in quadrature. Thus the Galactic disk provides only about 50 percent of the total Galactic potential well at the solar Galactocentric distance. The nature of the mass which generates the other half of the potential remains unknown. (It is useful to appreciate that the evidence for extended dark matter in the Galaxy which is dynamically significant interior to the Sun's orbit is derived from comparison of the local circular velocity with the circular velocity implied by the local surface mass density of the disk, and not directly from stellar-dynamical analyses.) There has also been continuing suggestive evidence that some substantial fraction of the 45 to $80 \, \mathcal{M}_\odot \, \mathrm{pc}^{-2}$ associated with the disk is also unidentified. Detailed local dynamical analyses provide the opportunity to measure the spatial distribution of the disk mass, and perhaps thereby to discover the nature of any unidentified contribution. Such analyses require both local and global measurements.

If we knew both the local *volume* mass density and the integral *surface* mass density of the Galactic disk, we could immediately constrain the scale height of any contribution to the local volume mass density that was not identified. For example, we might suspect that some fraction of the local volume mass density was unidentified (*i.e.*, a local "dark mass" problem), but we might nevertheless calculate a surface density which is effectively fully explained by observed mass. Then the unidentified contribution to the local volume density would have to have a small scale height, in order that its integral contribution to the surface density be small. In view of the very small scale height on which it must be distributed, it would then be plausible to deduce that any "local" dark mass unidentified in the volume mass density near the Sun was not the same "dark" mass which dominates the extended outer parts of galaxies.

8.1 MEASUREMENT OF THE GALACTIC POTENTIAL

Calculation of the volume mass density and the integral surface mass density near the Sun requires similar observational data, namely distances and velocities for a suitable sample of tracer stars, but rather different analyses.

All calculations of the mass distribution in the Galactic disk require a solution of the collisionless Boltzmann equation. In view of the inconvenience of general solutions of this equation derived from real data, in practice we use its vertical velocity moment, the vertical Jeans equation (Eq. 7.46), which we repeat here for convenience:

$$K_z = \frac{1}{\nu}\frac{\partial}{\partial z}(\nu\sigma_{zz}^2) + \frac{1}{R\nu}\frac{\partial}{\partial R}(R\nu\sigma_{Rz}^2), \tag{8.2}$$

where $\nu(R, z)$ is the space density of the stars, and $\vec{\sigma}_{ij}^2(R, z) = \langle v_i v_j \rangle - \langle v_i \rangle \langle v_j \rangle$ their velocity dispersion tensor. (Although Chapter 7 used moment notation for these quantities, it is more appropriate here to emphasize their tensor nature.)

The first term on the right-hand side of Eq. (8.2) is dominant, and contains a logarithmic derivative of the stellar space density $\nu(z)$, and a derivative of the vertical velocity dispersion, σ_{zz}. Since the stellar population in the solar neighborhood is, to a quite good approximation, tolerably well described by an isothermal stellar population, the part of this term containing the derivative of the space density dominates $K_z(z)$ near the Sun. This point is not often appreciated adequately, but means that we should measure stellar density profiles with even greater care than that required for the velocity dispersions.

The second term in the Jeans equation (8.2) describes the tilt of the stellar velocity ellipsoid away from the local cylindrical-polar coordinate system in which velocity dispersions are measured. We therefore need the R-gradients of σ_{Rz} and of ν. There are no general analytical solutions for this term, because it depends on the unknown third integral of the motion (see Chapter 7). Estimates of its importance may be derived by numerical integration of orbits in potentials which are thought to be realistic approximations to that of the Galaxy, or by analytic studies of idealized approximations to the Galactic potential.

There is no general method for constructing distribution functions which reproduce a particular density in a particular potential. Consequently we do not know exactly what the velocity ellipsoid does once the plane-parallel approximation breaks down. However, there are a few theoretical results that are useful. In Stäckel potentials (sometimes also known, in the Galactic context, as Eddington potentials) analytic solutions do exist. In these potentials, the Hamilton-Jacobi equation (which yields the equations of motion in generalized coordinates) separates in confocal ellipsoidal coordinates, so that knowing the coordinate system corresponding to a particular Stäckel potential automatically yields the alignment of the velocity ellipsoid, though necessarily in a model-dependent way. A special case is a spherical galaxy, in which the equations separate in spherical polar coordinates, so that one axis of the velocity ellipsoid always points toward the galactic center. Nonseparable potentials do not have this nice property: the

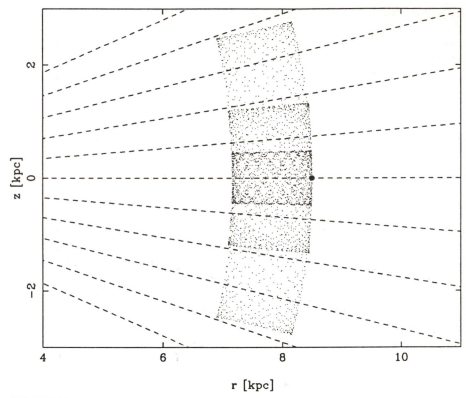

r [kpc]

FIGURE 8.1
The projection in the (R, z)-plane of three orbits calculated in a plausible Galactic potential. The dashed lines pass through the Galactic center. The envelopes of these orbits are well aligned with the Galactic radius vector in the plane, but diverge slightly from it at large z, providing an estimate of the orientation of the stellar velocity ellipsoid far from the plane. (Royal Astronomical Society)

velocity ellipsoids of different solutions to the collisionless Boltzmann equation do not have aligned axes.

To illustrate the significance of this nonseparability in determining the σ_{Rz} term, Fig. 8.1 shows the projections in the (R, z)-plane of three orbits computed in the disk–halo potential described by Carlberg and Innanen (1987). The envelopes of these orbits agree in the plane, but diverge slightly at large z. Consequently different orbits fill out boxes which are not part of a single coordinate grid, and this potential is not separable. Note that the envelopes of these orbits tilt toward the Galactic center, but not as strongly as they would have in a spherical potential, when they would have been aligned with the dashed lines in Fig. 8.1. Thus we may use numerical orbit integrations to estimate the likely range of values for the σ^2_{Rz} term in the Jeans equation.

We may therefore derive a realistic upper limit on the importance of the second term in the Jeans equation by considering velocity ellipsoids which are oriented toward the Galactic center. In this situation, if the disk of the Galaxy is self-gravitating and radially exponential, and has a constant vertical scale height, as is seen in external disk galaxies, vertical balance implies (for disk surface density μ) $\sigma_{zz}^2 \propto \mu$, and hence

$$\sigma_{zz}^2 \propto \mu \propto \nu \propto e^{-R/h_R}. \tag{8.3}$$

Thus we obtain

$$\frac{1}{R\nu}\frac{\partial}{\partial R}(R\nu\sigma_{Rz}^2) = 2(\alpha^2 - 1)\sigma_{zz}^2 \left[\frac{\alpha^2 z^3}{(\alpha^2 z^2 + R^2)^2} - \frac{Rz}{h_R(\alpha^2 z^2 + R^2)} \right] \tag{8.4}$$

as the tilting term for a radially exponential population of constant vertical scale height with a velocity ellipsoid of constant axis ratio $\sigma_{RR} : \sigma_{zz} = \alpha$ which points at the Galactic center (Kuijken and Gilmore 1989a). Since this term is proportional to σ_{zz}, inserting it into the Jeans equation (8.2) gives a linear equation in σ_{zz}, from which we can deduce K_z. The tilt term is clearly unimportant near the plane, as $z \Rightarrow 0$, and may legitimately be ignored in analyses restricted to local calculations of ρ_0.

8.2 FROM GRAVITY TO MASS

Given a measurement of the gravitational field $\vec{K}(R, z)$ in an axisymmetric galaxy, the total density ρ of gravitating matter follows from Poisson's equation:

$$\nabla \cdot \vec{K} = -4\pi G\rho. \tag{8.5}$$

For a disk galaxy we can express the R-gradient in $\nabla \cdot \vec{K}$ in terms of the observed circular velocity at the Sun, V_c, or in terms of the Oort constants of Galactic rotation A and B:

$$\rho = -\frac{1}{4\pi G}\left[\frac{\partial K_z}{\partial z} + \frac{1}{R}\frac{\partial}{\partial R}(RK_R) \right]$$

$$= -\frac{1}{4\pi G}\left[\frac{\partial K_z}{\partial z} + \frac{1}{R}\frac{\partial(V_c^2)}{\partial R} \right]$$

$$= -\frac{1}{4\pi G}\left[\frac{\partial K_z}{\partial z} + 2(A^2 - B^2) \right]. \tag{8.6}$$

For a disk galaxy with an approximately flat rotation curve, the second term is small within a few kpc of the disk plane (Kuijken and Gilmore 1989a provide a more exact calculation; for an exactly flat rotation curve $A^2 - B^2 \equiv 0$ at $z = 0$), so we can integrate in z to obtain the total column density $\Sigma(z)$ between heights $-z$ and z relative to the disk plane $z = 0$:

$$\Sigma(z) = \int_{-|z|}^{|z|} \rho(z)\, dz = \frac{|K_z|}{2\pi G} - \frac{(A^2 - B^2)}{\pi G}|z|. \tag{8.7}$$

The physical interpretation of a calculation of the Galactic $K_z(z)$ force law can now be seen from inspection of Equations (8.2), (8.6), and (8.7). In effect, we measure the pressure–gravity balance of the *collisionless* stellar "fluid." The hydrodynamic analogy that follows from the description of the collisionless Boltzmann equation as the equation of stellar hydrodynamics is particularly appropriate here. The dominant first term on the right-hand side of Eq. (8.2) contains a logarithmic derivative of the stellar spatial density $\nu(z)$, and the stellar velocity dispersion σ_{zz}. The spatial-density term plays the role of a scale height, the velocity dispersion is analogous to a temperature, and the product $\nu\sigma_{zz}$ is a pressure. Were there such an instrument as a stellar manometer, the analogy would be complete. However, some complexity is introduced by the fact that the stellar fluid is collisionless. This means that nondiagonal terms in the velocity dispersion tensor exist (the second term on the right-hand side of Eq. [8.2]) because pressure is not isotropic, and we have no equation of state to close the series of moment equations. It also means that we must measure the temperature-density balance locally by point-by-point sampling of many stars, imposing some inconvenience on the observational techniques.

It is evident from the preceding equations that the values for the local volume mass density ρ_0 depend on the square of any distance-scale errors in the tracer population, since they are derived from the second derivative of the stellar space-density distribution, whereas values for the surface mass density are linearly proportional to the distance scale, being based on the first derivative. Since the stellar space density is itself a derivative from the basic star-count data, it is evident that values of $K_z(z)$ and particularly of ρ_0 are very sensitive to sampling noise (and systematic errors) in the data. Direct analyses of combined density-law and velocity-dispersion data which proceed by substituting observed values into Equations (8.2) and (8.6) therefore tend to produce rather erratic answers.

Perhaps paradoxically, in spite of their highly divergent and unphysical force laws away from the plane, almost all calculations of the local *volume* mass density produced values which agreed to within better than a factor of 2. The reason for this follows from the structure of the Jeans equation in the form discussed below. That is, the derived force law and value of ρ_0 depend almost entirely on the adopted stellar density law. Since previous investigations tended to concentrate on intrinsically bright stars, all of which have similar velocity dispersions, only the density law was really being calculated, and since in practice this was rarely calculated for the specific investigation of interest, but tended to be adopted from a small number of published surveys, we can see why we can perceive the similarity, though not necessarily the correctness, of the results.

8.3   ISOTHERMAL DECOMPOSITIONS

Almost all calculations of ρ_0 model the distribution function of the tracer stars as one or more *isothermal* components. An isothermal component by definition has constant velocity dispersion, independent of z. Then we can easily see that the velocity distribution has to be Gaussian: by Eq. (8.2), if σ_{zz} is independent of z,

$$\nu(z) \propto e^{-\psi(z)/\sigma_{zz}^2}, \tag{8.8}$$

and hence the Abel inversion described below (Eq. 8.17) yields

$$f_z(E_z) \propto e^{-E_z/\sigma_{zz}^2} \propto e^{-\frac{1}{2}v_z{}^2/\sigma_{zz}^2}. \tag{8.9}$$

This method is especially simple if the tracer appears to be a single isothermal component and the analysis is restricted to stars near the plane, for then the Jeans equation simplifies to

$$K_z = \sigma_{zz}^2 \frac{\partial}{\partial z}\left[\ln \nu(z)\right] \tag{8.10}$$

and only the density gradient of the tracer population is needed.

The reliable decomposition of nonisothermal populations into isothermal components is neither trivial nor robust. First, there is the question of how physically meaningful it really is to separate a stellar sample into discrete isothermal subsamples. Given continual star formation and the continuous diffusion processes thought to be active in heating the disk, this can at best be only an *ad hoc* approximation to the true distribution. A very large number of discrete isothermals is clearly necessary for a close approximation to a real galactic disk. Second, as we shall see, such decompositions are far from unique, and lead to nonunique force laws. There is, however, no difficulty *in principle* with such decompositions. The difficulties lie entirely in the constraints on the data necessary to make the decomposition reliable.

At distance z_j, a superposition of N_{iso} isothermal components, each with squared velocity dispersion $\sigma_{zz,i}^2$ and density ν_i at $z = 0$, has density

$$\nu(z_j) = \sum_{i=1}^{N_{\text{iso}}} \nu_i e^{-\psi(z_j)/\sigma_{zz,i}^2}. \tag{8.11}$$

This makes clear that at different distances from the plane, the relative densities of the components are not the same, but those of highest velocity dispersion dominate increasingly at higher z. If we were able to specify the velocity dispersions and the spatial density normalization at $z = 0, \nu_i(0)$ of all components, we would have a direct solution for the potential from the density profile. However, in practice we have to derive the number density of the high-velocity-dispersion components at $z = 0$ from high-z velocity data, for only at greater heights are these components sufficiently in evidence to allow reliable calculation of their density. However, for $z \neq 0$, the relative densities of the isothermal components depend on the potential as well as the ν_i: thus we are faced with a self-consistency problem involving both the density profile and the potential far from the plane, even if we wish to solve for the potential only near the plane.

At height z_j, the mth velocity moment of a superposition of isothermals is

$$\langle |v_z|^m \rangle (z_j) = \left(\frac{m-1}{2} \right)! \frac{\sum_i \nu_i e^{-\psi(z_j)/\sigma_{zz,i}^2} \left(\frac{2\sigma_{zz,i}^2}{\pi} \right)^{m/2}}{\sum_i \nu_i e^{-\psi(z_j)/\sigma_{zz,i}^2}}. \qquad (8.12)$$

We could thus proceed by specifying a set of N_{iso} velocity dispersions $\{\sigma_{zz,i}^2\}$, and fit the densities and velocity dispersions observed at each of N_z heights z_j, solving for the potential and for the density normalizations of the individual components. However, without parameterizing the potential in some way, there are many (highly correlated) degrees of freedom in such fits: Eqs. (8.11) and (8.12) give $2N_z$ constraints for $(N_z+N_{\text{iso}}-1)$ fit parameters (the -1 arises from the free zero point of the potential). Typically, in past applications of this method, $N_z \approx 10$, and N_{iso} would have to be at least 3 to be able to describe the data adequately. With a small number of components, it turns out (not surprisingly) that the derived potential is quite sensitive to the dispersions of the individual model components. Since we require the derivative of this potential, we really have to (and also have the freedom to) adjust the dispersions of the model components until we obtain a potential which has some expected sensible form.

8.4   SELF-CONSISTENT SOLUTIONS FOR ρ_0

An almost invariable finding in early studies was that a *maximum* was found in the $K_z(z)$ law a few hundred pc from the plane. Such a result is physically impossible, corresponding to a layer of negative mass. A solution to this inconvenient situation was found by Oort (1960), who imposed consistency with the Poisson equation (8.5) on his solution of Eq. (8.10). In effect, he assumed that errors in the data rather than in the assumptions underlying the analysis are at fault, and that the theoretically reasonable solution which is most consistent with the (defective) data will produce an answer which is close to the "true" answer. Assignment of a realistic uncertainty to the resulting answer is, of course, somewhat problematic in this case. In Oort's (1960) reanalysis of Hill's (1960) K-giant data, he assumed the form of $\rho(z)$ and solved only for the scale factor ρ_0, rather than for all the quantities independently. Oort used three isothermal components in his fit to the K-giant data, so that a consistent solution was possible.

More recently, Bahcall (1984a,b,c) has extended this idea, and has developed a substantial improvement in the theoretical methods with which to calculate the local volume density of matter ρ_0, by deriving a joint solution of the Poisson and collisionless Boltzmann equations. This solution requires the derived $K_z(z)$ law to be physically possible, which is a highly desirable and quite reasonable constraint. The analytical techniques developed by Bahcall (1984a,b,c) represent a great improvement over those applied previously, and for the first time allow a derivation of ρ_0 which is limited by the quality of the available observational data, rather than by the approximate nature of the analysis.

In this technique, the solar neighborhood is divided into different isothermal components, each of which responds to the potential ψ *via* its velocity dispersion:

$$\rho_i = \rho_{i,0}\, e^{-\psi/\sigma_{zz,i}^2}. \tag{8.13}$$

Self-consistency of the potential and the total matter density in these components then requires that Poisson's equation be satisfied; *i.e.*,

$$4\pi G\rho = 4\pi G \sum_i \rho_{i,0}\, e^{-\psi/\sigma_{zz,i}^2} = \frac{d^2\psi}{dz^2}. \tag{8.14}$$

Here the density includes a constant "effective halo" density that arises from the halo mass and the radial gradients of the global gravitational field of the Galaxy (this term will be discussed further below). With the boundary conditions $\psi = \psi' = 0$ at $z = 0$, Eq. (8.14) can easily be integrated forward to obtain $\psi(z)$. A variety of dark-matter components can be added in, using the same prescription as for the visible components, and the resulting potentials calculated.

A critical problem is to analyze the gas and stars into isothermal components; this can be done to reasonable precision very near the plane, but at higher z the calculated potential becomes increasingly sensitive to the precise $z = 0$ velocity dispersions and densities. The modeling of the gas is also a problem, because it accounts for about half the locally identified volume density, but its precise density and spatial distribution are poorly known. Nevertheless, in the absence of detailed knowledge about the distribution of stars and gas within a few hundred pc from the plane, these models are very useful as a means of interpolating ψ over this distance range.

In Bahcall's (1984b,c) analyses of F-stars and of K-giants, in which he model-fitted potentials of this type, many different dark-matter components were considered. Here we shall restrict ourselves to two parameters, P_{ISM} and P_{stars}; these specify the amounts of dark matter which are distributed like the observed ISM and the observed stars, respectively, expressed as a fraction of the observed mass. In principle, of course, the dark matter can be distributed over an infinite number of isothermal components, whose relative proportions can be varied almost at will. However, varying just these two parameters covers most of the plausible types of potential that we wish to consider, and illustrates the essential features of the discussion.

Bahcall used his algorithm to reanalyze the available F-dwarf and K-giant high-Galactic-latitude data with new models which are self-consistent in the sense that the matter which generates the gravitational field itself responds to it in a manner described by the collisionless Boltzmann equation. Bahcall found that:

1. the gravitational field due to the $0.10\,\mathcal{M}_\odot\,\mathrm{pc}^{-3}$ of stars and gas that are identified in the solar neighborhood is inconsistent with the gravitational fields derived from the data;

2. depending on its scale height, a further 0.06–$0.14\,\mathcal{M}_\odot\,\mathrm{pc}^{-3}$ of unidentified matter is required. This unidentified matter is not part of a very extended halo, though that must also exist and have a local volume mass density of $\sim 0.01\,\mathcal{M}_\odot\,\mathrm{pc}^{-3}$, so that there is sufficient mass in the Galaxy to generate the potential required to explain the local circular velocity. Hence this result implies significant amounts of disk-like, dissipational dark matter in the solar neighborhood. The only significant assumptions required by Bahcall's analysis are that the available data are reliable, and that the tracer stellar populations analyzed obey the assumptions underlying the derivation of the collisionless Boltzmann equation.

8.5   UNCERTAINTIES IN THE LOCAL VOLUME MASS DENSITY

Because the existence of large amounts of dissipational dark mass would have major consequences, we should examine the uncertainties in the calculation of the Oort limit. The sensitivity of values for the local volume mass density ρ_0 to uncertain data lies in the modeling of the stellar velocity distribution near the Galactic plane, and in the calculation of the stellar density distribution with distance from this plane. Both F-dwarf and K-giant tracer samples have been analyzed to find a value for ρ_0, with both producing a result of $\rho_0 \sim 0.20\,\mathcal{M}_\odot\,\mathrm{pc}^{-3}$, where the identified mass provides $\rho_{0,\mathrm{obs}} = 0.10\,\mathcal{M}_\odot\,\mathrm{pc}^{-3}$ (Bahcall 1984c).

The effect of *random* errors on calculations of the Oort limit can be investigated by using Monte Carlo simulations of the data acquisition and analysis. Random errors can produce both random and systematic effects on a dynamically calculated quantity, as a result of effects similar to Malmquist bias. It is therefore important to understand the effects of such errors. Analyses of this type have been undertaken, but unfortunately disagree in their conclusions. Gilden and Bahcall (1985) conclude that random errors produce an unbiased uncertainty of ~ 12 percent in ρ_0, while Bienaymé *et al.* (1987) and Crézé *et al.* (1989) conclude that random errors produce both an uncertainty of ~ 50 percent, and a bias toward an erroneous detection of unidentified mass. The difference in these results comes from different techniques for handling observational errors in the simulations, suggesting that the appropriate uncertainty to apply to calculations of the Oort limit is is not yet well quantified.

An alternative, and perhaps more objective, method for generating an error bar is to do the experiment with two different but supposedly similar samples of stars, and to compare the resulting answers. Fortunately for this purpose, the F-star sample analyzed is the sum of two subsamples (F5 and F8; Hill *et al.* 1979), with no evidence for a difference between their velocity distributions (Adamson *et al.* 1988). For steady-state stellar populations, two tracer populations with the same kinematics in the same gravitational potential must follow the same spatial density distribution. For the F5 and F8 samples this is not true (Fig. 8.2). One or both of the data or the assumptions underlying the modeling of the F-star kinematics is thus clearly in error. The amplitude of the resulting uncertainty

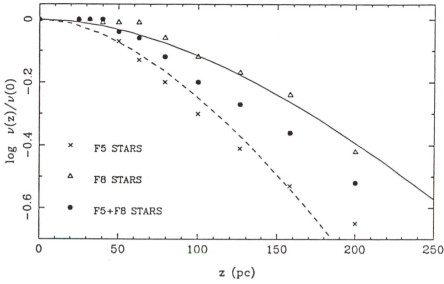

FIGURE 8.2

The Hill *et al.* (1979) F-star samples. The difference between the density profiles of the F5 and the F8 samples is evident. The curves show separate model fits calculated using the algorithm devised by Bahcall (1984a) to the F5 and the F8 subsets of the data defined by Hill *et al.* Only the averaged sample (solid points) was analyzed by Bahcall (1984b). The models shown have local volume mass densities: $\rho_0 = 0.11 \, \mathcal{M}_\odot \, \mathrm{pc}^{-3}$, *i.e.*, with no missing mass (solid line), and $\rho_0 = 0.29 \, \mathcal{M}_\odot \, \mathrm{pc}^{-3}$ (dashed line). (Royal Astronomical Society)

can be found by deducing ρ_0 from each of the three F-star samples, F5, F5+F8, and F8, using the algorithm derived by Bahcall (1984a). The resulting values of ρ_0 are $0.29 \, \mathcal{M}_\odot \, \mathrm{pc}^{-3}$, $0.185 \, \mathcal{M}_\odot \, \mathrm{pc}^{-3}$ (reproducing the result derived by Bahcall 1984b exactly), and $0.11 \, \mathcal{M}_\odot \, \mathrm{pc}^{-3}$, respectively. Thus we may deduce that there is twice as much mass missing as is observed in the local volume density, or that there is just as much missing mass as observed mass, or that there is no missing mass at all, depending on which sample of stars we choose to analyze. Clearly, the available F-star data cannot provide any evidence either for or against the concept of missing mass near the Sun.

The sample of K-giants which has been analyzed previously has been shown to have a velocity distribution which is consistent with a single isothermal, with a velocity dispersion of $\sim 20 \, \mathrm{km \, sec}^{-1}$ (Bahcall 1984c). Thus, unlike the F-stars, in this model the K-giants consist entirely of old disk stars, with neither young-disk nor thick-disk star representatives. Since stars of a wide range of masses become K-giants, including the present F dwarfs, this model is inherently implausible. Remember also that Oort (1960) found it necessary to model the same K-giant sample as the sum of three isothermal components, with (crudely) one component each for the young disk stars, the old disk stars, and the high-velocity halo stars. A further complication follows from a feature of Bahcall's analysis that

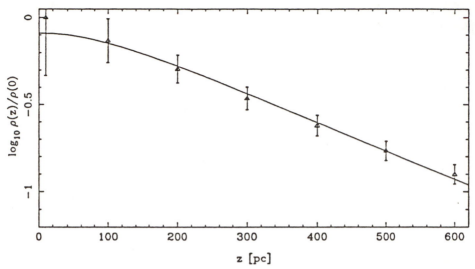

FIGURE 8.3

Weighted fit to the Upgren (1962) K-giant data, using the velocity distribution measured by Hill (1960). The model shown contains no dark matter in the Galactic disk, and has $\rho_0 = 0.10 \, \mathcal{M}_\odot \, \mathrm{pc}^{-3}$.

assigns high weight to the density profile near the plane, where the number of stars counted is smallest. Reanalysis of published data, including weighting of the density data by its Poisson noise and using the detailed fit to the local K-giant velocity data derived by Hill (1960), leads to a value of $\rho_0 = 0.10 \, \mathcal{M}_\odot \, \mathrm{pc}^{-3}$ (Fig. 8.3; see Kuijken and Gilmore 1989c). The previously derived value from the same data using the same analysis technique was $\rho_0 = 0.21 \, \mathcal{M}_\odot \, \mathrm{pc}^{-3}$ (Bahcall 1984c).

    Available values for the volume mass density near the Sun—the Oort limit—remain limited by systematic and random difficulties with the available data. We may deduce a local unexplained mass density of up to a factor of two larger than that mass density identified with stars and the interstellar medium near the Sun from some samples of (young) F-stars. Other samples of (older) F-stars and of K-giants, when analyzed using velocity distributions consistent with the structure of the local Galactic disk, provide no evidence for any unexplained mass near the Sun. Now that reliable analytical methods (Bahcall 1984a,b,c) are available for the calculation of ρ_0 from stellar density-distribution and velocity data, we may look forward to robust calculation of this important parameter as soon as adequate data can be obtained.

8.6   CALCULATION OF THE SURFACE MASS DENSITY

Because high-energy stars are present at all heights above the Galactic plane, measurements of the potential very close to the plane still require knowledge of the high-energy tail of the distribution function. Therefore either the tail of the velocity distribution at low z, or the density *and* potential at high z, are required to measure the potential at low z, and hence to deduce the local volume density of matter ρ_0. The phase-space distribution function we discuss below, however, depends on the density only at points farther from the plane than the height at which data are being analyzed. It is possible to capitalize on this insensitivity to the detailed shape of the potential (equivalently, the detailed mass distribution) near the plane to derive the potential at large distances from the plane from high-z data alone. Since a measurement of K_z at any height relates directly to the total surface density integrated to that height, this is extremely useful, allowing us to obtain meaningful values for the surface mass density of the Galactic disk from high-z data alone.

In most K_z studies, the density $\nu(z)$ is known to better precision than the velocity distribution. Instead of fixing the parameters of the latter, and then using these to model the density gradient, we should therefore work in the other direction, and predict the velocity distribution of a tracer in different model potentials, given its density. These velocity-distribution models can then be compared with the observed velocity data by using maximum-likelihood techniques.

Given a distribution function $f_z(E_z)$ and a potential $\psi(z)$, we can calculate the density $\nu(z)$, which is just a moment of f_z:

$$\nu(z) = \int_{-\infty}^{\infty} f_z(z, v_z) dv_z$$

$$= 2 \int_{\psi(z)}^{\infty} \frac{f_z(E_z)}{\sqrt{2\left(E_z - \psi(z)\right)}} dE_z. \tag{8.15}$$

Reparameterizing the z-height in terms of the potential ψ, we have

$$\nu(\psi) = 2 \int_{\psi}^{\infty} \frac{f_z(E_z)}{\sqrt{2(E_z - \psi)}} dE_z. \tag{8.16}$$

This equation is an Abel transform, which has the well-known inversion (see, *e.g.*, Binney and Tremaine 1987):

$$f_z(E_z) = \frac{1}{\pi} \int_{E_z}^{\infty} \frac{-d\nu/d\psi}{\sqrt{2(\psi - E_z)}} d\psi, \tag{8.17}$$

so that there is a unique relation between $\nu(\psi)$ and $f_z(E_z)$. Because of this equivalence of $\nu(\psi)$ and $f_z(E_z)$, there is a triangular mathematical relationship between the three functions $\psi(z)$, $\nu(z)$, and $f_z(E_z)$: any one of them can be deduced from the other two. Abel inversions are somewhat unstable, but not as unstable as taking a direct derivative of the data; they are in some sense "half-derivatives," in that any Fourier component $e^{i\omega\psi}$ of $\nu(\psi)$ corresponds to one $\propto \omega^{\frac{1}{2}} e^{i\omega E_z}$ in $f_z(E_z)$.

It is important to note that Eq. (8.17) shows that $f_z(E_z)$ depends on the density only at points where the potential exceeds E_z, *i.e.*, beyond the point

$z = \psi^{-1}(E_z)$. It is this property which allows the derivation of $K_z(z)$ at large z independently of the poorly known distribution of mass near the plane.

When starting from a set of (z, v_z) data for the tracer population, we know only the first of the three quantities $\nu(z)$, $f_z(E_z)$, and $\psi(z)$, since we need the potential to be able to convert $f_z(z, v_z)$ into $f_z(E_z)$. Therefore, before being able to make an inversion such as that given in Eq. (8.17), we have to make some *Ansatz* about the potential, or about $f_z(E_z)$. Assuming that the tracer is isothermal, $(f_z \propto e^{-E_z/\sigma_{zz}^2})$ is an example of the latter, leading to a trivial inversion to Eq. (8.11).

An analysis technique based on Eq. (8.17) has been devised by Kuijken and Gilmore, for the calculation of $K_z(z)$, and $\Sigma(z)$. The essential feature of that analysis is that we avoid the assumption of isothermality, by instead postulating a range of potentials $\psi(z)$, and for each of them calculating $f_z(z, v_z)$ from $\nu(z)$. The range of model distribution functions can then be compared to the observed distribution function of velocity–distance data, and used to select the best-fitting model potential.

This is not a direct measurement of the potential, but rather a modeling of it; so it is important to make sure that the models are sufficiently general. On the other hand, constraints can be built into the model potentials which direct derivations from data have more difficulty coping with. For example, the analysis by Hill (1960) of a K-giant sample found a $|K_z(z)|$ which started to decrease above a few hundred parsecs, which leads to negative dynamical masses—a physical absurdity. The same data were shown by Oort (1960) to be consistent with a K_z-law which did not show such a turnover. This illustrates the fact that some force laws can be ruled out on physical grounds, and so should not be included among the possible solutions (provided, of course, that acceptable fits to the data can be obtained from the restricted solution set). Consistency with the Galactic rotation curve can also be built into the potentials: if we were to find a very heavy disk mass, for example, which can generate most of the local circular speed by its gravity, we would not expect also to find evidence for a massive halo in K_z.

8.7 A SIMPLE PARAMETERIZATION OF PLAUSIBLE K_Z FUNCTIONS

As demonstrated above, $K_z(z)$ is related directly to the surface density to height z, $\Sigma(z)$. This means that the detailed distribution of the matter in the disk (whether it has a high or low scale height, whether its distribution is close to exponential or more like a sech^2 law,...) does not affect the high-z potential. Hence we can derive the disk surface density without needing to know how the mass inside it is distributed, and can investigate simpler model potentials than those required for measurements of ρ_0.

The total mass density along a slice perpendicular to the disk for a generic disk–halo system is shown schematically in Fig. 8.4. (By "halo" we here mean any mass component which is not distributed like the dominant mass in the old disk. Nothing is implied about the nature of this mass; it includes the luminous

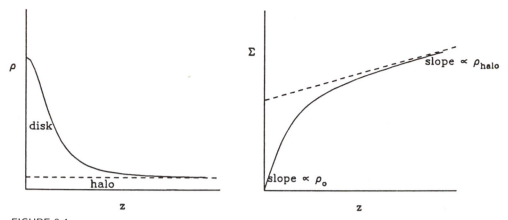

FIGURE 8.4
The volume density (a, left) and surface mass density, or equivalently $K_z(z)$, (b, right) as a function of vertical distance z from the disk plane in a simple disk-halo model. (Royal Astronomical Society)

"spheroid" or "bulge," as well as any roughly spherical distribution of dark matter.) At low z, the density is mostly due to the disk component, whereas further away from the plane the density of the halo, which at large Galactocentric radii is essentially constant over a few disk scale heights, dominates. The surface density, or equivalently $|K_z|$, for such a system is also shown in Fig. 8.4. This force law has the following generic features:

1. At low z, smaller than the scale heights of the dominant disk components, $|K_z|$ rises almost linearly, with slope $4\pi G\rho_0$.

2. At large z, beyond most of the mass of the disk, $|K_z|$ is again linear, but with a much reduced slope, equal to $4\pi G\rho_{\text{halo}}$.

3. The extrapolation of this latter linear portion back to $z = 0$ has an intercept of $2\pi G\Sigma_{\text{disk}}$.

Clearly, an accurately measured K_z-profile over a wide range in z yields the disk surface density, as well as the volume density of the halo and that of the disk at $z = 0$.

The essential features of the behavior of the K_z-function may be quantified in terms of a disk scale height D, a disk surface density K, and a halo of local density F, as follows:

$$- K_z = 2\pi GK\frac{z}{\sqrt{z^2 + D^2}} + 4\pi GFz, \tag{8.18}$$

corresponding to the potential

$$\psi(z) = 2\pi GK\left(\sqrt{z^2 + D^2} - D\right) + 2\pi GFz^2. \tag{8.19}$$

These functions reproduce all the generic behavior of a disk-halo system, except for the details of the disk-mass distribution near the plane, which are unimportant for calculations of $K_z(z)$ at large z.

The parameter F deserves further comment. It measures the large-z quadratic term in the potential, which is predominantly caused by the halo density (taken to be constant over a few disk scale heights). A part of it is due to the $(A^2 - B^2)$ term, however, discussed above. Because F behaves just like a constant halo density in the z-restriction of Poisson's equation, it has been termed the "effective halo density," ρ_{eff}. Although this name suggests that it is a function of the halo component only, this is misleading, because the quadratic part of the disk potential also contributes to the difference $\rho_{\mathrm{halo}} - \rho_{\mathrm{eff}}$.

Potentials of the form given by Eq. (8.19) can be combined with a measured density distribution $\nu(z)$ and Eq. (8.17) to predict a set of distribution functions of velocities as a function of distance. With appropriate care in handling observational errors, these can be compared with distance–velocity data using maximum-likelihood methods to find out what parameters in Eq. (8.19) are most likely to represent the Galactic potential. We may then derive the integral surface mass density of the Galactic disk. This technique has been applied by Kuijken and Gilmore (1989a,b), to calculate a dynamical total surface mass density $\Sigma_\infty = 46 \pm 9 \, \mathcal{M}_\odot \, \mathrm{pc}^{-2}$. These same authors integrate the local observed volume mass density in stars through their derived potential, and add the directly observed mass in the interstellar medium, to derive an observed integral surface mass density of $\Sigma_{\infty,\mathrm{obs}} = 48 \pm 8 \, \mathcal{M}_\odot \, \mathrm{pc}^{-2}$. There is thus direct evidence that there is no significant unidentified mass associated with the Galactic disk, if they are correct.

Calculating the appropriate uncertainties to apply to analyses of this type is extremely complex, for many of the parameters are correlated in subtle ways. Similarly, the amplitude of possible systematic errors, resulting for example from our limited knowledge of the gradient in the local rotation curve and of the shape and orientation of the stellar velocity ellipsoid at large distances from the Galactic plane, is extremely difficult to quantify. As with the calculation of ρ_0, the most reliable calculation of the uncertainty will come from the comparison of the results described here with the results of a similar measurement based on an independent sample of stars, when such is available.

The calculation of a low and fully identified total surface mass density for the Galactic disk has many implications of relevance here. It confirms that at least in our Galaxy a "maximal disk" fit to explain extended flat rotation curves is not viable (see Chapter 10). It disproves available models of chemical (and luminosity) evolution which require preferential formation of high-mass stars (e.g., by a bimodal stellar initial mass function) in the early evolution of the Galactic disk. Extrapolations of the stellar initial-mass function which result in significant mass in substellar-mass "brown dwarfs" are ruled out. It implies that the disk is comfortably stable against axisymmetric local perturbations (the Toomre Q parameter has a local value of ~ 2.1, where $Q > 1$ implies stability), and just stable against global bar mode instabilities (see Chapter 9). Severe constraints can also be set on most available alternative theories to Newtonian gravity, with the local gravitational potential gradient deduced from recent K-dwarf data being inconsistent with the most recently suggested modifications to the inverse-square law dependence of the gravitational force.

REFERENCES

Adamson, A. J., Hill, G., Fisher, W., Hilditch, R. W., and Sinclair, C. D. 1988. *Mon. Not. Roy. Astron. Soc.*, **230**, 273.

Bahcall, J. N. 1984a. *Astrophys. J.*, **276**, 156.

———. 1984b. *Astrophys. J.*, **276**, 169.

———. 1984c. *Astrophys. J.*, **287**, 926.

Bienaymé, O., Robin, S. C., and Crézé, M. 1987. *Astron. Astrophys.*, **180**, 94 (see also *erratum* in **186**, 357).

Binney, J., and Tremaine, S. 1987. *Galactic Dynamics*. Princeton, NJ: Princeton Univ. Press.

Carlberg, R. G., and Innanen, K. A. 1987. *Astron. J.*, **94**, 666.

Crézé, M., Robin, A., and Bienaymé, O. 1989. *Astron. Astrophys.*, **211**, 1.

Gilden, D. L., and Bahcall, J. N. 1985. *Astrophys. J.*, **296**, 240.

Hill, E. R. 1960. *Bull. Astron. Inst. Neth.*, **15**, 1.

Hill, G., Hilditch, R. W., and Barnes, J. V. 1979. *Mon. Not. Roy. Astron. Soc.*, **186**, 813.

Kuijken, K., and Gilmore, G. 1989a. *Mon. Not. Roy. Astron. Soc.*, **239**, 571 (Paper I).

———. 1989b. *Mon. Not. Roy. Astron. Soc.*, **239**, 605 (Paper II).

———. 1989c. *Mon. Not. Roy. Astron. Soc.*, **239**, 651 (Paper III).

Oort, J. H. 1932. *Bull. Astron. Inst. Neth.*, **6**, 249.

———. 1960. *Bull. Astron. Inst. Neth.*, **15**, 45.

Upgren, A. R. 1962. *Astron. J.*, **67**, 37.

CHAPTER 9

THE DYNAMICS OF DISKS

Ivan R. King

We now move to a more global view of the Galactic disk, examining its local oscillations and its large-scale dynamical behavior. As a first step, we derive the natural frequency with which stars oscillate in distance from the Galactic center. Since their motions with respect to a circular-velocity reference point turn out to be epicycles, this frequency is called the epicycle frequency.

9.1 EPICYCLE ORBITS

At the same time, in the 1920's, that Oort was developing the consequences of Galactic rotation, by using the collisionless Boltzmann equation, Lindblad was pursuing the same astronomical questions by the quite different technique of investigating how neighboring stars move relative to each other, as they all go around the Galactic center.

The first step is to restate the Newtonian equations of motion in polar coordinates with respect to the Galactic center. As shown in any standard text on dynamics, they take on the form

$$\frac{d^2 R}{dt^2} - R\left(\frac{d\theta}{dt}\right)^2 = -\frac{\partial \psi}{\partial R},$$

$$R\frac{d^2\theta}{dt^2} + 2\frac{dR}{dt}\frac{d\theta}{dt} = 0, \qquad (9.1)$$

$$\frac{d^2 z}{dt^2} = -\frac{\partial \psi}{\partial z}.$$

Lindblad's approach was to use an orthogonal but noninertial coordinate system whose origin is in uniform circular motion around the Galactic center, but whose ξ-axis always points away from the Galactic center. The motion of the origin is given by the equations

$$R = R_0 = \text{const.},$$

$$\theta = \theta_0, \qquad (9.2)$$

where θ_0 is such that

$$\frac{d\theta_0}{dt} = \omega_0 = \text{const.},$$

ω_0 being the angular velocity of circular motion at R_0. For differential coordinates we choose

$$\xi = R - R_0,$$
$$\eta = R_0(\theta - \theta_0). \tag{9.3}$$

(The coordinate lines are thus radial straight lines and circles around the Galactic center.)

In transforming from R, θ to ξ, η we will limit relationships to the first order, by considering that

$$\xi \ll R - R_0,$$
$$\frac{d(\theta - \theta_0)}{dt} \ll \omega_0. \tag{9.4}$$

(Note that η itself can be large; the restriction is only on its derivative.) We also expand the Galactic force field to first order around R_0; with a little manipulation this gives

$$\left(\frac{\partial^2 \psi}{\partial R^2}\right)_0 = \omega_0^2 - 4\omega_0 A, \tag{9.5}$$

where A is the first Oort constant, so that

$$\frac{\partial \psi}{\partial R} = \left(\frac{\partial \psi}{\partial r}\right)_0 + (\omega_0^2 + 4\omega_0 A)\xi. \tag{9.6}$$

With the substitution for R, θ in terms of ξ, η, the approximations indicated, and the subtracting off of the zero-order equation, the equations of motion (9.1) become (after some toil)

$$\frac{d^2\xi}{dt^2} - 2\omega\frac{d\eta}{dt} - 4\omega A\xi = 0,$$
$$\frac{d^2\eta}{dt^2} + 2\omega\frac{d\xi}{dt} \qquad = 0, \tag{9.7}$$
$$\frac{d^2\zeta}{dt^2} \qquad + \psi_{zz}\zeta = 0.$$

Note two liberties taken here. (1) We have dropped the zero subscript from ω, since the only angular velocity that remains is the local one. (2) We have also changed the z-equation into a differential one in ζ, using the notation ψ_{zz} for $\partial^2\psi/\partial z^2$. These, then, are the equations of motion of a particle with respect to a reference point that moves in a circular orbit.

Lindblad used an approach like this to follow the relative orbits of stars in a rotating galaxy, and he was able in that way to derive many of the same results that Oort found by such a different method. We shall discuss Lindblad's orbital solution here partly to demonstrate it as an independent approach to Galactic dynamics, partly because it demonstrates a natural oscillation frequency of the Galactic disk, and partly because of the intrinsic value of the resulting orbits in following the motions of young stars.

What Lindblad considered was motions parallel to the Galactic plane, as described by the first two of our equations. (The ζ-equation can be solved separately, and simply gives harmonic motion up and down through the Galactic plane, with a frequency that is controlled by the local density of the disk.)

Instead of going through the details of the solution of the ξ and η equations (which is not difficult), we shall simply quote the result:

$$\xi = c_1 + c \cos \kappa(t - t_0), \tag{9.8}$$
$$\eta = -2Ac_1(t - t_1) - \beta c \sin \kappa(t - t_0), \tag{9.9}$$

where

$$\kappa = \sqrt{4\omega(\omega - A)}$$
$$= \sqrt{-4B(A - B)} \tag{9.10}$$

and

$$\beta \equiv \frac{2\omega}{\kappa} \tag{9.11}$$
$$= \sqrt{\frac{\omega}{\omega - A}}$$
$$= \sqrt{\frac{A - B}{-B}}. \tag{9.12}$$

The two second-order equations required four integrations, yielding the constants of integration c, c_1, t_0, and t_1. The "epicycle frequency" κ and the related axial ratio β are, on the other hand, properties of the Milky Way.

The orbit that is parameterized by Eqs. (9.8) and (9.9) is a retrograde ellipse with semi-axes c and βc. Its center is at $\xi = c_1$ and drifts with respect to our circular-velocity reference point. The drift is backward for positive values of c_1 and forward for negative c_1. The constants t_0 and t_1 are time zero-points for the epicycle motion and the drift of the center. Common to all epicycles, however, are the frequency κ and the related value of the axial ratio of the epicycles, β. With the values of the Galactic rotation constants found in the discussion by Kerr and Lynden-Bell (1986), $A = 14$ and $B = -12$ km sec^{-1} kpc^{-1}, the epicycle period $2\pi/\kappa$ is 1.74×10^8 years, and the axial ratio of the velocity ellipsoid (or of the epicycle) is 1.47.

It may seem strange to modern students of astronomy to embrace a description that depends on epicycles, which are a mathematical device that recalls the Ptolemaic System. But in fact these epicycles give a valid description of the relative motions of stars.

Epicycle motion in the Galaxy is even simpler than our specific solution would imply. The constant c_1 does not give the star an orbit that is essentially

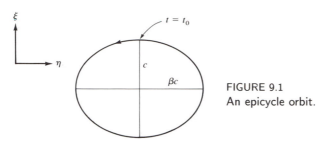

FIGURE 9.1
An epicycle orbit.

different; it merely allows for the drift of that star's orbit relative to the circular reference orbit with which we began. That drift is easily shown, however, to be a mere artifact of the choice of reference point. If we had instead chosen the reference point at $\xi = c_1$, then there would be no drift, and furthermore the center of the epicycle would have the local circular velocity. This fact becomes clear if we consider the circular velocity at $\xi = c_1$; relative to our local coordinate system it has an angular velocity $\Delta\omega = c_1 \, d\omega/dR$. Since $A = -\frac{1}{2}R \, d\omega/dR$, the angular differential velocity is $\Delta\omega = -(2A/R)c_1$; converted to a linear speed this is $-2Ac_1$, which is indeed the relative speed of the epicycle center.

Thus our general result is that every star follows an elliptical epicycle orbit around a center that moves in a circular orbit (Fig. 9.1). The amplitude of the epicycle is a characteristic of the star, but its axial ratio is determined only by the rotation curve of the Galaxy. The amplitude of the epicycle measures, from another point of view, how much the actual orbit of a star differs from a circular orbit. And from a global point of view the epicycle frequency κ is the characteristic vibrational frequency of local stellar motions, in directions parallel to the Galactic plane. (In this approximation, of course, the z-motions are decoupled and have a different oscillation frequency.)

9.2   APPLICATION TO LOCAL STELLAR MOTIONS

The epicycle ellipse is elongated in a direction perpendicular to the velocity ellipsoid. Thus it would appear that it predicts peculiar motions that are larger in the V direction rather than in the U direction. Such would certainly be the result if we calculated the mean square values of $\dot{\xi}$ and of $\dot{\eta} - \dot{\eta}_0$ (where $\dot{\eta}_0$ is the average η velocity in the epicycle). That result, however, would not apply to the velocity ellipsoid as it is normally defined. The fallacy here is that averaging a star's motion relative to a fixed, distant center is not the same as assessing its contribution to the *local* velocity ellipsoid. To do the latter, we must express the velocity of the star at every point with respect to the circular velocity at that same point. To do this, we again note that the velocity, with respect to the origin, of a circular-velocity standard at the place where the star finds itself is $-2A\xi$; thus its velocity relative to the local circular standard is

$$v = \dot{\eta} - (-2A\xi)$$
$$= 2Bc \cos\kappa(t - t_0). \tag{9.13}$$

The other local-velocity component is $\dot{\xi}$, which we will rename u, just for symmetry of notation:

$$u = -\kappa c \sin \kappa(t - t_0). \tag{9.14}$$

For these velocity components it is clear that the ratio of the mean squares is

$$\frac{\langle v^2 \rangle}{\langle u^2 \rangle} = \frac{4B^2}{\kappa^2}$$

$$= \frac{4B^2}{-4B(A - B)}$$

$$= \frac{-B}{A - B}, \tag{9.15}$$

which is indeed the correct result. The trouble we had in arriving at it, however, illustrates the kind of problem that we encounter when trying to go from the characteristics of individual orbits to the statistics of a velocity distribution.

In this vein, we shall not attempt to derive the asymmetric-drift equation from the statistics of epicycle orbits. We merely note the relevant factors: at a given point, such as the present location of the Sun, a glance at the diagram of an epicycle shows that the forward-moving stars are those whose epicycle centers are farther from the Galactic center than we are; similarly, the backward-moving stars have epicycles such that their characteristic place is closer to the Galactic center. Thus the backward-moving stars must predominate, by an amount that depends on the density gradient and also, obviously, on the general sizes of epicycles, *i.e.*, on the velocity dispersion.

9.3 EPICYCLE ORBITS OF INDIVIDUAL STARS

One of the unique advantages of the epicycle approach is that it gives us an easy way to convert a star's local velocity components into the characteristics of its Galactic orbit. The derivation of the transformation is straightforward, as follows: at $t = 0$, let $\xi = \eta = 0$, while $u = u_0$ and $v = v_0$. From the equations above for u and v, with the sine and cosine terms expanded into functions of κt and κt_0, expressions can be found for $c \cos(\kappa t_0)$ and $c \sin(\kappa t_0)$. This allows us to get c from the sum of the squares and t_0 from the arctan of the ratio. Putting $\xi = 0$, $t = 0$ in the equation for ξ then gives an expression for c_1, and similar treatment of the η equation gives t_1. The resulting form of the equations, derived by Blaauw (1952), is

$$\xi = \frac{1}{\kappa} \left[u_0 \sin \kappa t + v_0 \beta (1 - \cos \kappa t) \right], \tag{9.16}$$

$$\eta = \frac{1}{2B} \left[u_0 (1 - \cos \kappa t) + v_0 (2At - \beta \sin \kappa t) \right]. \tag{9.17}$$

Blaauw's equations conveniently allow us to go from any star's observed velocity components to its epicycle orbit. For the Sun, for example, the basic solar motion (whose inverse transfers the reference standard from the Sun to a point that moves with circular velocity) is approximately $u_0 = -10$ km sec^{-1}, $v_0 = +10$ km sec^{-1}. With the standard Galactic-rotation constants adopted earlier, we then get $c = 504$ pc, $\beta c = 742$ pc; these are the semi-axes of our own epicycle in the radial direction and in the direction of rotation. Our epicycle is centered $c_1 = 417$ pc outward from our present location, and the epicycle center drifts backward from our present location at $2Ac_1 = 11.7$ km sec^{-1}. Our present phase in our orbit is $\kappa(-t_0) = 146°$ (where the zero-point of time is, by inspection of the epicycle equations, at the outer end of the minor axis).

9.4   DISPERSAL OF AN ASSOCIATION

A very instructive application of Blaauw's equations is to follow the dispersal of a newly born stellar association, in which the stars have started in life together but are gravitationally unbound and will drift apart. Blaauw derived his equations, in fact, to study just this problem. In his 1952 paper he followed the configuration of a group of stars that begin expanding from a single point, in all directions, with relative velocities of 1 km sec^{-1}. His results are reproduced in Fig. 9.2. The numbers would be different now, because of different notions of the values of the rotation constants, but the qualitative picture remains the same: the group skews into a continually elongating ellipse. After one epicycle period the stars have returned to a tangential line, but the secular term in the η equation, $(A/B)v_0 t$, has spread them out over $(A/B)(2\pi/\kappa)$ parsecs (since Blaauw had assumed an expansion velocity of 1 km sec^{-1}). Blaauw's spread of 660 pc, which is proportional to A/B, is reduced by modern values of the rotation constants to 270 pc, but it is nevertheless impressive to see how rapidly the shear of Galactic rotation (which is what the secular term really expresses) spreads a newly formed star group along the tangential circle, for even a single kilometer per second of expansion velocity.

9.5   INSTABILITIES: THE JEANS LENGTH AND THE TOOMRE CRITERION

Long ago Jeans (1929,[1] p. 340) took up the question of a self-gravitating gas and found that under certain conditions it could be unstable enough to collapse under its own self-gravitation. His analysis considered an adiabatic gas with $p \propto \rho^\gamma$. We will do the equivalent thing here for a stellar system.

(Before taking up any details, we should clarify some terminology. A configuration is said to be stable if it does *not* form local condensations, and unstable if it does. This can be confusing, since the newly formed contractions themselves are often stable bodies. Nevertheless, this is the way the standard terminology goes, and it should be remembered carefully.)

---

[1] In its original form the work was published many years earlier.

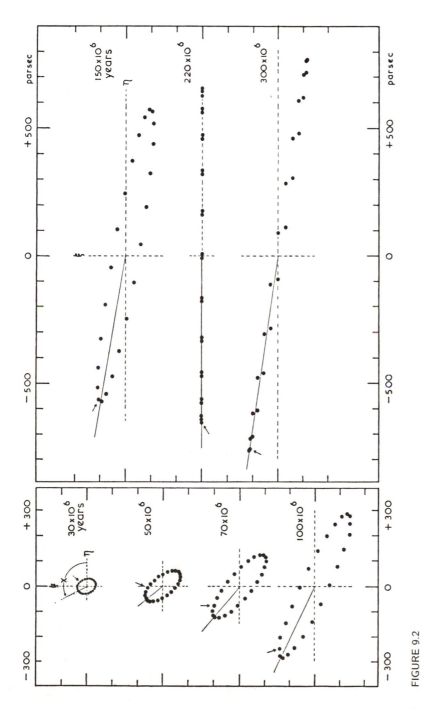

FIGURE 9.2

The dispersal of an association that begins expanding with relative velocities of 1 km sec^{-1}, as calculated by Blaauw (1952). Arrows mark the star that originally was moving in the direction of the $+\eta$ axis. These diagrams are calculated with the values $A = 20$ km sec^{-1} kpc^{-1}, $B = -7$ km sec^{-1} kpc^{-1}.

Conceptually we ask a simple question: will gravitation cause a configuration to collapse faster than velocity dispersion would cause it to fly apart? To answer this question we calculate a characteristic time for each process and simply assert that the faster one will win out. As might be expected, over small dimensions the velocities win; over larger scales the gravitation will predominate. Thus we will find a critical length above which gravitational instability asserts itself.

Consider a spherical region of radius L and density ρ, so that its mass is $\frac{4}{3}\pi L^3 \rho$. If a particle at the surface had nothing to hold it up, it would fall to the center in half the period of an infinitely narrow, elliptic orbit whose semi-major axis is $L/2$. (Everything beneath it would fall equally fast, since inside a homogeneous spherical distribution all periods are identical.) The collapse time would thus be

$$
\begin{aligned}
T_{\text{coll}} &= \frac{1}{2}\frac{2\pi(L/2)^{\frac{3}{2}}}{\sqrt{GM}} \\
&= \frac{\pi}{2^{\frac{3}{2}}} \cdot \frac{L^{\frac{3}{2}}}{\sqrt{G \cdot \frac{4}{3}\pi L^3 \rho}} \\
&= \sqrt{\frac{3\pi}{32G\rho}}.
\end{aligned}
\tag{9.18}
$$

On the other hand, if there were no gravitation, a star would run out to radius L in $T_{\text{esc}} = L/\langle v^2 \rangle^{\frac{1}{2}}$. In the critical case these two times are equal, and the radius for which equality holds is

$$
L_{\text{J}} = \sqrt{\frac{3\pi}{32}\frac{\langle v^2 \rangle}{G\rho}}.
\tag{9.19}
$$

This is called the *Jeans length*; if $L > L_{\text{J}}$, then stars cannot run fast enough and collapse will occur. Or, to put it in a different way, in a stellar medium that is characterized by given values of $\langle v^2 \rangle$ and ρ, all lengths $> L_{\text{J}}$ are gravitationally unstable. (Note: Jeans' analysis, for gas, had $\frac{5}{9}$ instead of $\frac{3}{32}$. The difference is not important, since we are merely finding an order of magnitude.)

The Jeans length is of great importance in discussions of the formation of galaxies, star clusters, stars, clusters of galaxies, and almost everything that might be imagined to condense in an originally more homogeneous Universe. It is usually expressed not as a length, but as the *Jeans mass*, which is the size of the entities that will be chosen out by the process of gravitational instability. Our application here, however, will be to the stability of a stellar disk, and we will need to deal only with lengths.

The situation in the Milky Way is somewhat different from the problem considered by Jeans. Not only do we have a flattened rather than a round configuration, but, more importantly, there is also differential rotation. Since the latter involves velocity differences that are proportional to ΔR, it might prevent collapses from taking place if the scale is too large. If so, we have velocity dispersion to protect against collapse on small scales and differential rotation to protect against it at large scales. But what happens in between? This is what Toomre (1964) analyzed at some length. He began with an approximate, intuitive discussion and then followed up with a more rigorous analysis. We will here follow the spirit of his intuitive introduction and merely quote his more accurate result at the end.

Since the Milky Way is disk-shaped rather than spherical, we repeat the previous analysis, but with $M = \pi L^2 \mu$, where μ is a surface density; but we still take the disk to attract (approximately) like a point mass. (This is not seriously in error; a sphere squashed down to infinite thinness has its equatorial attraction increased by a factor of less than 2.) Thus we have for the collapse time, as before,

$$T_{\text{coll}} = \frac{\pi}{2^{\frac{3}{2}}} \frac{L^{\frac{3}{2}}}{\sqrt{G \cdot \pi L^2 \mu}}$$

$$= \sqrt{\frac{\pi}{8} \frac{L}{G\mu}}. \tag{9.20}$$

When we set this equal to $T_{\text{esc}} = L/\langle v^2 \rangle^{\frac{1}{2}}$, the result is

$$L_{\text{J}} = \frac{\pi}{8} \frac{\langle v^2 \rangle}{G\mu}; \tag{9.21}$$

note the different form from the spherical case. [Toomre simply asserts that what we are calling T_{coll} "can be estimated to be of the order of $(L/G\mu)^{\frac{1}{2}}$ "; so he does not have the $\pi/8$.]

Toomre's important contribution was to consider Jeans instability in the presence of differential rotation. As already indicated, the problem is now a tripartite one; the new element added is that we have introduced a new fly-apart mechanism that is effective on *large* scales. What we shall do first is to analyze the balance between differential rotation and self-gravitation. Differential rotation manifests itself physically through the fact that a contracting region conserves angular momentum. This spins it up and creates a centrifugal force that might be able to inhibit further contraction. First, note from Oort's analysis of local differential rotation that the average angular velocity (with respect to a fixed system) is B. Now let us consider a region of original radius L_0. Its angular momentum at the edge, per unit mass, is $L_0^2 B$; if it contracts to radius L, the conservation of angular momentum requires that $L^2 \Omega = L_0^2 B$, so that its angular velocity will then be $\Omega = L_0^2 B/L^2$. Corresponding to this angular velocity there is a centrifugal acceleration, directed outward, whose size is $F_r = L\Omega^2 = L_0^4 B^2/L^3$. There is also a gravitational acceleration, directed inward, whose size is $F_g = G \cdot \pi L_0^2 \mu/L^2$. The question is, if these accelerations are initially in balance and a small contraction occurs, which one increases faster in size and wins out? If the centrifugal acceleration wins, the configuration is stable.

Let us suppose that the two are balanced at $L = L_0$ and ask what happens if there is a small contraction $-dL$. Then

$$dF_r = \frac{3L_0^4 B^2}{L^4} dL, \qquad\qquad dF_g = \frac{2G\pi L_0^2 \mu}{L^3} dL. \tag{9.22}$$

The critical condition that these be equal is (at $L = L_0$)

$$3B^2 = \frac{2G\pi\mu}{L_0}.$$ (9.23)

If we give this critical value of L_0 the name $L_{\rm rot}$, then

$$L_{\rm rot} = \frac{2\pi G\mu}{3B^2}.$$ (9.24)

We thus find that differential rotation will stabilize a disk against gravitational collapse for length scales $L > L_{\rm rot}$.

A disk can therefore be gravitationally unstable only in the intermediate range of lengths $L_{\rm J} < L < L_{\rm rot}$—if this intermediate range exists at all, $i.e.$, if $L_{\rm rot} > L_{\rm J}$. This was Toomre's great recognition: that the minimum condition for stability of a disk is that $L_{\rm rot} = L_{\rm J}$, or

$$\frac{2\pi G\mu}{3B^2} = \frac{\pi}{8}\frac{\langle v^2\rangle}{G\mu},$$

$$\langle v^2\rangle^{\frac{1}{2}} = \frac{4}{\sqrt{3}}\frac{G\mu}{B}.$$ (9.25)

If the stars of a disk have a lower velocity dispersion than this, the disk will be unstable.

Toomre's more elaborate (and complicated) analysis gave, in his notation,

$$\sigma_{u,\min} = 3.36\frac{G\mu}{\kappa},$$ (9.26)

where κ is the epicycle frequency. This is called the *Toomre criterion*. (In the solar neighborhood it is $\sim$ 30 km sec^{-1}.) If σ_u were lower, then presumably condensations would occur, and encounters between stars and these massive condensations would pump σ_u up to $\sigma_{u,\min}$.

The ratio of the actual velocity dispersion to $\sigma_{u,\min}$ is usually denoted by Q. It is an important quantity in the study of the dynamics of disks, and particularly in theories of spiral structure.

Note, however, that Toomre's analysis applies only to axisymmetric instabilities and that the critical length $L_{\rm T}$, where $L_{\rm J} = L_{\rm rot}$, is many kiloparsecs. Thus violation of the Toomre criterion does not provide directly for the formation of spiral arms. Note also that disks can be subject to other instabilities, such as the formation of a central bar.

9.6   BAR INSTABILITIES

In the 1970's, numerical simulations of galactic disks were plagued by a tendency to form a central bar, when the intent of the researchers had been to study the behavior of a disk that was close to axial symmetry. The situation was described most strikingly by Ostriker and Peebles (1973), who suggested that the bar instability would occur in a stellar system unless the rotational kinetic energy was less than 28 percent of the total kinetic energy. (The quantity that they actually quoted, the ratio of rotational to potential energy, is just half of this.) They were able to prove that this is so for fluid spheroids, but for stellar systems they advanced it only as an assertion that was supported quite well by a large range of numerical simulations, done by many techniques. The existence of massive halos in galaxies was just becoming recognized, and their specific suggestion was that a disk could be "cold," *i.e.*, have a low velocity dispersion, only if a massive halo was also present, with its kinetic energy in the form of random motions.

Later, Efstathiou *et al.* (1982) studied a set of numerical simulations designed to look as much like real galaxies as possible, with exponential disks but a flat rotation curve, for which they had to add a massive halo. Again they found that disk stability required that the majority of the total mass be in the massive (and in real galaxies, dark) halo. More recently, however, Athanassoula and Sellwood (1986) showed that disk stability could be achieved by increasing velocity dispersions by a reasonable amount in the central part of the disk; much less of a massive halo was then needed for stability.

9.7   THE GALACTIC WARP

The most distant hydrogen observed at 21 cm does not conform at all longitudes to the Galactic plane. Around $\ell = 90°$ the distant gas is north of the plane, while near 270° it is below the plane. (For details of the observations, see the review by Burton, 1988, and references therein.) Although out to the Sun's distance from the Galactic center the plane of the gas is very flat, somewhat beyond the solar circle this systematic trend begins, with a bending that increases with distance. The result is a systematic warp of the edge of the disk.

The warp has never been seen at optical wavelengths, because its material is at low latitude at a large distance, so that too much interstellar absorption intervenes. It has recently been observed in the distribution of IRAS point sources, however (Djorgovski and Sosin 1989).

As soon as the warp was discovered, more than three decades ago, it was noticed that it pointed in the direction of the Magellanic Clouds. The mass of the Clouds, however, is far too small to have such a striking effect on the potential field of the Milky Way; if the Magellanic Clouds are the cause of the warp, the interaction must operate in a more forceful, or else a more subtle way. Indeed, Hunter and Toomre (1969) were able to develop a theory in which a warp was induced by a recent close passage of the Clouds.

A warped edge is not unique to the Milky Way, however; numerous other galaxies have since been found to show warps (see, *e.g.*, Bosma 1981, and the discussion in Chapter 10). Some of these galaxies have no obvious perturber. It would appear that a warped disk is a frequent occurrence, and not necessarily transient. Thus it seems appropriate to seek for ways in which a galaxy can generate its own warp in a natural way. A new range of possibilities became available when it was discovered that the Milky Way has a large outer mass in its dark halo.

The bending modes of a self-gravitating disk, inside an additional potential, pose a difficult mathematical problem. Hunter and Toomre (1969) had found that for an isolated disk any discrete modes that could constitute a warp were accompanied by continuous modes that cause it to dissipate. In the presence of an oblate halo, however, the situation changes. Dekel and Shlosman (1983) and Toomre (1983) proposed such an origin for the warp, and Sparke and Casertano (1988) have worked out a model for the effect. They find that two types of bending are possible, depending on the core radius of the halo, thus raising the intriguing possibility that (if this theory is right) the shape of a warp can tell us otherwise unobservable facts about the dark halo.

9.8   THE DYNAMICAL EVOLUTION OF THE STELLAR DISK

Ever since velocity dispersions of stars were first measured, it has been clear that different types of stars have different velocity dispersions. The early tendency was to compare stars by spectral type; it was clear that the late-type stars had higher velocity dispersions than did the stars of early type. That recognition led to a strange aberration: since the mass of main-sequence stars decreases with advancing spectral type, the difference in velocity dispersions was taken by some astronomers as an indication of an approach to equipartition of kinetic energies. Since the time scale for such a process was known to be very long, the velocity dispersions were used as an argument for an age of more than 10^{12} years for the Milky Way!

The discovery of the energy-generating reactions in stars soon made it clear that such a time scale was nonsense, and also made it possible to assign ages to the stars of different types. The crucial step in interpreting the velocities was the recognition that the G, K, and M stars actually have quite similar velocity dispersions. Thus velocities increase from the O stars up to the G stars and then level off. This break, often called Parenago's discontinuity, occurs near the place where the main-sequence lifetime equals the age of the Galactic disk.

Now all becomes clear. What the velocity dispersions correlate with is the mean age of the group of stars. Our task, then, is to understand this; what dynamical mechanism causes velocity dispersions to increase with age?[2]

---

[2] This discussion applies only to the stars of the (thin) disk. The halo stars clearly had a different origin, which endowed them with their high velocities. Although less is known about the stars of the thick disk, a similar caveat probably applies.

9.9   THE SPITZER–SCHWARZSCHILD MECHANISM

Several decades ago, Spitzer and Schwarzschild (1951, 1953) suggested that the velocities of old stars might be increased over time by encounters with massive interstellar complexes. (In the modern context, we would tend to think of giant molecular clouds.)

Their basic idea becomes clear if we examine Chandrasekhar's (1960; see Eq. 2.379) formula for the time of relaxation as a result of stellar encounters:

$$T_E = \frac{1}{16}\sqrt{\frac{3}{\pi}} \frac{v_1^3}{N_2 G^2 m_2^2 \ln[D_0 v^2 / G(m_1 + m_2)]}. \tag{9.27}$$

Subscripts 1 and 2 refer to the objects followed and the objects encountered, respectively. The v's are velocity dispersions; the unsubscripted one is a typical relative velocity in an encounter. The encounter cutoff length D_0 can be taken to be the scale height of the disk; its exact value is unimportant, because the argument of the logarithm is so large that the value of the logarithm is quite insensitive to it. N_2 is the number density of objects encountered.

For ordinary encounters between stars in the solar neighborhood relaxation is far too slow to matter; for stars encountering stars, T_E is 10^{13} to 10^{14} years. But when the objects encountered are massive, the situation changes, because of the $N_2 m_2^2$ factor in T_E; remember that "2" refers to the *other* object. If we note that $N_2 m_2^2$ can be written as $m_2 \cdot N_2 m_2 = m_2 \rho_2$ (where ρ_2 is the mean spatial mass density of objects of type 2), then it is clear that the velocity-diffusion effect due to objects with a given average density is *proportional to the mass of an individual object*. Thus interstellar material collected into clouds of size $10^5 \, \mathcal{M}_\odot$ will be 10^5 times as effective in changing stellar velocities as the same amount of material would be if it were in chunks of one solar mass each. Note, however, that the presence of the v_1^3 factor makes it increasingly hard to pump up a velocity dispersion once it has begun to rise.

Spitzer and Schwarzschild worked out a theory based on solving the Fokker–Planck equation (which describes the diffusion in velocity space that results from random stellar encounters) and following the increase in the velocity dispersions of the stars. They concluded that if the existing interstellar material were organized into complexes of a million solar masses each, their theory could account for the velocity dispersions of old disk stars. It also predicted that the velocity dispersions of old stars would go as their age, t, to the 1/3 power (corresponding to the presence of the v_1^3 factor in the time scale set by Eq. [9.27]).

For a star in the Milky Way, it is necessary to integrate the diffusive effects over the motion of the star in its Galactic orbit. An easy way of doing this is to use Lindblad's epicyclic approximation and to calculate the changes in the constants that describe the epicycle orbit. (This refinement was introduced in the second Spitzer–Schwarzschild paper.)

Spitzer and Schwarzschild considered only the U and V velocity components. Lacey (1984) made a calculation that also included the W component. The result was disappointing; the theory, as applied to encounters with massive gas clouds, did not predict the correct shape for the velocity ellipsoids of older stars. In a numerical simulation, Villumsen (1985) came somewhat closer to predicting the observed value of $\langle W^2 \rangle / \langle U^2 \rangle$; but unfortunately a simulation does not give immediate insight into why it came out that way.

Another aspect of Lacey's discussion is worth noting. He pointed out that at late times the rate of "heating" of the stars would decrease for an additional reason, when their z-amplitudes of motion reached a size such that they no longer spent all their orbital time in the cloud layer.

9.10   THE PROBLEM OF THE DIFFUSION COEFFICIENTS

Earlier, however, Wielen had introduced a new difficulty. He had calculated the actual dependence of stellar velocity dispersions on age (Wielen 1974), and had found reasonable agreement with the Spitzer–Schwarzschild theory. But then he did a recalculation (Wielen 1977) and concluded that it did not agree well after all. Rather than increasing as $t^{1/3}$, as that theory would predict, velocities appeared to go more as $t^{1/2}$. A later study by Carlberg et al. (1985), however, gave a different result: they found that velocity dispersions increased with age up to 6 Gyr, but then remained constant. It is thus not clear what the dependence really is.

Wielen's conclusion in 1977 was not that the Spitzer–Schwarzschild theory was wrong, but that the nature of the process that increases stellar velocities was not well enough understood. Noting that their Fokker–Planck equation is a diffusion equation in velocity space, he retreated to a more pragmatic view of what kind of diffusion process could be taking place. Writing a Fokker–Planck equation that contained generalized, unspecified diffusion coefficients, he then tried coefficients of various forms, in order to see what kind of diffusion process would fit the observed dependence of velocity dispersion on age.

In the simplest model of a diffusion process, where the diffusion coefficient is a constant, it turns out that the average v^2 increases uniformly with time, so that v goes as the desired $t^{1/2}$. Wielen found that by assuming equal, constant diffusion coefficients for each of the three velocity components, he also got a velocity ellipsoid whose axis ratios agreed with the observed ones for the oldest stars. (More correctly, this is a statement about the ratio of the W dispersion to the others; as we saw earlier in this chapter, the ratio of the U and V dispersions is determined by the shape of the epicycles, which depends in turn on the ratio of the Oort A and B constants.)

Wielen thus achieved a good agreement between theory and observation—yet his theory is somewhat unsatisfactory, because it arbitrarily introduces a diffusion coefficient that is chosen to fit but has no real physical basis. The Spitzer–Schwarzschild theory did, after all, have a plausible a priori basis; can something be done to salvage it and bring it into agreement with the observations? The discrepancy, Wielen notes, is that the velocity-dependent diffusion coefficient that corresponds to their theory is unable to accelerate the oldest stars enough. Why not then postulate that the diffusion coefficient depends on time and that it was larger when these stars were young? This is quite a plausible behavior; the Milky Way has, in fact, continuously been converting interstellar material into stars, so there *must* have been more of the former in the past. What Wielen did, then, was to make a new calculation, in which the diffusion coefficient depends on velocity but decreases with time. With a suitable decay constant he was able to achieve agreement with observation.

Another possibility is that the diffusion coefficients really *are* independent of the velocity of the star encountered, as they would if the objects responsible for the velocity diffusion were high-velocity massive objects. Thus if the dark halo were made of massive black holes, they would produce velocity-independent diffusion coefficients (Lacey and Ostriker 1984; Wielen 1990). The dynamical effect of such a population on moving groups could lead to difficulties, however.

As an alternative perturber, it was proposed by Barbanis and Woltjer (1967) that the gravitational influence of transient spiral patterns might be responsible for velocity diffusion. A numerical simulation by Carlberg and Sellwood (1985) achieved reasonable agreement with a growth rate that is proportional to $t^{1/2}$; they did not follow the W component, however. In a later set of simulations Carlberg (1987) studied all three components. He found that in order to heat up the W component he had to introduce clouds in addition to the changing spiral arms. He also found a new complication: the shape of his velocity ellipsoid depended on the physical size of the clouds, which limits the effectiveness of close encounters.

In a recent paper, Binney and Lacey (1988) tried to pull together all the aspects of velocity diffusion. First, they showed (as did Lacey in his 1984 paper) that the Spitzer–Schwarzschild $t^{1/3}$ resulted from their having solved a problem in two dimensions rather than three. Next, they showed that in the three-dimensional problem encounters with clouds give a $t^{1/4}$ growth rate, as well as a wrong distribution of the magnitudes of the velocities. Abandoning clouds for changing spiral patterns, they found to their distress that only patterns of an improbably long wavelength will do. This left them with only rapidly moving massive objects, either the dark-halo black holes referred to previously or else dwarf galaxies plunging through the disk.

This is how the situation stands at present. It is clear that we do not yet fully understand this intriguing problem.

9.11 THEORIES OF SPIRAL STRUCTURE

Although spiral galaxies are so common, and although it was realized many decades ago that the Milky Way is a spiral galaxy, the details of spiral structure have been perversely stubborn in falling into place. It was 1952 before the locations of spiral arms in the Milky Way began to emerge, and they are still seen only in a murky and confused way. But our understanding of the dynamics of spiral structure has evolved even more slowly.

Lindblad devoted a large part of his career to the search for dynamical explanations of spiral structure, following the orbits of stars and their collective interactions with each other. Although he developed some ideas that were seminal to later work, his quest for spiral-arm theories was basically unsuccessful. The difficulty, certainly, was that he was trying to make spiral arms out of ordinary stars. An examination of any well-resolved spiral galaxy shows, on the contrary, that the spiral arms are marked by H II regions and bright, young stars. In other words, the spiral arms are regions of star formation and of recently formed stars. Indeed, the old stars of a disk that has no gas (as, for instance, in an S0 galaxy) show no tendency toward spiral structure, which is characteristic only of star-forming galaxies.

This suggests strongly, then, that it is to the interstellar material and to the young stars that we should look for the origin of spiral structure. The nature of a theory is still not obvious, however, because these components differ from the older stars in several different ways.

1. The gas plays an important role; is it possible that spiral structure is a phenomenon of gas dynamics rather than stellar dynamics?

2. The young stars have not yet fully mixed into the general field; could spiral structure arise from the particular way in which their original spotty distribution changes to a smooth one? Remember the way in which an association develops a trailing shape during much of the early dispersal time of its stars.

3. Interstellar gas and young stars both have lower velocity dispersions than old stars; does velocity dispersion play an important role in the nature of spiral structure? This question, in particular, is reminiscent of the Toomre criterion, and the way in which low velocities can lead to instability.

9.12   THE WINDING DILEMMA

In addition to looking at the population makeup of spiral arms, we must also consider their kinematics. We cannot, of course, watch them move, but we can predict how they ought to move, and a serious difficulty immediately arises. If, in fact, we take a simple-minded view of the motion of spiral arms, it is hard to see how differential rotation can allow them to persist at all. We noted, when studying epicycles, that two regions whose distances from the Galactic center differ by ΔR have a relative drift rate of $-2A\Delta R$. Since $(2A)^{-1} \cong 35$ Myr, a material structure such as a spiral arm is subject to such a strong shear that the arm would become tightly wound in only the time of one galactic rotation or so. This difficulty is called the "winding dilemma." It shows that material arms cannot persist. If an arm consists of the same material throughout its life, it must soon wind up hopelessly. The prevalence of spiral structure in so many galaxies could then be explained only by postulating a mechanism for continually generating new arms. But then we would expect to find, in the same galaxy, arms at all stages of winding; nor would we see the well-observed correlation between galaxy type and the openness of the arms.

One way out of the winding dilemma is to give up the idea that a spiral arm always consists of the same material. It might rather be the locus of a phenomenon that moves through the material of the disk, keeping the same spiral shape as it moves. Such a progressive wave has been imagined in two possible ways. One possibility is that a compressional wave of density moves through the material, inducing star formation as it goes. The other is that the regions of star formation persist and migrate (along with their recently formed stars) in such a way that at any given time they show a spiral pattern.

9.13   THE DENSITY-WAVE THEORY—AND OTHER MECHANISMS

Lindblad, and then especially Lin and Shu, envisioned a quasistationary spiral structure, in which a *density wave* moves through the stellar and interstellar material as a configuration whose shape stays fixed, in a framework that rotates with "pattern speed" Ω_p, which is less than the material speed of galactic rotation. Thus a particular point in the material sees density waves go by (in a direction opposite to galactic rotation, because the pattern speed is slower) at an angular rate $2(\Omega - \Omega_p)$, where 2 is the number of arms in a "grand design" spiral.

Of particular importance to this linear theory are the regions where resonances occur, because there the theory blows up, and (as shown by much more complicated, nonlinear calculations) there could be amplification, reflection, absorption, or even creation of waves. The most important of these are *corotation* (rather far out, in most density-wave fits to a galaxy), where $\Omega = \Omega_p$ and the density wave remains stationary with respect to the material, and the *inner Lindblad resonance*, where $\Omega_p = \Omega - \kappa/2$. At the latter point, density waves pass through in synchronism with the local epicycle frequency κ.

The basic density-wave theory of Lin and Shu assumes small-amplitude sinusoids and fits them so that the density perturbations, the gravitational perturbations, and the motion perturbations are all mutually consistent. Lin (1966) describes the process with a logical diagram showing the interconnections that are needed if the theory is to be self-consistent (Fig. 9.3). Note that the quantities referred to are not the total gravitation or density but rather the increments that are associated with the density wave. Lin starts with the gravitation (labeled 0). It has two aspects. First, it corresponds, through Poisson's equation, to the density (1). Second, it induces density changes, which are in two parts. The gas response to the gravitational perturbation (2) is calculated, in a reasonably straightforward way, by perturbing the hydrodynamic equations of a fluid. The

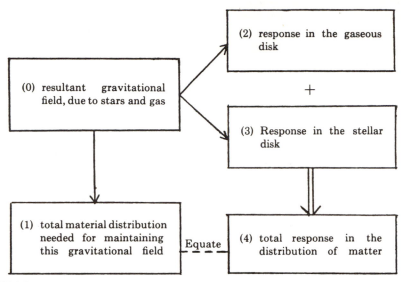

FIGURE 9.3
Logical outline of the relationships in the Lin–Shu density-wave theory. (From Lin 1966.)

stellar response (3) is found from a perturbed form of the collisionless Boltzmann equation; it is a rather complicated calculation. The stellar response can be quite important, and its amount is sensitive to how close to the Toomre criterion the stars are. The final step is to insist that the total induced density (4) be equal to the density (1) that causes it.

The mathematical details of the density-wave theory are much too compli-cated to present here. Suffice it to say that they have been worked out in some detail by Lin, Shu, and others. (See, *e.g.*, Lin 1966.)

At this point it should be clearly understood what the basic linear density-wave theory contributes to an understanding of spiral structure and what it does not contribute. It successfully describes a set of waves that are dynamically self-consistent, in that they arouse density perturbations that are just right to provide the gravitational forces that are needed in order to cause those pertur-bations. The theory does not, however, predict the amplitude of the waves; in a linear theory such as this, the amplitude is arbitrary (as long as it is small enough to justify the approximations that are made). Nor does the theory pre-dict that such waves must arise. The waves are also neutral waves, in the sense of a mathematical stability analysis; they neither grow nor decay. In short, the linear theory of density waves says neither that the waves must arise, nor how strong they should be, nor how they were generated, nor what can maintain them.

To attack these problems, we must investigate the natural modes of vibration of a disk that is composed of stars and gas. Immediately, however, three serious questions arise, on whose answers there has by no means been unanimity. First, *what* disk? The behavior of a disk depends on its density distribution and on the gravitation of any other mass that may be present. Second, should the spiral structure arise spontaneously or does it need to be excited by some other mechanism, such as a central bar or an external tidal force? Third, are spiral arms persistent, or do they continually dissipate and reform?

Lin and his associates have for many years sought modes that will grow spontaneously into "grand-design" spiral patterns (see, *e.g.*, Lin and Bertin 1985). The most promising mechanism that they have studied is amplification of a wave as it is reflected from the corotation circle. (Conversely, at the inner Lindblad resonance—if a galaxy has one—density waves are absorbed.)

Others have investigated mechanisms by which a more local irregularity can grow and be sheared into a spiral form. The most interesting of these is "swing amplification" (see, *e.g.*, Toomre 1981). Insofar as it is understood, this process is based on a combination of self-gravity, shearing, and the tendency for radial structures to grow (and therefore to be strongest when they are trailing).

The eventual truth might well draw on more than one process, in fact. After all, the galaxies themselves show different forms of spiral pattern (Elmegreen and Elmegreen 1982). The neat two-armed "grand-design" spirals are a minority; a more typical spiral has epidemic spirality but no clear, continuous arms. The name given to spirals of this type is "flocculent." The extreme of this tendency is exhibited by galaxies that are surrounded by almost-Medusa-like sets of arms. (See Figs. 9.4 and 9.5.) It has frequently been suggested (see, *e.g.*, Kormendy and Norman 1979) that grand-design spirals are produced by bars or tidal encounters, and that a galaxy left to itself will produce flocculent spiral structure.

The density-wave theory has some observational consequences that are different from those of other theories of formation of spiral structure. One consequence is the noncircular velocities that must occur as the wave passes. These have been seen in some grand-design spirals. (See, *e.g.*, Visser 1980.) Another consequence is that where there are density waves, the regions of star formation progress through the material; thus in a region where the wave has already passed, there should be rather recently formed stars but no *very* young ones. Across the wave there ought to be a recognizable difference in the distribution of stellar ages and therefore in the integrated colors of the regions. In other theories, although the regions of star formation are sheared by differential rotation, they do not move with respect to the material contained in them. Thus there should be no such differential color effect. Schweizer (1976) did a study of colors, in several grand-design spirals, but they did not clearly indicate such an effect (which may in fact be too small to observe). What he did find, however, were undulations of brightness that suggested a strong density response of the underlying stellar disks to a density wave. Elmegreen and Elmegreen (1984), in a study of many types of spirals, suggested that grand-design spirals tend in general to show this effect, whereas flocculent spirals do not. Such a difference might suggest that one class of galaxies derives its spiral structure from density waves, whereas the other is dominated by some other mechanism.

Dynamical problems apart, however, the basic density-wave theory can at least be looked upon as giving a kinematic description of any quasistationary

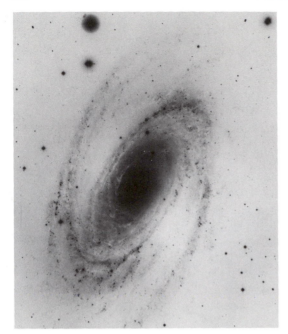

FIGURE 9.4
NGC 3031 (M81), a grand-design spiral.

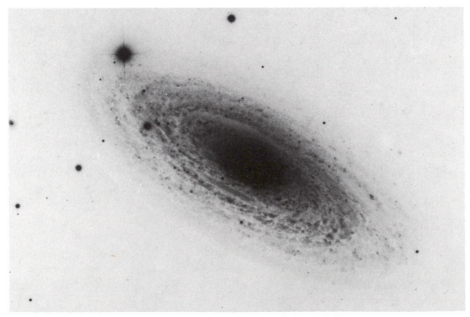

FIGURE 9.5
NGC 2841, a flocculent spiral.

spiral structure. If arms are to exist without winding up, they will move in this way and will set up the patterns of density and velocity perturbations that are described by the density-wave theory. For plausible values of the density contrast between arms and interarm regions, the local streaming velocities in the gas are of the order of 5 to 10 km sec^{-1}. There should also be similar local perturbations in the velocities of low-velocity stars, depending on where they are in relation to a spiral arm. The stars of higher velocity should be affected less.

Since the Oort constants, particularly A, are calculated from stars of low velocity, their values might be perturbed by the streaming associated with a density wave, and perhaps do not correspond to the shape of the Galactic rotation curve in the way that they would in an axisymmetric galaxy—if, indeed, the Milky Way is a density-wave spiral.

In addition to the strength of the density wave, the basic density-wave theory has another unspecified quantity. This is the pattern speed Ω_p, which is an arbitrarily adjustable parameter. It is possible, however, to make a virtue of this freedom by using the pattern speed to fit the density-wave theory to actual galaxies in a way that has an elegant astronomical plausibility. Noting that the theory works best in the zone between the inner Lindblad resonance and corotation, and using the empirical fact that spiral arms are the major loci of star formation, we can argue that the outermost H$\textsc{ii}$ region in a galaxy marks the outer limit of the spiral pattern, and therefore the location of corotation. The observed rotational velocity at that radius then determines the pattern speed, and from it, everything else about the geometry of the density wave. For some galaxies this procedure gives a good fit to the observed spiral structure (Roberts *et al.* 1975).

Another important application of the density-wave theory has come from the fact that if the amplitude of a density wave is large enough, the velocities can be supersonic. (The "speed of sound" is rather close to the atomic speed in an ionized medium, or the cloud speed in a cool medium, where the clouds are the "particles" that carry the density information. A rough value is 10 km sec^{-1}.) It is widely believed that shocking by the density wave can set off star formation in the arms.

9.14 STOCHASTIC STAR FORMATION

An extreme rival picture of the formation of spiral structure has been developed by Seiden (1983) and his associates. They call it stochastic self-propagating star formation. Without recourse to any dynamical processes, they simply follow the progress of star formation in a differentially rotating galaxy, under certain hypotheses about how star formation proceeds. They postulate a mixture of spontaneous and stimulated star formation. The spontaneous probability is taken to be low; but where stars have been forming, more are likely to form, as a result of supernova-induced shocking of the interstellar medium, for example. Thus stimulated star formation is represented by a second and larger probability co-efficient, which is applied to any cell that already has star formation and to the immediately adjacent cells of the representation as well. Because the coefficient is less than unity, it is possible for star formation to die out in any region where it exists; but it usually persists for some length of time. During this time, differential rotation draws out the star-forming patterns into spiral-shaped arcs. With lower probabilities of stimulated star formation, computer simulations produce flocculent spirals; with higher stimulated probabilities, they can sometimes even produce, temporarily, spirals with a grand design.

REFERENCES

Athanassoula, E., ed. 1983. *Internal Kinematics and Dynamics of Galaxies.* I.A.U. Symposium no. 100. Dordrecht: Reidel.

Athanassoula, E., and Sellwood, J. A. 1986. *Mon. Not. Roy. Astron. Soc.,* **221**, 213.

Barbanis, B., and Woltjer, L. 1967. *Astrophys. J.,* **150**, 461.

Binney, J., and Lacey, C. 1988. *Mon. Not. Roy. Astron. Soc.,* **230**, 597.

Blaauw, A. 1952. *Bull. Astron. Inst. Neth.,* **11**, 405 (no. 433).

Bosma, A. 1981. *Astron. J.,* **86**, 1825.

Burton, W. B. 1988. In Verschuur and Kellerman, p. 295.

Carlberg, R. G. 1987. *Astrophys. J.,* **322**, 59.

Carlberg, R. G., Dawson, P. C., Hsu, T., and VandenBerg, D. A. 1985. *Astrophys. J.,* **294**, 674.

Carlberg, R. G., and Sellwood, J. A. 1985. *Astrophys. J.,* **292**, 79.

Chandrasekhar, S. 1960. *Principles of Stellar Dynamics.* New York: Dover.

Contopoulos, G., ed. 1974. *Highlights of Astronomy.* Vol. 3. Dordrecht: Reidel.

Dekel, A., and Shlosman, I. 1983. In Athanassoula, p. 187.

Djorgovski, S., and Sosin, C. 1989. *Astrophys. J. (Letters),* **341**, L13.

Efstathiou, G., Lake, G., and Negroponte, J. 1982. *Mon. Not. Roy. Astron. Soc.,* **199**, 1069.

Elmegreen, D. M., and Elmegreen, B. G. 1982. *Mon. Not. Roy. Astron. Soc.,* **201**, 1021.

———. 1984. *Astrophys. J. Supp.,* **54**, 127.

Fall, S. M., and Lynden-Bell, D., eds. 1981. *The Structure and Evolution of Normal Galaxies.* Cambridge, Eng.: Cambridge Univ. Press.

Hunter, C., and Toomre, A. 1969. *Astrophys. J.,* **155**, 747.

Jeans, J. H. 1929. *Astronomy and Cosmogony.* Cambridge, Eng.: Cambridge Univ. Press.

Kerr, F. J., and Lynden-Bell, D. 1986. *Mon. Not. Roy. Astron. Soc.,* **221**, 1023.

Kormendy, J., and Norman, C. A. 1979. *Astrophys. J.,* **233**, 539.

Lacey, C. G. 1984. *Mon. Not. Roy. Astron. Soc.,* **208**, 687.

Lacey, C. G., and Ostriker, J. P. 1985. *Astrophys. J.,* **299**, 633.

Lin, C. C. 1966. *Lectures in Applied Mathematics,* **9**, pt. 2, 66.

Lin, C. C., and Bertin, G. 1985. In van Woerden *et al.*, p. 513.

Ostriker, J. P., and Peebles, P. J. E. 1973. *Astrophys. J.,* **186**, 467.

Roberts, W. W., Roberts, M. S., and Shu, F. H. 1975. *Astrophys. J.,* **196**, 381.

Schweizer, F. 1976. *Astrophys. J. Supp.,* **31**, 313.

Seiden, P. E. 1983. *Astrophys. J.*, **266**, 555.

Sparke, L. S., and Casertano, S. 1988. *Mon. Not. Roy. Astron. Soc.*, **234**, 873.

Spitzer, L., and Schwarzschild, M. 1951. *Astrophys. J.*, **114**, 385.

———. 1953. *Astrophys. J.*, **118**, 106.

Toomre, A. 1964. *Astrophys. J.*, **139**, 1217.

———. 1981. In Fall and Lynden-Bell, p. 111.

———. 1983. In Athanassoula, p. 177.

Verschuur, G., and Kellerman, K., eds. 1988. *Galactic and Extragalactic Radio Astronomy*. New York: Springer.

Villumsen, J. V. 1985. *Astrophys. J.*, **290**, 75.

Visser, H. C. D. 1980. *Astron. Astrophys.*, **88**, 159.

Wielen, R. 1974. In Contopoulos, p. 395.

———. 1977. *Astron. Astrophys.*, **60**, 263.

———. ed. 1990. *Dynamics and Interactions of Galaxies*. Heidelberg: Springer.

Woerden, H. van, Allen, R. J., and Burton, W. B., eds. 1985. *The Milky Way Galaxy*. I.A.U. Symposium no. 106. Dordrecht: Reidel.

CHAPTER 10

KINEMATICS AND MASS DISTRIBUTIONS
IN SPIRAL GALAXIES

Pieter C. van der Kruit

In this chapter I will address questions of kinematics in spiral galaxies. I will spend some time describing the observational techniques used to obtain velocity fields and velocity dispersions, and then review the results. This will lead to a discussion of local and global stability and mass distributions in spiral galaxies, and a comparison of them to our Galaxy. However, before doing all this I will first give a historical introduction to kinematic observations. Again I will use distances based on a Hubble constant of 75 km sec^{-1} Mpc^{-1}.

The rotation of spiral nebulae was recognized first by Wolf (1914) in M81 and Slipher (1914) in M104 from the inclination of the stellar absorption lines on spectra of the central regions, although the identification of these nebulae as disk galaxies was at that time controversial. That the discovery was made first with absorption lines instead of the more readily observable emission lines can be attributed to observational selection: the central regions of the nebulae were brightest, and the presence there of absorption lines, resembling that of an F- or solar-type star, had been known for a long time. The first observations to result in a plot of radial velocity versus radius were made by Pease (1916, 1918) for M31 and M104. It was a tedious business, requiring exposure times of about 80 hours for each absorption-line spectrum, but the results on M104 showed that the radial velocities relative to the center increased linearly with radius, reaching more than 300 km sec^{-1} at a distance of 2.5 arcmin. Pease's measurements along the minor axis of M31 indicated that the radial velocity was nearly constant at all positions, showing that the variation observed along the major axis "is without doubt to be attributed to the rotation of the nebula." The results on M31 are reproduced here in Fig. 10.1.

Emission lines from H$_{II}$ regions in the spiral nebulae were discovered about the same time as nebular rotation. Pease (1915) recorded [O$_{III}$], Hβ, and Hγ in a 34.5-hour exposure of NGC 604 in M33. For galaxies of large angular size, the use of emission lines to study the rotation is hampered by the fact that a single exposure gives only a single point on the rotation curve, because each H$_{II}$ region is discrete. In his classic study of the rotation of M31, Babcock (1939) derived 44 velocity measurements in the inner regions from 236 hours of exposure on the absorption lines; an additional 56 hours of exposure on emission lines was

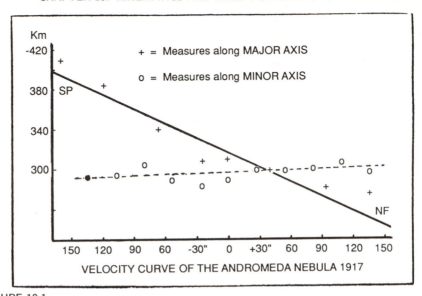

FIGURE 10.1
Observed velocities in the central part of M31 from absorption-line spectra obtained along the major (79 hours) and minor (84 hours) axes with the Mount Wilson 60-inch telescope. These observations provided the first evidence for the rotation of spiral galaxies. (From Pease 1918.)

required to provide four more points. These four points were, however, very important, for they permitted extension of the rotation curve by a factor of 3 further out into the main part of the disk.

The first detailed observations of the H I at 21-cm wavelength in M31 were made by van de Hulst *et al.* (1957), using the 25-m Dwingeloo radio telescope with an angular resolution of 0.6 degrees. The measurements resulted in a rotation curve from 0.6 to 2.5 degrees from the center, a significant extension of the 1.5-degree last point on Babcock's curve. With a similar angular resolution but improved receiving equipment, Argyle (1965) measured H I profiles over the whole image of M31, and was the first to plot a complete radial-velocity field for any galaxy (see Fig. 10.2). The resulting "spider diagram" is now in common use. It connects positions of equal radial velocity and gives a good overview of the rotation pattern and possible deviations from circular motion. Various parameters, such as systemic velocity, position angle of the (kinematical) major axis, inclination, and the position of the rotation center, can be derived from it, in addition to the rotation curve. This will be discussed in more detail later.

10.1   OBSERVATIONAL METHODS FOR STELLAR KINEMATICS

The derivation of the stellar kinematics in an external galaxy is done from absorption-line spectra, where the position and width of a line or a set of lines is

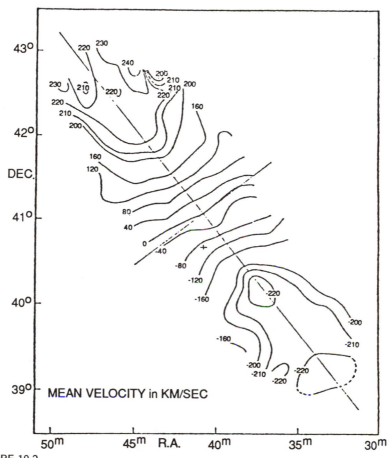

FIGURE 10.2
The first "spider diagram" to illustrate the observed radial velocity field over the image of a galaxy (here for H I in M31). Velocities are with respect to the systemic velocity and have been multiplied by 1.03 to correct for the inclination. The latter is not the usual practice anymore. (From Argyle 1965.)

compared to that of a Galactic star of the approximate type and luminosity class that dominates the integrated galactic light and therefore the spectrum. Such stars are late G or early K giants. The technique used to find the radial velocity and velocity dispersion is nowadays exclusively by Fourier methods (following the fundamental discussion of Simkin 1971). The basic point is that the galaxy spectrum is the convolution of that of the Galactic giant ("template") and a (often assumed Gaussian) velocity distribution function (also called broadening function). The procedure then is to express the observed galaxy and template spectra as a function of $\log \lambda$, so that Doppler shifts correspond to linear displacements, and to divide out low-order polynomial fits in order to suppress power at low Fourier frequencies. From this stage on there are in general three different methods, which will be discussed in turn.

The first (Illingworth and Freeman 1974) makes use of the peak in the cross-correlation function of the galaxy and template spectra to calculate the radial velocity, and uses the slope of the power spectrum to calculate the velocity dispersion (larger dispersions result in steeper slopes). With discrete Fourier techniques we calculate the Fourier transforms of the galaxy and template spectra, and from this the cross-correlation function between the two and the individual power spectra. The power spectrum of the template is also calculated after its observed spectrum has been broadened by a set of velocity dispersions, and the best-fitting one is then selected. The template data are in practice noise-free (taken from bright stars), but the galaxy data are not, and at high frequencies the galaxy's power spectrum is dominated by the noise in the observations. Usually this noise power spectrum is flat, and we can therefore find out over which frequency range in the power spectra the comparison of galaxy to template needs to be done. A useful way of estimating the noise power spectrum is to subtract two observations of the same object (or sky) from one another, and use this as the typical noise distribution. A disadvantage of this power-spectrum scheme is that the comparison of two power spectra is not easily made objective or automatic, and errors have to be estimated also by eye.

The second method is related, and makes an exclusive use of the cross-correlation function (Tonry and Davis 1979), where position and width of its peak are used for radial velocity and dispersion. In this method the width is compared to that of the cross-correlation function of preferably two different templates, or else two different observations of the same template star. This method works very efficiently in an automatic way, and a routine to estimate formal errors can be incorporated easily. Bottema (1988) has discussed this in detail, and has concluded that it is to be preferred in galactic disks where the surface brightness is faint, and the dispersions are low. For the practical case of a disk we must model the effects of integration along the line of sight on the shape of the broadening function, which in general may result in non-Gaussian functions. The template spectrum can then be convolved with the model function, and the resulting cross-correlation peaks can be fitted by a least-squares minimization technique to obtain a best value for the dispersion. The procedure has been illustrated in Fig. 10.3.

The third method, which is in widespread use, in particular for bright centers of elliptical galaxies and spiral bulges, is the Fourier quotient method, first introduced by Schechter (see Sargent *et al.* 1977). This assumes a Gaussian broadening function, and makes use of the fact that the Fourier transform of the broadening

FIGURE 10.3 (Facing page)
Cross-correlation peaks between the spectrum of a K0 III star with the same spectrum after convolution with Gaussian broadening functions with dispersions from 18.2 to 182 km sec^{-1} and addition of artificial noise. At the top there is no noise; in the middle panel the noise is that expected for a count level of (on average) 400 in each of the 600 spectral elements; in the lower it is 100. Note that the peaks have the same maximum at each dispersion independent of the count level, but that they become more irregular for noisier data. (From Bottema 1988.)

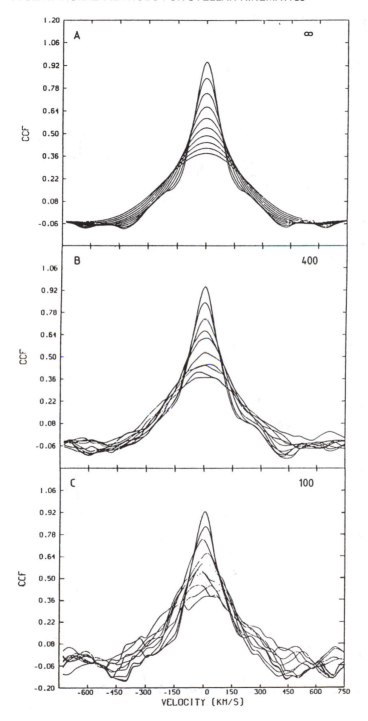

function is equal to the quotient of the Fourier transforms of the galaxy and template spectra. This quotient is then calculated from the observations and fitted to the Fourier transform of a Gaussian, which is also a Gaussian (but now complex). The problem is, of course, that division of noisy data is risky (but there are objective ways of finding out which wavenumbers to restrict yourself to), and this method therefore does not work very well for disks.

The Fourier quotient method, in the form developed by Schechter, also strengthens the weight of the velocity dispersion by noting that the Doppler shift consists, in the Fourier domain, of a systematic shift of the complex phase of each component, proportional both to the Doppler shift and to the component number. This phase shift can be removed from each component, which is then real and has only half the noise of the unshifted complex component. The weight of the velocity dispersion is thereby doubled.

I will say a few words about stellar velocity dispersions in spiral bulges before I turn to stellar disks in the next section. There are two aspects. The first is the V_m/σ versus ε diagram, which measures the importance of rotation relative to the flattening. This diagram has been used extensively to study general aspects of the dynamics of elliptical galaxies, and shows that in these systems rotation is not dynamically important enough to provide oblate flattening and that ellipticals are probably triaxial. The most extensive diagram with observed values is the one by Kormendy and Illingworth (1982), which shows that, contrary to ellipticals, bulges generally do follow the oblate, isotropic line. This means that a typical spiral bulge is probably flattened by its rotation (as well as possibly by the disk potential), and that the velocity distribution is more or less isotropic (at least in the central parts). This general relation can be used with values for the bulge of the Galaxy to estimate its flattening, and in Chapter 5 I derived an axis ratio of 0.7 ± 0.15.

The second relation is that between the velocity dispersion and the bulge luminosity (Kormendy and Illingworth 1983). This relation is interesting, because it can be used to derive information on how prominent the bulge of our Galaxy would appear from an external viewpoint. As before I use a velocity dispersion of 110 ± 10 km sec^{-1}. Then it follows that M31 (150–160 km sec^{-1}) and M81 (160–170 km sec^{-1}) have bulges that are at least 5 and at most 10 times more luminous than the bulge of the Galaxy. On the other hand, at least the disk of M31 has a radial scale length that is very similar to that of the disk of the Galaxy, so that these bulges are also brighter relative to the disk than is that in the Galaxy. This is important to realize, because in the semipopular literature and in elementary textbooks one of these two galaxies is often used as an example of what the Galaxy would look like as seen from outside. I will return to this point later in this book, when I discuss the possible Hubble type of the Galaxy.

The data given above using the V_m/σ–ε relation seem to suggest that the velocity distribution in the central parts of bulges is not too far from isotropic. This can be checked for the early type spiral NGC 7814, where a formal estimate of the velocity anisotropy can be made (van der Kruit and Searle 1982b). This galaxy has a luminosity distribution that is dominated by the bulge (it contains

93 percent of the light), but it still contains H$_I$ in the disk, so that the circular velocity can be measured and combined with stellar rotation and velocity dispersion in the bulge. This unique situation usually does not occur in ellipticals (when H$_I$ is seen, it is unclear whether it is in circular rotation), S0 galaxies (where H$_I$ is observed at larger radii than where the stellar kinematics can be measured), or later-type spirals (where the disk potential is a major contributor).

If we assume that one of the principal axes of the velocity ellipsoid is perpendicular to the plane and that the stellar space density goes as $R^{-\beta}$, it follows that in the plane we have

$$V_{\text{rot}}^2 = V_t^2 - R\frac{\partial \langle U^2 \rangle}{\partial R} + (\beta - 1)\langle U^2 \rangle + \langle (V - V_t)^2 \rangle. \tag{10.1}$$

V_{rot} is the circular velocity related to the radial force $(-K_R = V_{\text{rot}}^2/R)$, V_t is the rotation velocity of the bulge stars, and $\langle U^2 \rangle$ and $\langle (V - V_t)^2 \rangle$ are the squares of the radial and tangential velocity dispersions. From H$_I$ kinematics, optical surface photometry, and stellar kinematics along a line just above the dust lane, we know all parameters, except the radial velocity dispersion (the tangential one is in the line of sight). It is observed that in the bulge between 40 and 90 arcsec (2.9 to 6.5 kpc) V_{rot}, V_t, and the tangential velocity dispersion are all constant with R, and it then follows that we must have

$$\langle U^2 \rangle = \frac{V_{\text{rot}}^2 - V_t^2 - \langle (V - V_t)^2 \rangle}{\beta - 1} + \alpha R^{(\beta - 1)}. \tag{10.2}$$

Here α is an integration constant. From the observed velocity dispersion on the minor axis and the numerical values of the other parameters, it follows for NGC 7814 that α must be positive, and that therefore the radial velocity dispersion increases with radius. In the inner regions the velocities are indeed roughly isotropic, but become increasingly anisotropic with the radial dispersion exceeding the tangential one. In the Galaxy Woolley (1978) finds that the velocity ellipsoid of field RR Lyrae stars in the solar neighborhood belonging to the halo population is also highly anisotropic, with the radial dispersion exceeding the tangential one.

10.2 OBSERVATIONS OF STELLAR KINEMATICS IN DISKS

Measurement of the stellar velocity dispersions of the old disk populations serves a number of purposes. The first is that it allows us to estimate the disk mass distribution independently from the rotation curve (see below), which provides useful constraints on the mass distribution in the dark halo. Second, the velocity dispersion plays a role in both the global and the local stability of galaxy disks. Gravitational instability has been a major subject of theoretical study ever since it was described originally by Newton in his famous laws. It actually was Newton himself who worried in his correspondence with Dr. Bentley about what kept the

Sun and the fixed stars from collapsing together in a static universe. He proposed that stars were distributed evenly in space and that therefore all gravitational pulls would compensate each other. From the distribution of stars on the sky as a function of apparent magnitude (taken to indicate distance) he inferred that at least for the brightest stars (to fourth magnitude) their numbers occurred in the required ratios to conform to spatial uniformity and homogeneity. But such a distribution is, of course, very difficult to bring about, and even if this were possible at all, the supposed equilibrium was necessarily very unstable. The solution is, of course, that the equilibrium is maintained by the relative random motions of the stars. The problem was early this century reformulated by Jeans in the form of the now-called "Jeans instability criterion" in an infinite, homogeneous medium. His criterion states that for a given mass density and temperature (or velocity dispersion), masses above a certain critical mass are unstable relative to their own gravity and will collapse. The larger the velocity dispersion and the smaller the density is, the larger this critical mass becomes ($\mathcal{M}_{\text{Jeans}} \propto \langle V^2 \rangle^{3/2} \rho^{-1/2}$).

It is therefore also not at all obvious why stellar disks are smooth and stable. Of course, in particular the younger populations display spiral structure, which will be some kind of instability, but not of a violent type that completely redistributes the matter in the disk. On the other hand, the old disk population is generally speaking smooth, as is clearly seen in the disks of S0 galaxies, where young populations are absent. It is known that rotation in galactic disks can stabilize only low-wavenumber modes, and that the finite thickness of disks may be relevant at small wavelengths (Goldreich and Lynden-Bell 1965). The local stability of thin stellar disks was addressed by Toomre (1964), who showed that the stellar velocity dispersion needs to exceed a certain critical value to ensure stability against local axisymmetric modes. It is usually translated in the parameter Q, which has to exceed unity for local stability. This parameter is given by

$$Q = \frac{\langle U^2 \rangle^{1/2} \kappa}{3.36 G \sigma},$$
(10.3)

where κ is the epicyclic frequency $\kappa = 2[B(B - A)]^{1/2}$, with A and B the local Oort constants, and σ the disk surface density. This criterion is the equivalent of the Jeans instability in an infinite homogeneous medium, reformulated for a flat stellar disk. Global (bar-like) modes can be suppressed by a hot component (of high velocity dispersion), which may be either a dark halo, such as the ones that may produce the observed flat rotation curves (Ostriker and Peebles 1973), or the old disk population itself, if it has a high velocity dispersion in its inner regions (Sellwood 1983; Athanassoula and Sellwood 1986). The dispersion of the old-disk population required corresponds to values of Q of the order of 2.

It has been known for a long time that the stellar velocity dispersion in disks is a result of secular evolution of the kinematics of the stars. This is evident, for example, from the observed relation between the age of the stars and their velocity dispersion in the solar neighborhood in the Galaxy (*e.g.*, Wielen 1977).

Spitzer and Schwarzschild (1951) were the first to propose a mechanism for this effect, namely, scattering of the stellar orbits by massive gas concentrations in the interstellar medium. Such concentrations are now known to exist in the form of giant molecular clouds, but we do not yet know whether there are enough of them to provide the observed change of kinematics for the solar neighborhood, or whether this process may give the actual axis ratios of the local velocity ellipsoid (*e.g.*, Lacey 1984). It may very well be that spiral structure (in the form of either density waves or transient mass-density disturbances) also is important in this respect, but I will not discuss this here in detail, since it has been covered in the previous chapter.

We have seen that edge-on disks show that the vertical scale parameter of the old-disk population is independent of galactocentric distance. For the isothermal sheet approximation the relation between the scale height z_0 in the density distribution $\mathrm{sech}^2(z/z_0)$, the surface density $\sigma(R)$, and the velocity dispersion is

$$\langle W^2 \rangle_* = \pi G \sigma(R) z_0. \tag{10.4}$$

If z_0 is not a function of R, and if $\sigma(R)$ is an exponential with scale length h, then it follows that

$$\langle W^2 \rangle_*^{1/2} = [\pi G \sigma(0) z_0]^{1/2} \exp(-R/2h). \tag{10.5}$$

So, if the old-disk populations have an $\mathcal{M}/L$ ratio independent of galactocentric distance, we expect that the vertical velocity dispersion drops with R as an exponential, with an e-folding length of twice the scale length of the surface-brightness distribution. This is open to observational confirmation. The first detailed check of this prediction was made by van der Kruit and Freeman (1986) on the face-on spiral NGC 5247. The data are illustrated in Fig. 10.4 and reach out to about two optical scale lengths. This is not very far out into the disk, but this work already required a total exposure of 8.8 hours with the 3.9-m Anglo-Australian Telescope, which indicates the difficulty of obtaining such observations as a result of the low surface brightness and small dispersions. The least-squares fit, indicated by the dashed line, has an e-folding length of 2.4 ± 0.6 optical scale lengths, and is in good agreement with the predictions and therefore with no change in $\mathcal{M}/L$ with radius. Note that the central z-velocity dispersion is about 60 km sec^{-1} and certainly not dynamically insignificant. Bottema (1988 and work in progress) has confirmed this conclusion for a few more spirals. More evidence that $\mathcal{M}/L$ for the old disk population is constant with radius will be discussed later, but in what follows I will always make this assumption. From edge-on galaxies we know mean values for the vertical scale parameter z_0, and from this we can estimate values for the surface density and $\mathcal{M}/L$ from the vertical velocity dispersion (and the surface brightness). The latter comes out around 6 in solar B-units for the old-disk population exclusively.

The constant vertical scale height of the old-disk population as a function of radius provides us with a prediction for the radial dependence of the vertical velocity dispersion. No such *a priori* prediction exists for the dispersions parallel

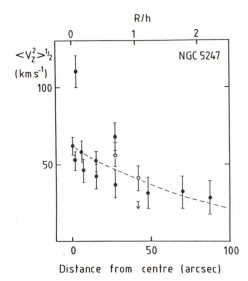

FIGURE 10.4
The observed velocity dispersion of the old disk population in the face-on spiral NGC 5247. The galactocentric distance R has been expressed at the top in units of the optical scale length h of the disk. The dashed line is a least-squares fit to the points and is consistent with a decline with an e-folding length of $2h$. (From van der Kruit and Freeman 1986.)

to the plane, although we know from fundamental galactic dynamics that the ratio of the radial and tangential dispersions is related to the local Oort constants as

$$\frac{\langle (V - V_t)^2 \rangle}{\langle U^2 \rangle} = \frac{B}{B - A}. \qquad (10.6)$$

There is, however, already evidence that these velocity dispersions must decrease with galactocentric distance from the observation of the sharp edges to the radial light distribution. At these radii the exponential decline is 1 kpc or less. At a typical radius of 20–25 kpc and a rotation curve of 200–250 km sec^{-1}, the stars will go through epicycles with half-axes in the R-direction of about 1 kpc, if their random motions are 10–15 km sec^{-1}. This is much lower than, *e.g.*, in the solar neighborhood, so that we may indeed expect these velocity dispersions to drop as well with increasing radius. It can also be shown for the Galactic disk, that unless $\langle U^2 \rangle^{1/2}$ decreases with radius, Toomre's stability parameter Q will become less than unity a few kpc inward from the solar position.

There are two independent ways in which observations can reveal values for parallel velocity dispersions. The first is by direct measurement in highly inclined galaxies, but then we must make corrections for the integration along the line of sight. Such corrections are in practice straightforward to make. The second method is by measuring the asymmetric drift. This is the effect that, in a disk in equilibrium, the mean tangential velocity V_t of a mass component is less than the circular velocity (corresponding to a centrifugal force necessary to compensate for the gravitational force) by an amount that depends on the velocity dispersion. The random motions provide a pressure support, so that a mean tangential velocity less than the circular velocity is required. The relevant equation is

$$-K_R = \frac{V_t^2}{R} - \langle U^2 \rangle \left\{ \frac{\partial}{\partial R}(\ln \nu \langle U^2 \rangle) + \left(\frac{1}{R}\right) \left[1 - \frac{\langle (V - V_t)^2 \rangle}{\langle U^2 \rangle}\right] \right\} \qquad (10.7)$$
$$+ \langle UW \rangle \frac{\partial}{\partial z}(\ln \nu \langle UW \rangle).$$

Here ν is the stellar density distribution, K_R the radial force, and V_t the mean stellar tangential velocity. The last term, which contains the cross term $\langle UW \rangle$, is small. If the velocity ellipsoid points toward the galactic center, the term reduces to $(1/R)(1 - \langle W^2 \rangle / \langle U^2 \rangle)$, and this is probably the largest value this term may take. There is even a compelling reason to ignore the term altogether if the galaxy has a flat rotation curve, as was first pointed out by van der Kruit and Freeman (1986). For an axially symmetric system, Poisson's equation is

$$\frac{\partial K_R}{\partial R} + \frac{K_R}{R} + \frac{\partial K_z}{\partial z} = -4\pi G\rho. \qquad (10.8)$$

Now, near the plane the first two terms on the left are equal to $2(A-B)(A+B)$. In areas away from the central regions, where we have flat rotation curves, $A = -B$ and the terms vanish. So, for a flat rotation curve, the plane-parallel case (with no cross terms involving U and W) happens to be an excellent approximation. The stellar dynamics is then independent of the radial gradients; there is no reason for the velocity ellipsoid to point toward the galactic center; and we expect it to be parallel to the plane over the range of z occupied by the old disk, at least over the relevant extent in z (most old disk stars are at $z < z_0$).

We then may represent the ratio of the two velocity dispersions in terms of the Oort constants as given above, and write $-K_R = V_{\text{rot}}^2/R$. For the exponential disk with constant thickness (and $\mathcal{M}/L$), the density distribution of the old-disk population ν gives rise to $\partial(\ln \nu)/\partial R = -1/h$. Then we have

$$V_{\text{rot}}^2 - V_t^2 = \langle U^2 \rangle \left[\frac{R}{h} - R\frac{\partial(\ln \langle U^2 \rangle)}{\partial R} - \left(1 - \frac{B}{B - A}\right)\right]. \qquad (10.9)$$

Now we can measure h from surface photometry, and V_{rot}, A, and B for the gas from optical emission lines or 21-cm H I mapping. The latter is so, because the gas has a very low velocity dispersion, and therefore for the gaseous component V_t is very close to V_{rot}. The only unknown in the equation above then is the radial velocity dispersion and its radial derivative.

For illustrative purposes we can calculate the radial dependence of the stellar velocity dispersion for a disk with a flat rotation curve. Then $B/(B - A) = 0.5$. First we may assume that the ratio of vertical to radial velocity dispersion is always the same (although there is no theoretical reason for this); then we have

$$\langle U^2 \rangle^{1/2} \propto \exp(-R/2h) \qquad (10.10)$$

and

$$V_{\text{rot}}^2 - V_t^2 = \langle U^2 \rangle(2R/h - 0.5). \qquad (10.11)$$

The other possibility would be that the disk will during its evolution heat up to an equilibrium value for Q that is roughly constant with galactocentric distance. For example, Sellwood and Carlberg (1984) find such a behavior in their numerical experiments at a value of Q around 1.7 or so. This implies for a flat rotation curve [when $\kappa = (2V_{\rm rot}^2/R^2)^{1/2}$] that

$$\langle U^2 \rangle^{1/2} \propto R \exp(-R/h) \tag{10.12}$$

and

$$V_{\rm rot}^2 - V_{\rm t}^2 = \langle U^2 \rangle (3R/h - 2.5). \tag{10.13}$$

Over the ranges where the stellar velocity dispersion can be measured, the difference is actually small (see also Martinet 1988). To illustrate this, Table 10.1 gives Q as a function of radius (arbitrarily normalized to unity at $R = 1.5h$) for the first case. For $R < h$ the rotation curve is generally not flat, and the approximation used must fail. We see that Q starts to increase significantly only beyond $R = 3h$, and there it is impossible to measure stellar kinematics. The conclusion, then, is that we cannot now distinguish between the two possibilities from stellar spectroscopy alone. On the other hand, at larger R there is a significant contribution from the gas to the disk surface density. The calculation of Q then has to involve a "mean" velocity dispersion between stars and gas, and a higher surface density than the exponential stellar disk. This may easily lower the value of Q by a factor up to two, so that the predicted increase can in actual galaxies very well be compensated for by the gas. With this in mind we may say that the two assumptions are likely to be very similar in actual galaxies.

The method then is to measure $V_{\rm rot}$ from the gas, $V_{\rm t}$ from the stars, and h from surface photometry and then to fit these data to a radial dependence of the radial velocity dispersion. This was first done by van der Kruit and Freeman (1986) for NGC 7184 and over the radial extent R/h between 1 and 2. The observed asymmetric drift between the stars and the gas is illustrated in Fig. 10.5. The data are consistent with the models just given, and imply a radial velocity dispersion of about 100–120 km sec^{-1} at $R/h = 1.0$ (NGC 7184 has a $V_{\rm rot}$ of 266 km sec^{-1}) and a value of Q of about 1.5 to 2, if the $\mathcal{M}/L$ of the old disk is 6 in solar B-units. Again Bottema (1988 and in preparation) has confirmed the conclusion of a roughly constant Q (of order 2) out to 2 or 3 scale lengths

TABLE 10.1
Q as a function of radius

R/h	Q
1.0	1.17
1.5	1.00
2.0	0.96
3.0	1.06
4.0	1.31
5.0	1.73

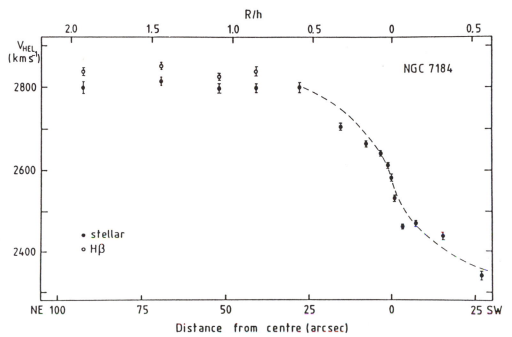

FIGURE 10.5

The asymmetric drift between the stars and the gas in the disk of NGC 7184 is evident from the displacement of the observed rotation curves for these two components. The data have not been corrected for inclination and derive from the same spectra. The dashed line shows a symmetric stellar rotation curve for the central areas. The top scale gives the radial distance in units of the photometric scale length h. (From van der Kruit and Freeman 1986.)

in other spirals, and Martinet (1988) has argued for this conclusion on general theoretical grounds.

Before leaving the subject of stellar kinematics, we need to say a few words on the origin of the observed radial variation of the velocity dispersions with galactocentric distance and on differences between galaxies. The equilibrium situation in disks is apparently such that the radial velocity dispersion of the stars is close to that necessary to keep Q constant and ensure marginal local stability, and the vertical one decreases with a radial e-folding length of two luminosity (and density) scale lengths in order to keep the old disk thickness constant. I have already shown that, when effects of the gas on the value of Q are taken into account, these two properties may in practice arise if there is a constant axis ratio of the velocity ellipsoid. The constant Q may arise from the fact that disks heat until this situation has been established, after which suppression of large instabilities then feeds back into less heating, as is indeed observed in numerical experiments (Sellwood and Carlberg 1984). Note that the central velocity dispersion is then so large that it also suppresses global stabilities (Sellwood 1983;

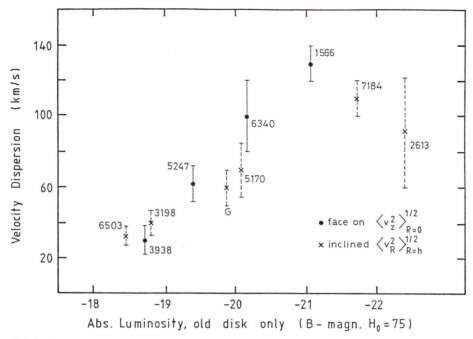

FIGURE 10.6

The stellar velocity dispersion in galactic disks as a function of the integrated magnitude of the old disk component. For inclined galaxies (crosses) the observed radial velocity dispersion at one photometric scale length from the center has been plotted; for face-on galaxies (dots) the vertical velocity dispersion at the center is entered. (From Bottema, private communication.)

Athanassoula and Sellwood 1986), and this may also be part of the feedback cycle.

    Bottema (1988) has found that the radial stellar velocity dispersion at one scale length from the center correlates with the rotation velocity of the flat rotation curve and therefore through the Tully–Fisher relation also with the integrated luminosity. An up-to-date version of his relation is shown in Fig. 10.6, where the horizontal axis is the luminosity of the old disk only. Face-on galaxies have been added, with the vertical velocity dispersion at the center. For the exponential decline described above, the dispersion at $R = h$ will be a factor $e^{-1/2} = 0.61$ lower than at the center, which is about the ratio of vertical to radial velocity dispersion of the stars in the disk of the Galaxy near the Sun; so this central value of the vertical dispersion should be comparable to the radial one at one scale length. The correlation can be understood qualitatively as follows. Assume again a flat rotation curve with rotation velocity V_{m} and an exponential disk with central surface brightness μ_0 and constant $\mathcal{M}/L$. The radial velocity dispersion at one scale length h can then be written as

$$\langle U^2 \rangle_h^{1/2} \propto \mu_0 (\mathcal{M}/L) Q h / V_{\mathrm{m}}. \tag{10.14}$$

Noting that the disk luminosity L is proportional to $\mu_0 h^2$, we get

$$\langle U^2 \rangle_h^{1/2} \propto \mu_0^{1/2} (\mathcal{M}/L) Q L^{1/2} / V_{\mathrm{m}}. \qquad (10.15)$$

Taking the Tully–Fisher relation as $L \propto V_{\mathrm{m}}^n$ replaces V_{m} by $L^{1/n}$. A reasonable value for n is about 4, so that over a range in L of 4 magnitudes, the velocity dispersion is predicted to increase by a factor 2.5, similar to what is seen in Fig. 10.6, if indeed μ_0, $\mathcal{M}/L$, and Q are constant from galaxy to galaxy. The constancy of μ_0 is "Freeman's Law," which has been presented already; its possible origin will be discussed later. It is then a remarkable property of galactic disks that $\mathcal{M}/L$ (resulting from the mass spectrum of stars at birth, the so-called initial mass function) and Q (resulting from dynamical heating) or at least the product of these two, are independent in first approximation of the mass of the galaxy. Note that the Galaxy is also included in the figure, using the data from Lewis (1986; see also Lewis and Freeman 1989) and that the point fits rather well with the external galaxies.

10.3   OBSERVATIONS OF THE KINEMATICS AND DISTRIBUTION OF THE GAS

The distribution and kinematics of the gas can be observed using optical emission lines, such as Hα, Hβ, and [O III], radio lines, such as the 21-cm line of H I, or molecular lines, such as from CO. The older method of optical slit spectroscopy will not be discussed here in detail, and I will concentrate on observations that map the velocity field over a large part of the galaxy. In principle a large number of slit positions can be used for this also, while at radio frequencies a velocity field can in principle be found from a large number of observations at individual positions with a single-dish instrument. The fastest method these days is the use of interferometric techniques. At radio frequencies this is done with the use of aperture-synthesis telescopes, such as at Westerbork or the VLA. At optical wavelengths the last few years have seen the advent of scanning Fabry–Perot instruments, such as TAURUS.

   The detailed workings of such instruments will not be discussed here, and I will limit myself to saying that we end up after observation and calibration (in position, intensity, and wavelength) with a set of so-called channel maps. These are maps of the intensity on the sky in a small range of wavelength or radial velocity. A set of maps contiguous in radial velocity make up the "data-cube" (two spatial and one velocity dimension), and it is this information that is used to derive the desired kinematics. One critical procedure that precedes all analysis is the subtraction of the continuum emission, which is derived by averaging all maps at velocities outside the range present in the galaxy.

   There are two methods for deriving the three maps of parameters that we wish to evaluate, namely, integrated intensity, radial velocity, and velocity dispersion. These are, of course, related at each position to the first three moments of the intensity distribution as a function of velocity, the so-called profile. The first method directly evaluates these moments. However, the velocity ranges outside the line emission from the galaxy contain noise that, especially in areas of low signal-to-noise ratio, have devastating effects on the first and second moments, and it is always unwise to add unnecessary noise in the calculation of the zeroth moment. We overcome this by defining a "window" in velocity at each position, over which emission from the galaxy is observed. This can be done automatically, but it is much better to do this interactively by displaying cuts through the data-cube on a television screen. These cuts have a spatial (*e.g.*, R.A. or Dec) and a velocity dimension, and are often referred to as "*l–V* diagrams." We then use a cursor to define at each sky position the velocity extent over which galaxy emission is present in this diagram. Only the regions within the window are then used to calculate the moments. This method is in principle sensitive to personal bias, and may depend on the (usually color) display levels on the monitor.

The second procedure is to fit a Gaussian to each individual profile and find values for the kinematic parameters in this way. The advantage of the window method is that it does not presuppose the profile to have any particular functional form, so that higher-order moments can in principle also be calculated. However, these are difficult to measure, and Gaussian profiles are usually assumed as soon as we start to interpret velocity dispersions. The fitting of Gaussians can be done automatically without human interference. The two methods have been compared in detail for calculation of the H$\textsc{i}$ velocity dispersion in face-on spirals from Westerbork maps by van der Kruit and Shostak (1982), using the data on NGC 3938. It turns out that for the window method, the personal bias is small, and that for the integrated emission and velocities both methods give very similar results. However, for calculation of velocity dispersions, the fitting of Gaussians is to be preferred, but we must always check how well the observed profiles are actually fitted by Gaussian functions. For H$\textsc{i}$ in undisturbed galaxy disks the fit is usually good. In areas of low signal-to-noise ratio the window method usually produces a superior estimate for the surface densities.

The observed velocity fields across disks of galaxies are almost always dominated by the pattern indicative of circular rotation. From this we then want to derive a rotation curve $V_{\rm rot}(R)$ that specifies the rotational velocity as a function of galactocentric distance. The usual procedure is to use a least-squares scheme to find values for five parameters: position of the rotation center (two numbers), systemic velocity $V_{\rm sys}$, inclination i of the normal to the galaxy plane with the line of sight (so $0°$ is face-on), and the position angle ϕ_0 of the line of nodes (major axis). These parameters are evaluated from different symmetry properties of the observed velocity field, as is illustrated in Fig. 10.7. Note, however, that a systematic pattern of radial expansion or contraction has an effect similar to a change in ϕ_0. Furthermore, deviations in the velocities owing, for example, to a density wave may perturb the derived rotation curve; if so, a detailed self-consistent model has to be used to extract an estimate of the unperturbed rotation curve. Finally, a transverse motion of the galaxy perpendicular to the line of sight produces errors in both ϕ_0 and $V_{\rm rot}(R)$, but these effects are only

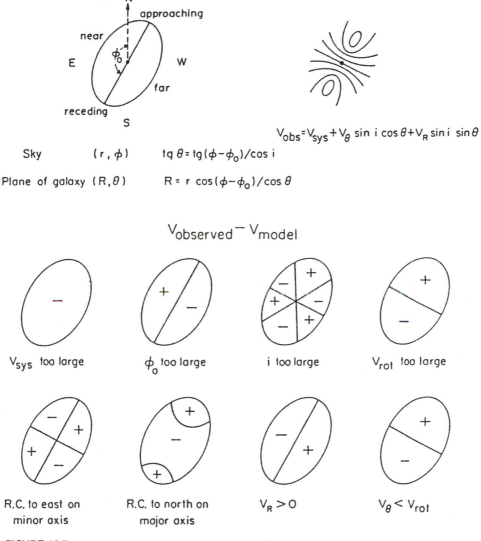

$$V_{obs} = V_{sys} + V_\theta \sin i \cos \theta + V_R \sin i \sin \theta$$

Sky (r, ϕ) $tg\, \theta = tg(\phi - \phi_0)/\cos i$

Plane of galaxy (R, θ) $R = r \cos(\phi - \phi_0)/\cos \theta$

$$V_{observed} - V_{model}$$

V_{sys} too large ϕ_o too large   i too large V_{rot} too large

R.C. to east on   R.C. to north on $V_R > 0$ $V_\theta < V_{rot}$
minor axis   major axis

FIGURE 10.7
Schematic representation of the rotation of disk galaxies and the calculation of dynamical parameters. A galaxy with the orientation at top left shows the pattern of line-of-sight velocities at top right, where lines of equal radial velocity are sketched. The contours close around the radius of maximum rotation velocity on the line of nodes; a flat rotation curve will fail to show this feature. The relevant equations are then given. The systematic pattern of residual velocities is then shown for the situation in which all but one of the dynamical parameters are chosen correctly. Note that the patterns are different, so that these parameters can be fitted independently to the observed velocity field. The last two examples show the effects of noncircular motion in the radial and tangential direction. (From van der Kruit and Allen 1978.)

appreciable for galaxies of large angular size. Lately, various refinements have been used in view of the observation that H I layers often deviate from a single plane beyond the optical extent of the disk; these have included evaluating i and ϕ_0 as a function of galactocentric distance by analyzing the velocity field in annular rings.

These methods of analysis can be applied to all observations of a data-cube as described, whether it comes from radio or from Fabry–Perot interferometry. I will illustrate some important results for H I synthesis. First I choose NGC 628 (Shostak and van der Kruit 1984) to show the derivation of a rotation curve, a warp, and the H I velocity dispersion. This is a rather face-on system, so that the velocity variation due to rotation over a resolution element (telescope beam) is smaller than the effect of velocity dispersion. First we have in Fig. 10.8 the integrated H I surface density. The gas is much more extended than the optical disk, and, although this is somewhat extreme in NGC 628, this is a general property of disks. A close inspection shows that the H I has maxima on top of the optical spiral arms, and that these gaseous spiral arms actually continue beyond the extent of the optical structure. Also, note the lack of H I in the inner regions, which is a feature common to many spirals.

A general feature of the radial H I distributions is that (at least beyond a central hole, if present) it declines much slower than the optical surface brightness. For Sb and Sc galaxies, typical values of this ratio are 10^{-1} $M_\odot/L_{\odot,B}$ in the inner regions to a few at the optical edges (see, *e.g.*, Wevers *et al.* 1986). In earlier types the general H I surface density is lower, but the amount of variation is similar. The largest H I surface density that occurs in late-type spirals is usually about 8.0 $M_\odot$ pc^{-2}.

The radial distribution of CO and therefore probably also of H_2 has been derived now in a fair number of spirals. A recent review has been given by Young (1987). In Sc galaxies the azimuthally averaged intensities of CO peak in the center and decrease with radius in the disk. The distributions of Hα, and of optical and radio-continuum surface brightness are similar, which can be interpreted to mean that the massive-star-forming efficiency is constant with radius, since massive stars appear to form exclusively in molecular clouds. This similarity also means that the ratio of molecular to atomic hydrogen decreases as fast as that given above for H I to optical light. In Sa and Sb galaxies at least 40 percent (improved resolution may increase this number) show a central hole in their CO distributions, similar to that observed in our Galaxy. At present none of the observed Sc galaxies have CO distributions resembling that in the Milky Way.

FIGURE 10.8 (Facing page)
Distributions of optical light and H I gas in NGC 628 represented on the same scale. The H I spiral structure in the inner regions fits onto the optical arms, but in the gas these extend out to much larger radii. The two sharp features in the H I map are processing artifacts, and the crosses indicate positions of field stars. (From van der Kruit and Shostak 1983.)

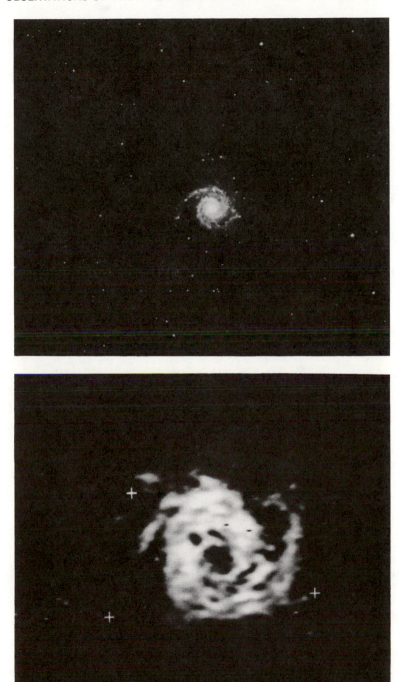

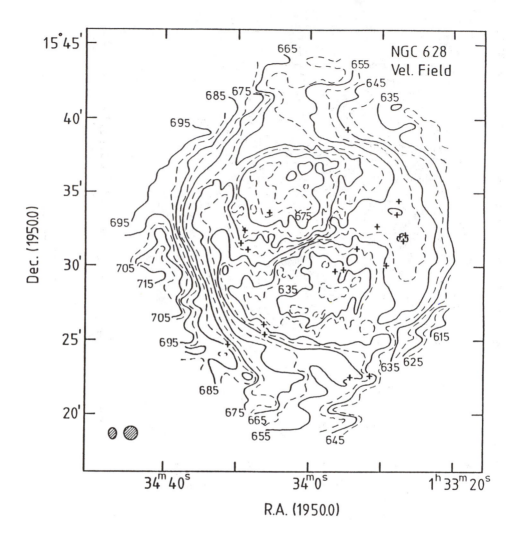

FIGURE 10.9

The observed radial velocity field of NGC 628. The resolutions used are indicated at the bottom left, and are 1 arcmin FWHM for the outer regions and 14 × 48 arcsec in the inner parts. Contours are shown every 5 km sec^{-1}. The velocity field is smooth down to this resolution, and continuous contours can be drawn. (After Shostak and van der Kruit 1984.)

The observed H I velocity field of NGC 628 is given in Fig. 10.9. At first glance this looks very disturbed, but some general features appear upon closer inspection. First, notice that over roughly the optical extent we can discern a pattern as drawn schematically in Fig. 10.7 with the dynamical major axis in a position angle of about 30°. This part can be fitted well by the procedure described above, although because of the low inclination the result here is insensitive to the value of i. The observed rotation curve has an amplitude of about 25 km sec^{-1} in the line of sight. From the integrated luminosity and the Tully–Fisher relation we can estimate the inclination-corrected rotation amplitude, and from this an inclination of 5° to 7° can be estimated. We can subtract the symmetric rotation field corresponding to the fit from the observed velocities and obtain a field of residual velocities. This field has an r.m.s. value of only 3.9 km sec^{-1} and displays no systematic pattern; this shows that there are essentially no vertical motions in excess of a few km sec^{-1}, unless there is an organized field of vertical motions, such that the overall pattern of its line-of-sight components mimics circular rotation. This is very unlikely, and the result therefore indicates an extreme flatness of the disk. For comparison, we can calculate the vertical amplitude of motion, if the H I layer in the solar neighborhood had a z-velocity of 4 km sec^{-1}. This is only 45 pc, which illustrates that the H I disk of NGC 628 has to be flat to within values of this magnitude.

Further inspection of Fig. 10.9 shows that beyond about 7 arcmin (20 kpc) the pattern remains organized, except that the kinematical major axis rotates to a position angle of about 100°. Furthermore, at this radius the contour at the systemic velocity of 655 km sec^{-1} essentially closes around the galaxy (allow for finite resolution effects). This means that the outer H I layer changes orientation with respect to the inner plane, and actually is in the plane of the sky at 20 kpc radius. The angle in space between the outermost ring and the inner plane is small, namely, no more than 9° or so. Such warps (except in very special situations) always produce effects of a changing kinematical major axis in the observed velocity field, although the effects on a contour diagram as in Fig. 10.9 are usually less pronounced (e.g., see Bosma 1981a,b). Note also the discussion of warps in Chapter 9.

Next I turn to the velocity dispersions, which are derived by fitting Gaussians to each profile. These fits were always satisfactory and show no evidence for superpositions of two Gaussians ("tails" to the profile). In Fig. 10.10 the values have been illustrated by the distribution in the (velocity dispersion, H I surface density) plane. The area of the symbol indicates the number of grid points at which the values occurred. For this diagram only the area between 1.5 and 4.5 arcmin (5 to 15 kpc) has been selected, and regions near the minor axis have also been deleted, because there the gradient of observed velocity across the telescope beam is too large to derive the velocity dispersion. At each vertical set of points a small horizontal line indicates the median, and we see that it increases from about 7 km sec^{-1} in low-surface-brightness regions (the "interarm" regions) to about 10 km sec^{-1} in the arms. However, there is no radial dependence of the H I velocity dispersion.

The velocity dispersion of the gas is likely to be isotropic as a result of the frequent collisions of gas clouds. There is no obvious reason why the H I velocity dispersions always come out about 7 to 10 km sec^{-1} in all galaxies observed so far (see van der Kruit and Shostak 1984). It is, however, interesting to note that neutral interstellar gas has a very sharp rise in the curve of cooling rate

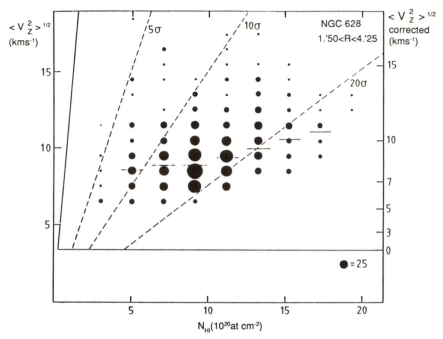

FIGURE 10.10

The distribution of points over the (H I surface density, velocity dispersion) plane in the inner disk of NGC 628. The area of each symbol is proportional to the number of grid points. The horizontal line is at the instrumental velocity resolution of 3.5 km sec^{-1}, and at the right are the corrected velocity dispersions. Points to the left of the full-drawn, slanted line have peak surface brightnesses in the profiles less than that where Gaussian fits can be attempted. The dashed lines indicate the peak surface brightnesses of the Gaussians expressed in units of the r.m.s. noise in the channel maps. Profiles in which this is less than 5 are rejected by the automatic routine, because then the formal error in the profile area would exceed 25 percent. The horizontal bars are the medians of the vertical distributions, and appear to increase with surface density. Generally speaking, $n(H I) > 10 \times 10^{20}$ H-atoms cm^{-2} correspond to the spiral arms. (From Shostak and van der Kruit 1984.)

versus temperature at about 10^4 K because of ionization of the hydrogen. For sufficient heat input the gas will thus always get warmer until it reaches this temperature and then will start to cool very efficiently. A kinetic temperature of 10^4 K interestingly corresponds to a one-dimensional velocity dispersion of 9 km sec^{-1}. The difference between arm and interarm may simply be the result of heat input from young O and B stars in the spiral arms.

I will now turn to an edge-on galaxy, NGC 891, to illustrate measurements of the thickness of the H I layer as a function of galactocentric distance. The data are from Sancisi and Allen (1979). In such a galaxy we have to deal with line-of-sight effects, and as a result the profiles are certainly not Gaussian. The basic data can now be given in an l–V diagram, where l is measured along the galaxy's major axis (see Fig. 10.11, top). Sancisi and Allen have pointed out how this diagram

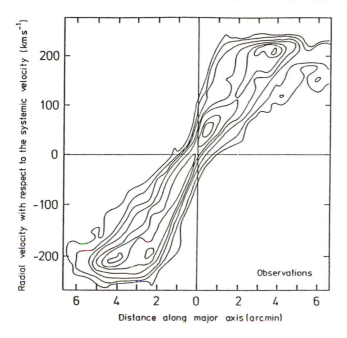

Observations

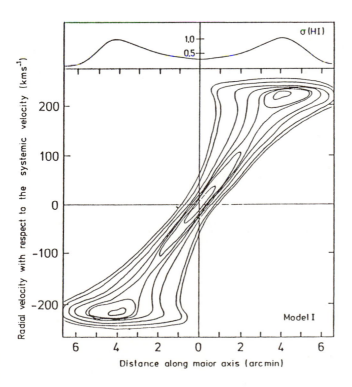

FIGURE 10.11
(Top) The observed Hɪ
intensities of NGC 891
in the *l–V* diagram.
(Adapted from Sancisi
and Allen 1979, where
l is measured along the
major axis.) Contour
levels are in relative
units 0.5, 1.0, and in
steps of 1.0 up to 6.0.
(Bottom) The Hɪ
intensities for the
model in the same
relative units. In this
model the derived
radial Hɪ distribution is
given at the top; the
peak value corresponds
to 3.7 $\mathcal{M}_\odot$ pc^{-2}, and
the radial distance of
this peak is 11 kpc.
The derived rotation
curve rises from the
center to 225 km sec^{-1}
at 0.25 arcmin (0.7
kpc) and remains
constant at this level
for larger radii. (From
van der Kruit 1981.)

maps the galactic plane. At a distance R along the major axis, the various parts of the line of sight map onto different observed velocities as follows. On the line of nodes we observe the full rotation, but moving away from this line a smaller part of the rotation velocity projects onto the line of sight, and therefore the observed velocity approaches more and more the systemic velocity. For example, at $l = 3$ arcmin, the line of nodes is seen at about 225 km sec^{-1} with respect to the systemic velocity, and the edge of the H<small>I</small> along this line of sight occurs at about 70 km sec^{-1}. The upper envelope of the H<small>I</small> distribution then corresponds to the rotation curve, except for an amount dependent on the velocity dispersion. This one-to-one mapping of the galaxy's plane onto the diagram is, of course, similar to the interpretation of the l–V diagrams in traditional H<small>I</small> studies in the Galaxy. The first step in interpreting Fig. 10.11 is to model the observed diagram in a radial distribution of the H<small>I</small> and a rotation curve, using circular symmetry and a value for the H<small>I</small> velocity dispersion. For this purpose we need to simulate the observational procedure (angular and velocity resolution) in the calculations. The best-fitting solution is illustrated in Fig. 10.11, bottom.

We can use the data-cube to derive the distribution of the thickness of the H<small>I</small> layer over the l–V diagram. For this purpose Sancisi and Allen derived the equivalent width, which is at each position in this diagram the sum of H<small>I</small> over all points perpendicular to the major axis, divided by the value in the plane. The result is given in Fig. 10.12, left. We can see there that the largest equivalent widths are at each l closest to the systemic velocity and therefore at the radial boundaries of the H<small>I</small> distribution. As Sancisi and Allen pointed out qualitatively, and as was subsequently confirmed quantitatively, line-of-sight effects will have a significant influence on this diagram if the inclination deviates slightly from 90°. Fortunately, the signature of inclination and layer-thickening are different across the diagram; the widths at the line of nodes (at each l the maximum deviation from systemic velocity) are not affected by inclination. Detailed modeling, again taking the observational resolutions into account, is necessary for a detailed interpretation.

Let us first see what to expect in an exponential disk with constant $\mathcal{M}/L$. In the isothermal disk with a vertical scale parameter z_0, the vertical distribution of a second component, moving in this force field with a velocity dispersion $\langle W^2 \rangle_{\mathrm{g}}^{1/2}$, will be

$$\rho_{\mathrm{g}}(z, R) = \rho_{\mathrm{g}}(0, R)\mathrm{sech}^{2p}(z/z_0), \tag{10.16}$$

where

$$p = \frac{\langle W^2 \rangle_*}{\langle W^2 \rangle_{\mathrm{g}}}. \tag{10.17}$$

In disks the velocity dispersion of the gas is considerably smaller than that of the stars, so that p is larger than unity and may possibly approach unity at the edges. Now we can calculate from this the full width of the H<small>I</small> layer at half-density points (FWHM) as

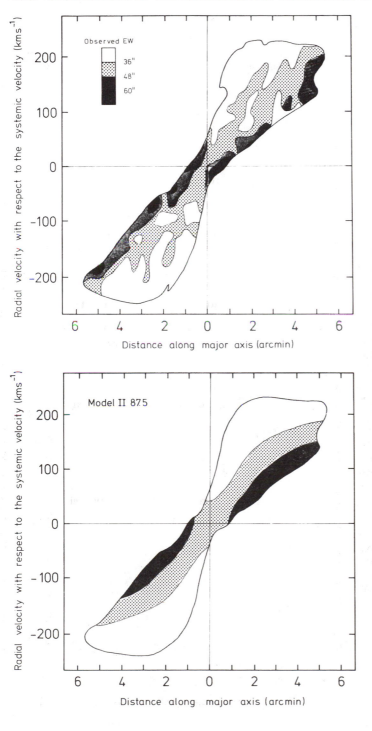

FIGURE 10.12
(Top) The observed
equivalent width of
the H I layer in the
disk of NGC 891
over the *l–V*
diagram. (Adapted
from Sancisi and
Allen 1979.)
(Bottom) The
best-fitting model
for the observed
equivalent width,
derived from
detailed modeling
as described in the
text. This involves
a gradual
thickening of the
H I layer with
galactocentric
distance and an
inclination of 87°5.
(From van der
Kruit 1981.)

$$\mathrm{FWHM_g} = 1.6625\, p^{-1/2} z_0 \qquad \text{for} \quad p \gg 1, \tag{10.18a}$$

$$\mathrm{FWHM_g} = 1.7628\, z_0 \qquad \text{for} \quad p = 1. \tag{10.18b}$$

With the equation for z_0 this becomes then, to within 3 percent,

$$\mathrm{FWHM_g} = 1.7 \langle W^2 \rangle_g^{1/2} \left[\frac{z_0}{\pi G (\mathcal{M}/L)\mu_0} \right]^{1/2} \exp(R/2h), \tag{10.19}$$

so that the thickness of the H I layer increases exponentially with an e-folding length of two optical scale lengths, if the gas velocity dispersion is constant with radius as observed in face-on systems.

When this thickening of the gas layer is taken into account, modeling of the diagram in Fig. 10.12 (left) results in a best fit, which is illustrated in Fig. 10.12 (right). As described above, this procedure involves calculating both the inclination and the parameters of the H I thickness. The fit is good and implies that in NGC 891 $\mathrm{FWHM_g} = (0.22 \pm 0.02) \exp(R/2h)$ kpc with $h = 4.9$ kpc, and that the inclination is about $87°5$. This can again be taken as evidence that $(\mathcal{M}/L)$ of the old-disk population is not a function of radius.

The thickening of the H I layer has also been observed in the Galaxy and possibly in the Andromeda galaxy. It is therefore interesting that Bottema *et al.* (1986) found that no such thickening is apparent in NGC 5023. It is true that this dwarf galaxy has a disk radius of only 7.8 kpc and an extent in H I of about 10 kpc. In this dwarf the H I seems at all radii to have the same thickness as the stellar disk, which also is rather blue, with $(U - B)$ about 0.0 and $(B - V)$ about 0.4. A possibility is that star formation here is so vigorous that especially in the inner parts the gas is heated much more efficiently than in larger spirals and that as a result of this the H I does have a variation in velocity dispersion with radius. This is consistent with the H I observations, but cannot be proven from these data.

Finally I will summarize the observations of warps in the H I layers. The presence of such deviations of the gas layer from the inner flat plane was first found in the Galaxy and was inferred from the change in kinematical major axis in M83 by Rogstad *et al.* (1974); they modeled it with the methods involving rings with changing orientations as described above ("tilted-ring model"). Another example of such a warp has already been described for NGC 628. The most direct way to look for warps is, of course, in edge-on galaxies. Sancisi (1976) has found warps in the H I distribution in four out of five systems, and work since then on both edge-on and moderately inclined galaxies has confirmed that warps are the rule rather than the exception (Bosma 1981a,b). One of these exceptions is NGC 891, the galaxy that has just been described. The largest regular warp has been identified by Bottema *et al.* (1987) in the edge-on galaxy NGC 4013. There the warp sets in at precisely the edge of the optical disk (9.5 kpc) and reaches a height of about 6 kpc from the inner plane at a distance of 20 kpc from the center. The warp is extremely symmetric between the two sides. The inner disk has a flat rotation curve at 195 km sec^{-1}, which suddenly drops to about 170 km sec^{-1} at the edge of the stellar disk and the start of the warp, after which it remains constant up to the last measured point.

10.4 THE DISTRIBUTION OF MASS IN SPIRAL GALAXIES

The distribution of mass in a spiral galaxy can be inferred from the rotation curve. Freeman (1970) has calculated what rotation curve to expect from an exponential disk with constant $\mathcal{M}/L$. The result for an infinitely flat disk is

$$V_{\text{rot}}(R) = (\pi Gh\sigma_0)^{1/2} \left(\frac{R}{h}\right) \left[I_0\left(\frac{R}{2h}\right) K_0\left(\frac{R}{2h}\right) - I_1\left(\frac{R}{2h}\right) K_1\left(\frac{R}{2h}\right)\right]^{1/2}.$$

(10.20)

I and K are modified Bessel functions. This curve has a maximum at about 2.2 scale lengths, and then declines steadily and eventually becomes Keplerian. The value of the maximum is

$$V_{\max} = 0.8796 \, (\pi Gh\sigma_0)^{1/2}.$$

(10.21)

Allowing for the finite thickness of the disk lowers this value by, e.g., 4.5 percent when $h/z_0 = 5$.

The discovery of the flat rotation curves in the H$_I$ out to large radii (Roberts 1975; Bosma 1981a,b) has immediately shown that the light distribution alone cannot account for the observed curves if a constant $\mathcal{M}/L$ is assumed. Bosma and van der Kruit (1979) observed six spirals in H$_I$ and in the optical, and inferred what change in $\mathcal{M}/L$ would be required if all the mass indicated by the rotation curve were in the disk. This gave local values of $\mathcal{M}/L$ of order ten (in solar B-units) in the inner regions, after which it increased steadily and reached values of a few hundred at the edges of the optical disks. Since the flat rotation curves continue beyond these edges, there must be material with even much higher values there.

It is now generally assumed that the material with high $\mathcal{M}/L$ is distributed in a more or less spherical "dark halo," the constitution of which remains to be discovered. This fits with the evidence for constant $\mathcal{M}/L$ in disks, and will be discussed in somewhat more detail below. Carignan and Freeman (1985) have proposed to model the rotation curves with a combination of two components, one from the observed distribution of surface brightness (if necessary composed of a disk and a bulge), and an isothermal sphere. They fitted the inner rotation curve to that expected from the light distribution, and added then an isothermal sphere to represent the outer part of the rotation curve. This worked satisfactorily in most cases.

Van Albada et al. (1985) have followed a similar scheme, except that they chose a somewhat different form for the density distribution in the dark halo (see below). In fitting the H$_I$ observations of NGC 3198, they found that satisfactory fits could be found for a large range of values for the $\mathcal{M}/L$ of the disk, all the way from zero (no disk!) to 3.6 in the B-band. The last solution (the "maximum-disk" model) gives an amplitude of the rotation curve from the disk alone that is almost equal to that of the outer flat rotation speed and consequently gives rise to a "disk–halo conspiracy." Kent (1986) followed the same procedure, and found that features in the rotation curve could be reproduced from features in the light distribution; he therefore argued that the disk must contribute dominantly to the inner rotation curve, and that therefore the solution to be preferred in practice would be the maximum-disk model or one very close to this.

Begeman (1987) has made sensitive H I observations of eight spirals in an attempt to trace out rotation curves to as large a distance as possible relative to the optical scale length h. His record came with NGC 2841, in which H I could be detected out to 17.8 h ($h = 2.4$ kpc in this galaxy; so the H I extends out to 42.6 kpc). The analysis proceeds as follows. The disk surface density $\sigma(r) = L(r)(\mathcal{M}/L)$ can for a flat distribution be translated into a circular velocity curve with (Casertano 1983)

$$V_c^2(R) = -8GR \int_0^\infty r \int_0^\infty \frac{\partial \rho(r, z)}{\partial r} \frac{K(p) - E(p)}{(Rrp)^{1/2}} \, dz \, dr, \qquad (10.22)$$

where $p = x - (x^2 - 1)^{1/2}$ and $x = (R^2 + r^2 + z^2)/(2Rr)$, and where K and E are complete elliptic integrals. For the z-dependence of $\rho(r, z)$ we may use the isothermal distribution $\operatorname{sech}^2(z/z_0)$.

The bulge is assumed to be spherical with an observed surface density distribution $\sigma(r) = \mu(r) \, (\mathcal{M}/L)$ (where $\mathcal{M}/L$ may, of course, be different from that of the disk) and then (Kent 1986)

$$\begin{aligned} V_c^2(R) = & \frac{2\pi G}{R} \int_0^R r\sigma(r) \, dr \\ & + \frac{4G}{R} \int_R^\infty \left[\arcsin\left(\frac{R}{r}\right) - \frac{R}{(r^2 - R^2)^{1/2}} \right] r\sigma(r) dr. \end{aligned} \qquad (10.23)$$

The halo density distribution is assumed to be

$$\rho(R) = \rho_0 \left[1 + \left(\frac{R}{R_c}\right)^2 \right]^{-1}, \qquad (10.24)$$

which gives a circular velocity curve

$$V_c^2(R) = 4\pi G \rho_0 R_c^2 \left[1 - \left(\frac{R_c}{R}\right) \arctan\left(\frac{R}{R_c}\right) \right]. \qquad (10.25)$$

This is indeed a flat curve for large R, and the asymptotic value is $(4\pi G \rho_0 R_c^2)^{1/2}$.

The contributions from these components have to be added in quadrature to calculate the "total" rotation curve.

The fitting procedure then is as follows. From the observed light distributions the shape of the circular-velocity curves of disk and bulge can be calculated; these

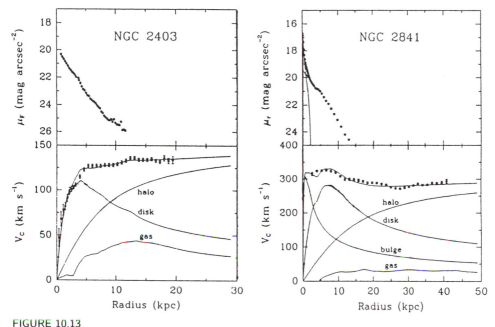

FIGURE 10.13
Analysis of the rotation curve and mass model for NGC 2403 and 2841. The upper panels show the photometry from Kent (1987) and a decomposition in truncated bulge and disk. The lower panels show the observed rotation curve (points and error bars), the circular-velocity curves for the four components, and the total model rotation curves for the best-fitting models. (From Begeman 1987.)

must be scaled by a linear factor that depends on the value of M/L. In the same way the contribution of the gas is calculated, but there is no necessity to scale this, since the surface density is measured (25 percent He is assumed). As a first guess for the disk and bulge M/L, the inner rotation curve is represented by the luminous components only (luminous includes the gas). Then the halo is added with a first guess at R_c (which determines the shape), and the amplitude is derived by least-squares fitting. The M/L's of disk and bulge need to be adapted (decreased), and then varied; for each combination a χ^2 is calculated, and the best fit is selected. This process is repeated for different values of R_c until the final best fit is obtained. Since this process starts out with the largest possible value for M/L of the disk, which then is decreased only to accommodate the dark halo, the final solution has still the largest possible M/L, and therefore this is a "maximum-disk" fit.

The results for NGC 2403 and 2841 are shown in Fig. 10.13. The photometry in the top panels is from Kent (1987) and is represented by a (truncated) bulge and a disk. The lower panels show the observations, the four contributing circular-velocity curves, and the total-model rotation curve. Note that the amplitude of the disk circular-velocity curve is similar to the asymptotic value for the dark halo. The M/L of the disk is 1.8 for NGC 2403 and for NGC 2841 9.4, and that of the halo 5.6 for NGC 2841. Within the optical boundaries (14 kpc) the mass of the halo constitutes 34 percent of the total mass, and within the outermost point (44 kpc) the dark mass constitutes 79 percent of the total mass, which is 5.9×10^{11} $\mathcal{M}_\odot$. For his other seven galaxies Begeman finds $\mathcal{M}/L$'s of the disk of 3.1 ± 1.2; so NGC 2841 is here somewhat exceptional. In NGC 5371 there is no need to introduce a dark halo in order to arrive at a fit, but here the rotation curve is observed out to only 4 scale lengths. For the remaining galaxies the dark mass makes up 44 ± 9 percent of the total mass within the optical boundaries. For the sample (except NGC 5371) the parameters of the dark halos have ranges of 0.008 to 0.036 $\mathcal{M}_\odot$ pc^{-3} for ρ_0 and 2.5 to 15 kpc for R_c. The ratio of dark to total matter (Λ) at the largest radius R_M where HI is observed varies between 0.84 and 0.64. The ratio of Λ and R_M (expressed in optical scale lengths h) is 0.072 ± 0.019, and Λ increases with R_M/h, as expected.

It should be stressed that the maximum-disk hypothesis implicitly assumed in the modeling above has some, although not very strong, observational support. The support mainly comes from the fact that there are features in some rotation curves that appear to correspond to features in the luminosity profiles. However, the correspondence is not always present, and often features in either the rotation curve or the luminosity profile are not accompanied by features in the other. There are also galaxies where the luminosity profile is affected by spiral structure and where it actually is the contribution from the HI that appears responsible for the structure in the rotation curve. It is therefore of great importance to see whether there is other evidence, and this could come from independent ways of measuring the disk mass. Before discussing these I will first look at the values of disk $\mathcal{M}/L$ that are implied by the maximum-disk hypothesis.

Begeman has, as described above, found that the maximum-disk solutions imply values for $\mathcal{M}/L$ in the rough range 2 to 5 in solar B-units, except in NGC 2841, where the value is about 9. It is of importance to see first what values are reasonable in view of information available on the solar neighborhood. The integrated surface density, according to the data from Pioneer 10 (van der Kruit 1986) is $20 \pm 2 L_{\odot,B}$ pc^{-2}. A lower limit to the surface density comes from just adding the contribution from known stars of 24 $\mathcal{M}_\odot$ pc^{-2} (Tinsley 1981) to that from stellar remnants and gas, for which a minimum estimate is about 10 $\mathcal{M}_\odot$ pc^{-2}. Then $\mathcal{M}/L$ is at least 1.5 in the usual units. The maximum comes from Bahcall's (1984) determination of the surface density from vertical dynamics as 80 $\mathcal{M}_\odot$ pc^{-2}, so that $\mathcal{M}/L$ is at most 4.5. The solar neighborhood is not in a region of spiral structure and many young stars, so that the upper limit appears to be a strong one. In any case the values seem "reasonable," except for that for the disk of NGC 2841 of 9.4.

One other way of independently estimating the disk mass and checking the maximum-disk hypothesis is to use the fact that disks appear to have sharp edges in their light and therefore probably in their mass distributions. If the latter is also true, then we may expect a feature in the rotation curve at this radius $R_{\max}$, which can be used to constrain the disk-density distribution. The effects of a sharp edge were calculated by van der Kruit and Searle (1982a) and by

Casertano (1983). Inside R_{max} the rotation curve from a truncated disk is flatter than expected from a pure exponential disk, because there is no outer material to exert an outward pull. At R_{max} a rather sudden decline in the disk rotation curve sets in, and this signature is more pronounced when the ratio R_{max}/h is smaller. Casertano found that the H<small>I</small> observations of NGC 5907 showed the signature of the disk truncation, and derived from this the relative contributions of the disk and halo to the total mass. Within the optical boundaries the dark mass constitutes 60 percent of the total mass, which is rather higher than the values found by Begeman in his maximum-disk models, which average for this ratio at 44 percent.

A second example where the signature of the disk truncation may be identified in the rotation curve is NGC 4013 (Bottema *et al.* 1987). This galaxy with its severe H<small>I</small> warp has already been described. At R_{max} the rotation velocity suddenly drops by 25 km sec^{-1} from an inner rotation velocity of 195 km sec^{-1} and then again remains flat. The interpretation is that the latter is due to the dark halo and that the drop is the truncation feature of the disk. The model that can fit these data has within the optical boundaries this time a dark halo with only 25 percent of the total mass, which is even less than what was found by Begeman. So we see that this method is not yet conclusive.

A more direct way is to use "vertical dynamics," which means the use of vertical density distributions and velocity dispersions to infer the gravitational field of the disk. In the disk the gravitation is almost entirely due to the disk itself. We have already seen some applications of this in which it was, for example, inferred that the disk $\mathcal{M}/L$ is constant with radius, but until now I have not so far estimated values for $\mathcal{M}/L$ from these data. The method is, of course, similar to classical estimates of the local surface density of the disk of our Galaxy, as pioneered by Oort (1932). (See also Chapter 8.) The relevant equations have already been given for the isothermal-sheet approximation; I will first extend these to different, but plausible distributions at low z, and then discuss the outcomes of these studies. These are important to consider, because the actual distribution of stars and mass at low z does have important effects on the numerical results, even though the distribution at larger distances from the plane, where we can interpret optical surface photometry, are the same. So I will first consider alternatives to the isothermal-sheet formulas that do share the property that the density distribution tends to an exponential at larger z. This is partly motivated by the fact that Wainscoat (1986) has found some evidence, from surface photometry in the near infrared in one edge-on galaxy, that the z dependence continues exponentially to low values of z, and does not flatten off as expected from the isothermal sech^2 function.

The isothermal-sheet description was originally introduced for two reasons. First, it has the property that at large distances from the plane it approximates an exponential, as indicated by existing observations of edge-on galaxies. Second, it is known from the kinematics of stars in the solar neighborhood (*e.g.*, Wielen 1977) that stars that have ages above a few Gyr all have roughly the same velocity dispersion, so that the mix of stars should at least moderately far from the plane be dominated by a stellar population with essentially a single velocity dispersion. Since this old-disk population indeed has most of the disk mass and appears to be self-gravitating, the isothermal sheet immediately suggests itself, at least for light distributions away from the central dust lane. At low z, however, deviations may be expected, because there the stellar mix may no longer be described as isothermal.

At this point it is not definitely known what distribution for the mass density to use, but it is reasonable to take the sech^2 and an exponential all the way to $z = 0$ as the two extremes. Van der Kruit (1988) has calculated the effects of using the family of density laws

$$\rho(z) = 2^{-2/n} \rho_e \, \text{sech}^{2/n}(nz/2z_e). \tag{10.26}$$

The isothermal is the extreme for $n = 1$, and the exponential is the other for $n = \infty$. Here ρ_e is the (extrapolated) density in the plane and z_e the exponential scale height at large z. For comparison we have for the isothermal sheet $\rho_0 = \rho_e/4$ and $z_0 = 2z_e$. I will now summarize the relevant equations for both these extreme cases and the intermediate case $n = 2$. The subscripts refer to the value of n. The density distributions are shown in Fig. 10.14. The differences are large only at such small z, where in general optical surface photometry cannot be attempted because of dust absorption.

Surface densities:

$$\sigma_1 = \rho_e z_e, \tag{10.27a}$$

$$\sigma_2 = (\pi/2)\rho_e z_e, \tag{10.27b}$$

$$\sigma_\infty = 2\rho_e z_e. \tag{10.27c}$$

Vertical force K_z, if the distribution is self-gravitating:

$$K_{z,1} = -2\pi G\sigma \tanh(z/2z_e), \tag{10.28a}$$

$$K_{z,2} = -4G\sigma \arctan[\sinh(z/z_e)], \tag{10.28b}$$

$$K_{z,\infty} = -2\pi G\sigma[1 - \exp(-z/z_e)]. \tag{10.28c}$$

Velocity dispersion (squared) at $z = 0$:

$$\langle W^2 \rangle_{0,1} = 2\pi G\sigma z_e, \tag{10.29a}$$

$$\langle W^2 \rangle_{0,2} = (\pi^2/2)G\sigma z_e, \tag{10.29b}$$

$$\langle W^2 \rangle_{0,\infty} = \pi G\sigma z_e. \tag{10.29c}$$

Velocity dispersion (squared) as a function of z:

$$\langle W^2 \rangle_1 = \langle W^2 \rangle_0, \tag{10.30a}$$

$$\langle W^2 \rangle_2 = \langle W^2 \rangle_0 \{1 - (2/\pi)^2 \arctan^2[\sinh(z/z_e)]\} \cosh(z/z_e), \tag{10.30b}$$

$$\langle W^2 \rangle_\infty = \langle W^2 \rangle_0 [2 - \exp(-z/z_e)]. \tag{10.30c}$$

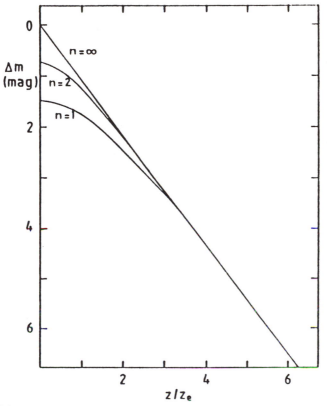

FIGURE 10.14

The density distribution as a function of distance z from the symmetry plane for the three density models in the text. The parameters have been chosen to give the same density at large z. The vertical scale is in magnitudes. (From van der Kruit 1988.)

In a face-on stellar disk the observed velocity dispersion follows from an integration; the values (squared) are

$$\langle W^2 \rangle_{\text{FO},1} = 2\pi G\sigma z_{\text{e}},\tag{10.31a}$$

$$\langle W^2 \rangle_{\text{FO},2} = (1.7051)\pi G\sigma z_{\text{e}},\tag{10.31b}$$

$$\langle W^2 \rangle_{\text{FO},\infty} = (3/2)\pi G\sigma z_{\text{e}}.\tag{10.31c}$$

A second nongravitating component with isothermal velocity dispersion $\langle W^2 \rangle_{\text{II}}^{1/2}$ can be calculated using the parameter

$$p = \frac{\langle W^2 \rangle_0}{\langle W^2 \rangle_{\text{II}}}.\tag{10.32}$$

The result is

$$\rho_{\text{II},1}(z) = \rho_{\text{II}}(0)\text{sech}^{2p}(z/2z_{\text{e}}),\tag{10.33a}$$

$$\rho_{\text{II},2}(z) = \rho_{\text{II}}(0) \exp[-(8/\pi^2)pI(z/z_e)], \tag{10.33b}$$

$$\rho_{\text{II},\infty}(z) = \rho_{\text{II}}(0) \exp\{-2pz/z_e + 2p[1 - \exp(-z/z_e)]\}. \tag{10.33c}$$

In the second equation the function I follows from

$$I(y) = \int_0^y \arctan(\sinh x)\, dx. \tag{10.34}$$

An analytical method to calculate this integral and various other useful approximations is given in van der Kruit (1988). In Fig. 10.15 are some properties plotted for comparison. In all these examples representative values have been chosen for the solar neighborhood, namely, $\sigma = 80\ \mathcal{M}_\odot\ \text{pc}^{-2}$ and $z_e = 325$ pc.

We now use these equations to derive values for $\mathcal{M}/L$ of the old-disk population. First we look at derivations using the observed velocity dispersion of the old-disk stars in face-on disks, combined with the observation in edge-on galaxies that $z_e = 0.35 \pm 0.1$ kpc. Van der Kruit and Freeman (1986) have discussed the data in terms of the isothermal-sheet approximation only and found for the old disk $\mathcal{M}/L = 6 \pm 2$ (as always below in solar B-units). The effects of the two additional models are that it will have to be increased by 17 percent for $n = 1$ and by 33 percent for the exponential model. On this basis we may take as the best value $\mathcal{M}/L = 7.5 \pm 2.5$, after allowance for the uncertainty about which model to choose.

As the bottom-left panel in Fig. 10.15 shows, the H I layer is severely affected by the choice of n, understandably because of the fact that the gas resides at small z and is therefore sensitive to the precise form of K_z near the plane. The width of the H I layer at half-density points FWHM_g can be conveniently expressed in terms of a length parameter d_g, which equals

$$d_g = \left(\frac{\langle W^2\rangle_g\, z_e}{G\sigma}\right)^{1/2} \tag{10.35}$$

Then for the three models we have for $p \gg 1$

$$\text{FWHM}_{g,1} = 1.33\, d_g, \tag{10.36a}$$

$$\text{FWHM}_{g,2} = 1.18\, d_g, \tag{10.36b}$$

$$\text{FWHM}_{g,\infty} = 0.94\, d_g. \tag{10.36c}$$

For p of order unity the numerical values need to be decreased by about 7 percent.

The effects of the different models are rather large, especially when we remember that σ follows from d_g squared. The effect on $\mathcal{M}/L$ is even more severe, because in edge-on galaxies the face-on surface brightness is calculated by being fit to the z profiles at large z, and the isothermal value is a factor two lower than if an exponential were assumed. For the fit to the H I layer in NGC 891 (van der Kruit 1981), the isothermal approximation gave for the old disk 8 ± 4; this becomes 4 ± 2 for $n = 2$, and 2 ± 1 for the exponential. Hence such H I measurements are useful only to monitor a change in $\mathcal{M}/L$ and are not very accurate in discovering its value.

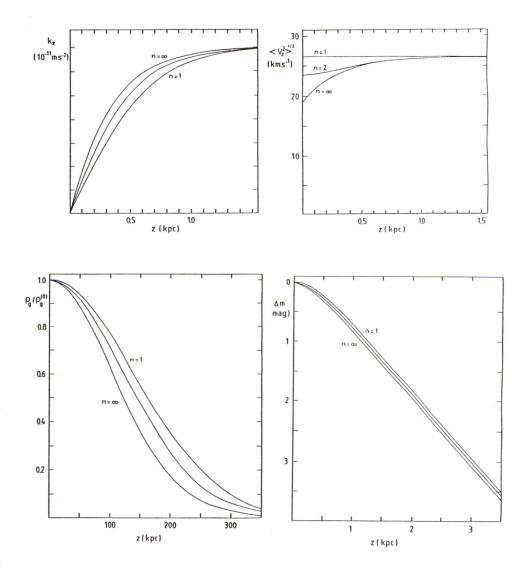

FIGURE 10.15
Distribution of some properties in the force field of various vertical density distributions,
indicated by external galaxies and given in Fig. 10.14. All three models have the same
integrated surface density of 80 $\mathcal{M}_\odot$ pc^{-2} and exponential density slope at large z of 325
pc, typical for the solar neighborhood. (Top left) The vertical force K_z. (Top right) The
stellar velocity dispersion of the self-gravitating disks. (Bottom left) The density
distribution of a gas layer with negligible surface density and a velocity dispersion of 8
km sec^{-1}. (Bottom right) The density distribution of a second isothermal component with
negligible surface density and velocity dispersion 45 km sec^{-1}. (From van der Kruit 1988.)

The conclusion from such work is that the best value for $\mathcal{M}/L$ of the old-disk population is 6 ± 2. This cannot yet be compared directly to Begeman's values, because we have to add in the young population. The disk colors can be used for this as a guide. For example, in the Bahcall and Soneira (1984) model for the Galaxy, the integrated colors of the contributions to the disk surface brightness are $B - V$ about 0.56 for old dwarfs, 1.30 for disk giants, and -0.03 for young Population I (see van der Kruit 1986). The old disk can on this basis be expected to have a $B-V$ of about 0.9, and from the observed colors we may then estimate in a galaxy disk the relative contributions of old disk and young Population I. From this it follows that for a typical Sb disk with a $B - V$ of 0.65, the $\mathcal{M}/L$ would be 4, if for the old-disk population only it is 6. For an Sc disk with $B-V =$ 0.4, the $\mathcal{M}/L$ becomes 2. So we can see that these estimates of disk masses result in $\mathcal{M}/L$'s in the range 1 to 5 for disks of Sb galaxies or later types, and fit with the maximum value of 4.5 quoted above for the Galactic disk. This shows that Begeman's values of 3 ± 1 (excluding NGC 2841) are indeed "reasonable." On the other hand, the uncertainties are large, so that this must not be taken as a strong confirmation of the maximum-disk hypothesis. At this point it therefore is not entirely clear yet that a disk-halo conspiracy indeed exists. This conspiracy expresses the fact that in maximum-disk fits the disk-alone and (dark) halo-alone rotation curves have essentially the same amplitude. It is not even clear from current observations whether the ratio of luminous to dark matter is closely the same in all galaxies.

Now that we have at least some indications of the $\mathcal{M}/L$ to be expected in a disk, it is of interest to return to the stability questions raised above. For the local stability I have already indicated that these results lead to values for Toomre's Q in the range of 1.5 to 2, and disks appear stable locally. For global stability Efstathiou *et al.* (1982) have derived from numerical experiments an empirical parameter Y, which should exceed 1.1 for stability:

$$Y = V_{\mathrm{m}} \left(\frac{h}{G\mathcal{M}_{\mathrm{d}}} \right)^{1/2}, \tag{10.37}$$

where $\mathcal{M}_{\mathrm{d}}$ is the disk mass and V_{m} the velocity in the flat part of the rotation curve. Estimates of this parameter for samples of galaxies with rotation curves and surface photometry give mean values somewhat below the critical value, but generally within the uncertainties. On the other hand, I have already indicated above that the observed central velocity dispersions in the disks are sufficiently high to contribute also to global stability. A rough calculation (van der Kruit and Freeman 1985) shows that the condition $Y = 1.1$ means that within the optical boundaries the dark halo must make up about 75 percent of the mass. In Begeman's sample this condition is not met, since for his galaxies the maximum-disk fits gave values of about 45 ± 10 percent. So it follows that the maximum-disk hypothesis in general implies that $Y < 1.1$, and that therefore the dark halo would not be sufficiently massive in the inner regions to provide global disk stability.

For NGC 891 the value for Y can be calculated more accurately than above, because $\mathcal{M}/L$ is not needed here. For the observed thickness of the H<small>I</small> layer, van der Kruit (1981) found $\text{FWHM}_g = 0.22 \exp(R/2h)$ kpc, and using $z_e = 0.49$ kpc the disk central surface density becomes for the three models 266 (isothermal), 210, and 133 (exponential) $\mathcal{M}_\odot$ pc^{-2}. With $h = 4.9$ kpc the disk mass becomes 4.0, 3.2, and 2.0×10^{10} $\mathcal{M}_\odot$, respectively, and the maximum in the "disk-alone" rotation curve 111, 99, and 79 km sec^{-1}. This is significantly lower than the observed rotation of 225 ± 10 km sec^{-1}. The total mass within $R_{\max} = 21$ kpc is about 2×10^{11} $\mathcal{M}_\odot$; so the disk does make up only a small fraction of the mass within the optical boundaries. The parameter Y also follows immediately as 1.20, 1.34, and 1.70, respectively. Hence NGC 891 does not conform to the maximum-disk hypothesis (and this is independent of any assumed value for $\mathcal{M}/L$), and its halo must be massive enough to stabilize the disk against global modes.

There is an interesting coincidence that I want to point out, finally, concerning dark halos and the maximum-disk hypothesis. In Chapter 5 I showed component separations in the light distributions of the two edge-on galaxies NGC 891 and NGC 7814. These are two extremes, in the sense that in NGC 891 82 percent of the light from old stars is in an exponential disk, whereas for NGC 7814 93 percent is in an $R^{1/4}$-bulge. In spite of these major differences in the light and therefore the luminous-mass distributions, the rotation curves of the two galaxies are very similar, both rising to about 225 km sec^{-1} in about 3 kpc and remaining flat thereafter out to at least 20 kpc. This seems to imply that the force field cannot be dominated by the gravitation of the luminous material and must be set by a dark component.

Finally, I note that optical and near-infrared searches have been made for material that could constitute the dark halos. I reviewed these recently (van der Kruit 1987), and concluded that H-burning main-sequence stars can be ruled out, especially by the measurements in the near infrared. The local $\mathcal{M}/L$ (in B) in the halo is in excess of 1100. IRAS data have not been able to rule out black (or brown) dwarfs; so low-mass objects that are below the H-burning limit are still possible.

10.5   THE MASS DISTRIBUTION IN OUR GALAXY

To conclude this chapter I will make a brief comparison to our Galaxy. In the first place I note that the Galaxy's disk velocity dispersion was included in Fig. 10.6, and that it appears to fit in with the trend noted there. Given the discussion in the preceding section, can we decide whether or not, according to the available data, the Galaxy has a "maximum disk"? Observations of the rotation curve and the local disk surface density are presented and discussed in Chapters 6 and 8, and I will not repeat this. As in essentially all other galaxies, the rotation curve is probably flat, and the amplitude will be taken in what follows as 230 ± 15 km sec^{-1}. The local disk surface density is currently in some dispute, as discussed in Chapter 8, and is either $80 \pm 20 \, \mathcal{M}_\odot \, \mathrm{pc}^{-2}$ following Bahcall (1984) or 45 ± 9 $\mathcal{M}_\odot \, \mathrm{pc}^{-2}$ in the Kuijken and Gilmore (1989) analysis. These two values are derived from completely different samples: Bahcall (B84) uses G and F dwarfs and K giants; Kuijken and Gilmore (KG) base their analysis on their own sample of K dwarfs.

First, let us look at some local properties that have been discussed above for external galaxies. The first is $\mathcal{M}/L$. The local disk surface brightness was estimated from the Pioneer 10 data in van der Kruit (1986) as 23.8 ± 0.1 B-mag arcsec^{-2} ($\simeq 20 \pm 2 \, L_\odot \, \mathrm{pc}^{-2}$), so that the total disk $\mathcal{M}/L$ becomes 4.0 ± 1.0 (B84) or 2.3 ± 0.5 (KG). It was estimated that the old-disk population contributes two thirds of the surface brightness, so that the old-disk $\mathcal{M}/L$ can estimated as 6.0 ± 1.5 (B84) or 3.5 ± 0.8 (KG). The first value is comparable to what was found in external galaxies. Another quantity that can be calculated is Toomre's Q, which for a radial stellar velocity dispersion of 45 ± 5 km sec^{-1} comes out as 1.5 ± 0.5 (B84) or 2.7 ± 0.9 (KG). Again, the value based on Bahcall's surface density is comparable to that estimated for external disks.

Schmidt (1985) has recently presented a mass model of the Galaxy consisting of three components: a "bulge + spheroid" (which in the nomenclature of this book would simply be called bulge), an exponential disk with a scale length of 3.5 kpc, and a "corona" or dark halo. A revised version of this has been presented in van der Kruit (1986) for a disk scale length of 5 kpc and has been reproduced in Fig. 10.16. The bulge has been left unaltered and constitutes a mass of about $10^{10} \, \mathcal{M}_\odot$. In this model the local disk surface density has been taken as $80 \, \mathcal{M}_\odot$ pc^{-2}, and its total mass then is $7 \times 10^{10} \, \mathcal{M}_\odot$. The dark halo, which Schmidt modeled as $\rho \propto (a^2 + R^2)^{-1}$, has been adapted to $a = 2.8$ kpc, and has a mass within the probable disk radius $R_{max} = 25$ kpc of $1.9 \times 10^{11} \, \mathcal{M}_\odot$. The local mass density in the dark halo is $0.009 \, \mathcal{M}_\odot \, \mathrm{pc}^{-3}$. The halo constitutes 70 percent of the mass within the optical disk radius, which is significantly higher than Begeman's mean value of 45 percent, which follows from his maximum-disk fits. The inferred value for the global stability parameter Y is 0.98 ± 0.25.

The model given has a less massive disk than in the maximum-disk hypothesis. Clearly, the local surface density from Kuijken and Gilmore would fail by an even larger margin to accomplish this. We must still see whether the uncertainties actually rule out a maximum-disk model for the Galaxy. From the maximum-disk fits in other galaxies, such a model would imply that the disk-alone rotation curve would need to have an amplitude of 195 km sec^{-1} or more. The disk in the model in Fig. 10.16 has a maximum rotation of about 135 km sec^{-1}. To have an amplitude of 195 km sec^{-1}, the disk would need to have a local surface density of about $140 \, \mathcal{M}_\odot$ pc^{-2}, if the disk scale length is 5 kpc, but this can be brought down to $100 \, \mathcal{M}_\odot$ pc^{-2}, if the scale length were 3.5 kpc. This in itself would imply that our Galaxy is probably not maximum disk, but we should really interpret maximum disk to mean that all the luminous material (disk and bulge) by itself should be able to explain the amplitude (or most of it) of the inner rotation curve. From Fig. 10.16 we see that at the solar position the bulge-alone rotation curve has an amplitude of about 80 km sec^{-1}, and therefore the amplitude

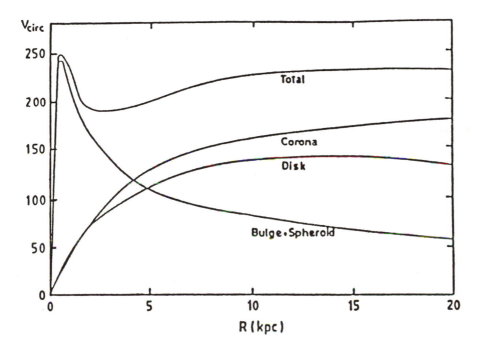

FIGURE 10.16

Mass model of the Galaxy represented by the rotation curves corresponding to the gravitational fields of the individual components indicated. This is an adaption of the model given by Schmidt (1985), allowing for a disk scale length of 5 kpc. The Sun is at 8.5 kpc from the center. Further parameters of the model are in the text. (From van der Kruit 1986.)

of the rotation curve from all the luminous material is in the model about 160 km sec^{-1}. This still falls short of the required amplitude in the maximum-disk hypothesis.

Sellwood and Sanders (1988) have tried to construct a maximum-disk model for the Galaxy by pushing all constraints to their limits, that is, by assuming a disk scale length of 4 kpc and a much more massive bulge of $(3-4) \times 10^{10}$ $\mathcal{M}_\odot$ (about 3 or 4 times more than in the model of Fig. 10.16). It is especially this bulge mass that allows their solution, but the problem is that the local bulge density then is $(4.6-6.3) \times 10^{-4}$ $\mathcal{M}_\odot$ pc^{-3}, which is at least twice the best observational constraints (Schmidt 1985). This bulge mass is also inconsistent with the observed drop in the Galactic rotation curve between 1 and 4 kpc, although it can be argued that the drop results from lack of gas or noncircular motions in this region. My conclusion still is that our Galaxy is probably not maximum disk. Sellwood and Sanders also note that the Kuijken and Gilmore local surface density would make the solution even more extreme, requiring a local circular velocity of 200 km sec^{-1} and a solar Galactocentric distance of 9.5 kpc.

REFERENCES

Albada, T. S. van, Bahcall, J. N., Begeman, K., and Sancisi, R. 1985. *Astrophys. J.*, **295**, 305.

Argyle, E. 1965. *Astrophys. J.*, **141**, 750.

Athanassoula, E., ed. 1983. *Internal Kinematics and Dynamics of Galaxies*. I.A.U. Symposium no. 100. Dordrecht: Reidel.

Athanassoula, E., and Sellwood, J. A. 1986. *Mon. Not. Roy. Astron. Soc.*, **221**, 213.

Babcock, H. W. 1939. *Lick Obs. Bull.*, **19**, 41.

Bahcall, J. N. 1984. *Astrophys. J.*, **287**, 926.

Bahcall, J. N., and Soneira, R. M. 1984. *Astrophys. J. Supp.*, **55**, 67.

Begeman, K. 1987. Ph.D. thesis, Univ. of Groningen.

Bosma, A. 1981a. *Astron. J.*, **86**, 1791.

———. 1981b. *Astron. J.*, **86**, 1825.

Bosma, A., and Kruit, P. C. van der. 1979.*Astron. Astrophys.*, **79**, 281.

Bottema, R. 1988. *Astron. Astrophys.*, **197**, 105.

Bottema, R., Shostak, G. S., and Kruit, P. C. van der. 1986. *Astron. Astrophys.*, **167**, 34.

———. 1987. *Nature*, **328**, 401.

Carignan, C., and Freeman, K. C. 1985. *Astrophys. J.*, **294**, 494.

Casertano, S. 1983. *Mon. Not. Roy. Astron. Soc.*, **203**, 735.

Efstathiou, G., Lake, G., and Negroponte, J. 1982. *Mon. Not. Roy. Astron. Soc.*, **199**, 1069.

Freeman, K. C. 1970. *Astrophys. J.*, **160**, 811.

Goldreich, P., and Lynden-Bell, D. 1965. *Mon. Not. Roy. Astron. Soc.*, **130**, 125.

Hayli, A., ed. 1975. *Dynamics of Stellar Systems*. I.A.U. Symposium no. 69. Dordrecht: Reidel.

Hulst, H. C. van de, Raimond, E., and Woerden, H. van. 1957. *Bull. Astron. Inst. Neth.*, **14**, 1.

Illingworth, G., and Freeman, K. C. 1974. *Astrophys. J.*, **188**, L83.

Kent, S. 1986. *Astron. J.*, **91**, 1301.

———. 1987. *Astron. J.*, **93**, 816.

Kormendy, J., and Illingworth, G. 1982. *Astrophys. J.*, **256**, 460.

———. 1983. *Astrophys. J.*, **265**, 632.

Kormendy, J., and Knapp, G. R., eds. 1987. *Dark Matter in the Universe*. I.A.U. Symposium no. 117. Dordrecht: Reidel.

Kruit, P. C. van der. 1981. *Astron. Astrophys.*, **99**, 298.

———. 1986. *Astron. Astrophys.*, **157**, 230.

———. 1987. In Kormendy and Knapp, p. 415.

———. 1988. *Astron. Astrophys.*, **192**, 117.

Kruit, P. C. van der, and Allen, R. J. 1978. *Ann. Rev. Astron. Astrophys.*, **16**, 103.

Kruit, P. C. van der, and Freeman, K. C. 1986. *Astrophys. J.*, **303**, 556.

Kruit, P. C. van der, and Searle, L. 1982a. *Astron. Astrophys.*, **110**, 61.

———. 1982b. *Astron. Astrophys.*, **110**, 79.

Kruit, P. C. van der, and Shostak, G. S. 1982. *Astron. Astrophys.*, **105**, 351.

———. 1983. In Athanassoula, p. 69.

———. 1984. *Astron. Astrophys.* **134**, 258.

Kuijken, K., and Gilmore, G. 1989. *Mon. Not. Roy. Astron. Soc.*, **239**, 605.

Lacey, C. G. 1984. *Mon. Not. Roy. Astron. Soc.*, **208**, 687.

Lewis, J. R. 1986. Ph.D. thesis, Australian Natl. Univ.

Lewis, J. R., and Freeman, K. C. 1989. *Astron. J.*, **97**, 139.

Martinet, L. 1988. *Astron. Astrophys.*, **206**, 253.

Oort, J. H. 1932. *Bull. Astron. Inst. Neth.*, **6**, 249.

Ostriker, J. P., and Peebles, P. J. E. 1973. *Astrophys. J.*, **186**, 467.

Pease, F. G. 1915. *Publ. Astron. Soc. Pacific*, **27**, 239.

———. 1916. *Proc. Natl. Acad. Sci. (U.S.)*, **2**, 517.

———. 1918. *Proc. Natl. Acad. Sci. (U.S.)*, **4**, 21.

Roberts, M. S. 1975. In Hayli, p. 331.

Rogstad, D. H., Lockhart, I. A., and Wright, M. C. H. 1974. *Astrophys. J.*, **193**, 309.

Sancisi, R. 1976. *Astron. Astrophys.*, **53**, 159.

Sancisi, R., and Allen, R. J. 1979. *Astron. Astrophys.*, **74**, 73.

Sargent, W. L. W., Schechter, P. L., Boksenberg, A., and Shortridge, K. 1977. *Astrophys. J.*, **212**, 326.

Schmidt, M. 1985. In van Woerden *et al.*, p. 75.

Sellwood, J. A. 1983. In Athanassoula, p. 197.

Sellwood, J. A., and Carlberg, R. G. 1984. *Astrophys. J.*, **282**, 61.

Sellwood, J. A., and Sanders, R. H. 1988. *Mon. Not. Roy. Astron. Soc.*, **233**, 611.

Shostak, G. S., and Kruit, P. C. van der. 1984. *Astron. Astrophys.*, **132**, 20.

Simkin, S. M. 1971. *Astron. Astrophys.*, **31**, 129.

Slipher, V. M. 1914. *Lowell Obs. Bull.*, **II**, no. 12.

Spitzer, L., and Schwarzschild, M. 1951. *Astrophys. J.*, **114**, 385.

Tinsley, B. M. 1981. *Astrophys. J.*, **250**, 758.

Tonry, J., and Davis, M. 1979. *Astron. J.*, **84**, 1511.

Toomre, A. 1964. *Astrophys. J.*, **139**, 1217.

Wainscoat, R. J. 1986. Ph.D. thesis, Australian Natl. Univ.

Wevers, B. M. H. R., Kruit, P. C. van der, and Allen, R. J. 1986. *Astron. Astrophys. Supp.*, **66**, 505.

Wielen, R. 1977. *Astron. Astrophys.*, **60**, 263.

Woerden, H. van, Burton, W. B., and Allen, R. J., eds. 1985. *The Milky Way Galaxy.* I.A.U. Symposium no. 106. Dordrecht: Reidel.

Wolf, M. 1914. *Vierteljahresschrift Astron. Gesell.*, **49**, 162.

Woolley, R. v. d. R. 1978. *Mon. Not. Roy. Astron. Soc.*, **184**, 311.

Young, J. S. 1987. *Star Formation in Galaxies.* NASA Conf. Proc., **2466**, p. 197.

CHAPTER 11

STELLAR POPULATIONS
IN THE GALAXY

Gerard Gilmore

Reliable taxonomy is an essential prerequisite for understanding in any complex system. The Linnaean achievement of imposing order and coherence on the apparently disparate body of observational data concerning the distribution of luminosity in external galaxies, and the distribution functions over space, color, and luminosity of stars in our Galaxy, is provided by Baade's (1944) concept of stellar populations (see Chapter 1). As with all great unifying concepts, Baade's image of galaxies as the sum of two discrete populations of stars survived a vast improvement in the quality and variety of observational data, though of course with increasing refinement and complexity. An explanation of the apparent differences between the stellar populations in terms of an evolutionary process requires the appreciation of kinematic, chemical-abundance, and age data for stars in the Galaxy, in addition to large-scale spatial distribution information.

The spatial distribution of stars in the Milky Way Galaxy, and the related distribution of luminosity in other disk galaxies, are described in earlier chapters (2 and 5), as are the dynamical concepts needed to relate these distributions to the kinematic properties of stars near the Sun (Chapter 7). We will now consider the variety of other data available describing the distributions of chemical abundances, kinematics, and ages of stars in the Galaxy, and relate these to the taxonomic concept of stellar populations. These data and their classification provide the essential basis for the explanation (see Chapter 13) of the observed properties of stars in the Galaxy as an evolutionary sequence beginning at the time of the formation of the Galaxy.

11.1 STELLAR POPULATIONS IN OTHER DISK GALAXIES

A detailed discussion of the photometric properties of the stellar components of disk galaxies is presented in other chapters of this volume. For now we need to consider just two aspects of available data: the general kinematic properties of the high-surface-brightness regions of spiral bulges, and the abundance distribution in the halo of M31, the only large disk galaxy other than the Milky Way in which studies of individual halo stars are available.

Unresolved Bulges

The kinematic properties of spiral bulges have been well studied in only five galaxies, four early types (S0/Sa) and one Sb (NGC 4565), which has, however, a pronounced box-shaped bulge (Kormendy and Illingworth 1982; Illingworth and Schechter 1982). These galaxies were chosen for study because they are edge-on, minimizing thin-disk light contamination, but also because they are both unusually luminous (mean $M_B \sim -20$) and unusually round (mean ellipticity $\sim$ E5, axial ratio $\sim$ 1:2). "Typical" spiral bulges are both less luminous and flatter than this (Kormendy and Illingworth 1982), complicating a comparison of their properties. A further complication is that reliable data do not exist fainter than surface brightnesses of ~ 24 B-mag arcsec^{-2}. This surface-brightness level corresponds to ~ 3 half-light radii in the galaxies studied, which is similar to the solar Galactocentric distance, though the Galactic-halo surface brightness near the Sun is some two magnitudes fainter than 24 B-mag arcsec^{-2}. Detailed modeling (Shaw and Gilmore 1989) of the surface-brightness data for the only galaxy studied which is similar in Hubble type to the Galaxy (the Sb galaxy NGC 4565) shows that those few data which extend well beyond the box-shaped isophotes contain a significant contribution from thin-disk light, and are thus not ideally suited to study of that galaxy's bulge. [The Fourier quotient analysis technique adopted in the spectroscopic studies has a well-known limitation, in that it is strongly biased toward narrow-lined contributions to the total light (*e.g.*, McElroy 1983), which makes interpretation of these data problematic.] At face value, however, the bulge of NGC 4565 a few thin-disk scale heights above the major axis has a rotation curve and a velocity-dispersion profile which are initially flat and then decline slowly with distance from the minor axis to at least 1.5 thin-disk scale lengths from its center. We might perhaps anticipate similar behavior in our Galaxy.

Interpretation of the data is less equivocal for the central few half-light radii of the luminous, unusually round bulges. Such systems rotate almost as rapidly as self-gravitating, oblate spheroidal models whose shape is dominated by angular-momentum support (see Binney 1982; though remember that spiral bulges are *not* self-gravitating). The rotation velocity decreases much more slowly than does the luminosity perpendicular to the plane. Thus the bulge luminosity is dominated by a stellar population which is rotating sufficiently rapidly that rotational support dominates over pressure support. This holds equally for bulges whose luminosity profile is adequately described by a single $r^{1/4}$ law,[1] such as NGC 7814, and those bulges whose luminosity distribution requires an exponential thick disk, such as NGC 3115.

---

[1] In other chapters that used spherical polar coordinates we have used the symbol R, but here that symbol has been preempted for our usual system of cylindrical polar coordinates. The spherical-polar coordinate here will be r.

Population II Stars in M31

M31, being the nearest large spiral, offers a unique opportunity for the study of individual stars in another galaxy like the Milky Way. Resolution of individual stars in the halo of M31 by Baade (1944) led (among other things) to the population concept (see Chapter 1), so that the field stars in the halo of M31 were and remain the paradigm of "Population II." Only a rather qualitative description of the properties of halo stars in M31 was, of course, possible with photographic techniques. Modern CCD detectors now allow a reappraisal of Baade's "Population II," by quantitative study of substantially fainter stars than was possible in the 1940's. A striking example of the efficacy of silicon technology in this field is provided by comparison of the color–magnitude diagrams for field halo stars in M31 published by Baade (1944) and by Pritchet and van den Bergh (1987; these authors also provide an excellent discussion of other relevant work). With the recent data it is possible for the first time to derive reliable values for the mean chemical abundance of field stars in the halo of M31.

The most detailed data are available for a field 12 kpc down the minor axis, 8.6 kpc from the center of M31. This field has negligible contamination from either the (thin) disk of M31 or from foreground stars in our Galaxy. We may calculate the equivalent position in the Galaxy of this field, from the fact that the halos of M31 and of our Milky Way Galaxy have similar effective radii, and the observed flattening of M31 ($c/a \sim 0.6$) is also very similar to that of the Milky Way (see Chapter 2). These values imply that the isophotes passing through the field studied by Pritchet and van den Bergh cross the major axis of M31 at ~ 15–20 kpc from the galactic center. This field is therefore well beyond the equivalent of the solar neighborhood, though since the halo of M31 is intrinsically brighter than that of the Galaxy, there are in fact not dissimilar surface brightnesses in the M31 field and for the Galactic subdwarf halo-star distribution at the solar Galactocentric distance. Since there are no strong radial gradients in the observed photometric properties of the M31 halo stars or in the subdwarf system in our Galaxy, we might have expected the M31 stars in the Pritchet and van den Bergh field to have properties like those of the subdwarf population in the solar neighborhood of our Galaxy. Contrary to expectation, the color–magnitude data, together with an earlier study of RR Lyrae stars by the same authors, show that the mean metallicity at this large distance from the center of M31 is similar to that of a somewhat metal-rich globular cluster in our Galaxy. Their best estimate of the mean stellar abundance is [Fe/H] ~ -1.0.

Although quite unlike that of the Galactic subdwarf system (which has mean metallicity ~ -1.5 dex, see below), the metallicity derived is, however, equal to that of the thick-disk component in the solar neighborhood of our Galaxy. Does this suggest that the halo stars are members of a thick disk in M31? This is extremely unlikely, since it would imply that the M31 thick disk would dominate at very large galactocentric distances, far beyond those where the thick disk in our Galaxy has been studied directly, and far beyond the distance at which thick-disk stars are expected to be found in significant numbers, based on available

models of local star-count data (see Chapter 2). The available evidence regarding
the existence of a thick disk in M31 similar to that in the Galaxy is rather sparse.
Some suggestive evidence for a population like the Galactic thick disk is provided
by the distribution of planetary nebulae. Nolthenius and Ford (1987) show the
M31 planetaries to be distributed in a thick disk with small asymmetric drift.
Their estimates for the thickness of this disk are very model-dependent, but are
$\sim$ 1–3 kpc, very similar to that in the Galaxy.

The high mean metallicity for the halo of M31 is in fact consistent with
available spectroscopic line-strength and color data for the bulges of other disk
galaxies (Chapter 14), although data on other galaxies exist only at smaller
galactocentric radii than the stellar data in M31. (Note that the existence of a
color–magnitude relation for spiral bulges is not important for these comparisons,
even if the relation is driven by metallicity, since the amplitude of the effect
over the narrow magnitude range relevant here is small; see Visvanathan and
Griersmith 1977.) There is a real spread in color evident in the M31 stellar data,
indicative of a dispersion in metallicity, with perhaps 10 percent of stars in the
field being as metal-poor as M92 ([Fe/H] $\sim$ −2.25). Other photometric data
show these more metal-poor stars to dominate only at $\gtrsim$ 20 kpc along the minor
axis (Mould 1986). It is of interest to note that the 10 percent of stars detected
in the tail of the color–magnitude distribution in the data of Pritchet and van
den Bergh is the only detection at measurable surface brightnesses of a stellar
population whose properties are consistent with those of the Galactic subdwarf
system in a large external galaxy.

Thus Baade's original "Population II" tracers in M31 are similar to the
Galactic metal-rich globular clusters, and are quite unlike solar-neighborhood
subdwarfs and Galactic metal-poor globular clusters.

Globular Clusters in M31

It is not obvious that the properties of the distribution of globular clusters in
a galaxy are a reliable guide to the properties of the distribution of all stars in
that galaxy. Still, in our Galaxy the globular clusters form a two-component
system, with a metal-poor, pressure-supported halo population, and a more
metal-rich, rotationally supported (thick) disk population, very similar to the
two-component distribution of stars more metal-poor than about −1 dex. It is
therefore of interest to see if such a distribution is evident in M31. The most detailed study of a large sample of clusters in M31 is that by Elson and Walterbos
(1988), who compiled all available data from earlier studies. Although there are
real and unexplained spectroscopic differences between clusters in the Galaxy
and in M31, with the M31 clusters having stronger CN and Balmer lines, there
is considerable similarity in the spatial and kinematic properties of the Galactic
and M31 cluster systems. M31 clusters tend to be bluer than the underlying
halo starlight, and the metal-poor clusters form a system which is also more
extended. There is either a two-component halo–disk distribution of clusters as
in the Galaxy, with the more metal-rich clusters forming a rotating disk-like

distribution with flattening similar to that of the halo light, or possibly there is a continuous change from an extended pressure-supported system for the most metal-poor clusters to a flattened rotationally supported system for the more metal-rich clusters. This similarity of the detailed properties of the cluster distributions in the Galaxy and in M31 supports models in which the clusters form as an integral part of the formation of the galactic halo.

11.2   DYNAMICS, KINEMATICS, AND DENSITY LAWS IN THE MILKY WAY GALAXY

Gravity provides a quantifiable relationship between stellar kinematics and stellar spatial distributions; hence a description of the large-scale properties of the Galaxy can be tested for self-consistency by combining spatial-distribution and kinematic data with dynamical analyses. The large-scale structural properties of stars in the Galaxy are reviewed in Chapter 2, and the dynamical methods are presented in Chapter 7. We will now discuss kinematic data for halo and old-disk stars in the Galaxy, and illustrate the dynamical relationships between these data and the large-scale spatial distribution of stars in the Milky Way.

The Shape of the Galactic Halo

Metal-poor stars ($[Fe/H] \lesssim -1$) in the Galaxy have a flattened oblate-spheroidal distribution, with minor:major axial ratio $c/a \approx 0.6$ within a few kpc of the Sun (see Chapter 2), although until recently it had been thought that the stellar distribution was more round ($c/a = 0.8^{+0.2}_{-0.05}$). Which if either of these values is consistent with dynamical analyses, and which should we have expected prior to analysis of the star-count data?

We may calculate the velocity dispersions in three orthogonal coordinates for stars near the Sun in two ways. The most direct is by straightforward analysis of each of the three components of velocity of a sample of stars. This is possible provided only that data on reliable proper motions, radial velocities, and distances for a kinematically unbiased sample of metal-poor field stars in the solar neighborhood are available—a nontrivial yet practicable task. Alternatively, we may analyze just radial velocity and distance data for a suitable stellar sample, in which case the stars must be distributed over a fairly large fraction of the sky to allow statistically valid averaging of the unknown transverse velocities. Norris (1986) provides an excellent description of the methods and their sensitivity to the realities of available data. The conclusion of the three available direct analyses of kinematically unbiased samples of metal-poor stars near the Sun is that such stars have an anisotropic velocity-dispersion tensor, with $\sigma_{RR} : \sigma_{\theta\theta} : \sigma_{zz} \sim \sqrt{2} : 1 : 1$. The specific values derived, in km sec^{-1}, are:

$$\sigma_{RR} : \sigma_{\theta\theta} : \sigma_{zz} = 128 \pm 8 : 96 \pm 8 : 93 \pm 5 \quad \text{(Carney and Latham 1986)},$$
$$\sigma_{RR} : \sigma_{\theta\theta} : \sigma_{zz} = 131 \pm 6 : 106 \pm 6 : 85 \pm 4 \quad \text{(Norris 1986)},$$
$$\sigma_{RR} : \sigma_{\theta\theta} : \sigma_{zz} = 133 \pm 8 : 98 \pm 13 : 94 \pm 6 \quad \text{(Morrison et al. 1990)},$$

giving a weighted average of

$$\sigma_{RR} : \sigma_{\theta\theta} : \sigma_{zz} = 131 \pm 7 : 102 \pm 8 : 89 \pm 5.$$

Since the velocity-dispersion tensor behaves like an anisotropic stress tensor in the equations governing stellar dynamics, we may expect this anisotropic "pressure" to result in an anisotropic shape, *i.e.*, a flattened metal-poor halo. Binney and May (1986) investigated this idea in more detail for various Galactic potential–distribution function pairs, and concluded that in the locally nonspherical potential felt by the subdwarfs, due to the presence of the disk, the observed velocity-dispersion anisotropy implies a substantially flattened halo, with shape $\sim$ E7 or axis ratio $\sim 1 : 4$. A spherical potential would imply shape $\sim$ E3, with axis ratio $c/a \sim 0.7$. White (1989) quantified the derivation of the shape of a set of isodensity contours from an asymmetric velocity ellipsoid using the tensor virial theorem for axisymmetric galaxies. This approach has the disadvantage of treating the two components of the velocity dispersions perpendicular to the vertical direction on an equal footing. That is, his analysis is valid if $\sigma_{RR} \equiv \sigma_{\theta\theta}$, which is not compatible with the observations. Thus the numerical values resulting from this analysis will not be valid. Nevertheless, the general results should be an adequate illustration of the amplitude of the flattening of isodensity contours to be expected from an anisotropic velocity ellipsoid. Using the velocity dispersions quoted above (insofar as we can, given their inconsistency with the assumptions underlying the analysis), we find from the flattening of the velocity ellipsoid that in a spherical potential the subdwarfs should have axis ratio $c/a \sim 0.4$, whereas in a potential due to a mass distribution that is also flattened we obtain $c/a \sim 0.3$. The ratio of the vertical velocity dispersion to the circular velocity also constrains the flattening, and predicts $c/a \sim 0.53$. The precise numerical value for this flattening is model-dependent, and hence uncertain. It is in fact possible to have a nearly spherical stellar system with a velocity ellipsoid as anisotropic as the values quoted above with a reasonable distribution function. Thus, although we may suspect flattening of the stellar halo from the simplest dynamical arguments, observation is required to resolve the uncertainty.

The dynamical and direct photometric studies of the shape of the halo stellar system within a few kpc of the Sun are therefore in good agreement. The kinematic data of Ratnatunga and Freeman (1985, 1989) for distant metal-poor K giants can also most easily be explained by allowing these stars to form a spatially flattened distribution. The most important feature of the Ratnatunga–Freeman data is that the line-of-sight velocity dispersion toward the south Galactic pole does not increase with distance. This is surprising, since as we increase the distance of a star from the Sun, the angular distance between the Sun and the

Galactic center as seen from the star becomes smaller. Thus an observer near the Sun sees an increasingly larger contribution of the star's Galactocentric radial velocity to that star's radial speed from the Sun. The assumption behind the expectation of a rising line-of-sight dispersion with distance is, then, that the distant metal-poor K giants have the same radially biased velocity-dispersion tensor as do local subdwarfs. It is not possible to construct self-consistent Galactic models using straightforward distribution functions which are consistent with a change as large as that seen by Ratnatunga and Freeman, basically since high-velocity stars travel such large distances that some of the stars which dominate distant data will also pass through the solar neighborhood. Adopting a global form of the distribution function in either a spherical (White 1985) or in an oblate (Levison and Richstone 1986) potential requires a flattened spatial distribution for the halo stars, again with axis ratio $\sim 1:4$.

Alternatively, we can depress the observed velocity dispersion at large distances by allowing suitable discontinuities in the stellar distribution function, essentially by assuming that all stars beyond a given Galactocentric radius are on circular orbits. This latter approach allows a fit which is consistent with a spherical spatial distribution for these distant stars, though at the expense of a somewhat contrived distribution function (Sommer-Larsen 1987; Sommer-Larsen and Christensen 1989; Dejonghe and de Zeeuw 1988). However, even these models are not consistent with the observed local-velocity anisotropy noted above, and provide a poor fit to data nearer the Galactic plane. Thus, local kinematic and dynamic analyses are consistent with direct star-count data for field stars and RR Lyraes in showing the Galactic halo to be flattened, with axial ratio $c/a \sim 0.6$. This flattening probably is also appropriate at much larger distances from the Sun (to ~ 20 kpc from the Sun), though the available data at such large distances are sparse and in poor internal agreement.

Axisymmetric Dynamics of the Galactic Disk

The collisionless Boltzmann equation describes the dynamical equilibrium of a stellar system, and is described in Chapter 7 above. The radial velocity moment of this equation provides the relationship between the relative amounts of pressure and angular-momentum balance to the gravitational potential gradients, and the physical size of some tracer stellar distribution. This relationship, the *asymmetric-drift* equation, is derived in Chapter 7. For convenience we repeat it here, in terms of observables in the Galactic plane ($z = 0$) near the Sun, with R the planar radial cylindrical coordinate, and θ the coordinate in the transverse (rotational) direction adopted in this section:

$$V_c^2 - \langle V_\theta \rangle^2 = \sigma_{\theta\theta}^2 - \sigma_{RR}^2 - \frac{R}{\nu}\frac{\partial(\nu\sigma_{RR}^2)}{\partial R} - R\frac{\partial\sigma_{Rz}^2}{\partial z}$$

$$= \sigma_{RR}^2\left[\frac{\sigma_{\theta\theta}^2}{\sigma_{RR}^2} - 1 - \frac{\partial\ln(\nu\sigma_{RR}^2)}{\partial\ln R} - \frac{R}{\sigma_{RR}^2}\frac{\partial\sigma_{Rz}^2}{\partial z}\right]. \qquad (11.1)$$

Here $\nu(R)$ is the planar radial spatial-density distribution of the stars, V_c is the circular velocity (*i.e.*, $V_c^2 = R\partial\psi/\partial R$, where we adopt a locally flat rotation curve with $V_c = 220$ km sec^{-1} for now), $\langle V_\theta \rangle$ is the mean rotation velocity of the relevant sample of tracer stars and is the only mean streaming motion, and the tracer stellar sample has velocity variances $\sigma_{RR}^2, \sigma_{\theta\theta}^2$, and covariance σ_{Rz}^2. The quantity $V_c - \langle V_\theta \rangle \equiv V_a$ is often called the *asymmetric drift* of a stellar population.[2]

Eq. (11.1) relates measurable local moments of the stellar distribution function to global properties of the Galaxy. Physically, the equation describes the various contributions to the support against the galactic potential gradient which are required from pressure and angular momentum to maintain some specified spatial length scale of a stellar distribution. In view of its importance in studies of the Galaxy, we briefly discuss the evaluation of each term.

$\dfrac{\sigma_{\theta\theta}^2}{\sigma_{RR}^2}$:   For thin-disk stars, the ratio of these velocity dispersions is well described by epicycle theory, and is easily shown to be expressible in terms of the local Oort constants. That is, this ratio is a measure of the local gradient in the Galactic potential. However, some care is required in measuring it. Kinematic properties of thin-disk stars are a function of the stellar age and its metallicity. These correlations and their significance are discussed in Chapters 9 and 13 (see also Delhaye 1965 and Fuchs and Wielen 1987). Since we are here discussing the time-independent collisionless Boltzmann equation, we restrict the discussion to values which are adequately representative of old-disk stars. The velocity dispersions at $z = 0$ of old-disk stars are most reliably adopted from the values for the nearby, spectroscopically selected K and M dwarfs with good parallax distances. These give, in km^2 sec^{-2}, $\sigma_{RR}^2 : \sigma_{\theta\theta}^2 : \sigma_{zz}^2 = 39^2 : 23^2 : 20^2$ (Wielen 1974). For low-metallicity field stars the relevant values derived above are $\sigma_{RR}^2 : \sigma_{\theta\theta}^2 : \sigma_{zz}^2 = 131^2 : 102^2 : 89^2$. It is important to appreciate that these values are derived from *nonkinematically selected* samples of metal-poor stars near the Sun. With these numerical values, the first term in Eq. (11.1) becomes $\sigma_{\theta\theta}^2/\sigma_{RR}^2 = 0.35$ for the old disk, and $\sigma_{\theta\theta}^2/\sigma_{RR}^2 = 0.61$ for the low-abundance field stars.

$\dfrac{\partial\,\ln(\nu\sigma_{RR}^2)}{\partial\ln R}$:   A product of a spatial density and a velocity dispersion is a pressure, so that this term is analogous to a radial pressure gradient in the stellar "fluid." The thin exponential disks of spiral galaxies are apparently self-gravitating (see Chapter 2) and are observed to have a constant thickness independent of radius (see Chapters 5 and 10). Thus, since the scale height of a self-gravitating disk varies as $h_z \propto \sigma_{zz}^2/\Sigma$ with $\Sigma \propto \exp(-R/h_R)$ being the disk surface mass density, we expect that the radial variation of the vertical velocity

---

[2] Chapter 7 denoted the quantities σ_{ij}^2 by $\langle U^2 \rangle$, $\langle UV \rangle$, etc. Here, as in Chapter 8, it is more appropriate to use the present notation, which emphasizes the nature of σ^2 as a tensor whose components may change from place to place.

dispersion will be $\sigma_{zz} \propto \exp(-R/h_R)$. Assuming also that the *shape* of the velocity ellipsoid is independent of position, or more specifically that $\sigma_{RR}^2 \propto \sigma_{zz}^2$ (some indirect evidence in support of this assumption is discussed by Freeman 1987), leads to

$$\frac{\partial \ln (\nu \sigma_{RR}^2)}{\partial \ln R} = 2 \times \frac{\partial \ln \nu}{\partial \ln R} = -2h_R^{-1} R. \qquad (11.2)$$

For simplicity we restrict discussion of spheroidal distributions to an isothermal halo, $\sigma_{RR}^2 = $ constant, so that

$$\frac{\partial \ln (\nu \sigma_{RR}^2)}{\partial \ln R} = \frac{\partial \ln \nu}{\partial \ln R}. \qquad (11.3)$$

$\dfrac{R}{\sigma_{RR}^2} \dfrac{\partial \sigma_{Rz}^2}{\partial z}$:   The term involving σ_{Rz} describes the orientation of the velocity ellipsoid, has no general analytic solution, and has no useful direct observational bases. It describes the relationship between the *measured* stellar radial velocity and the true space motion, some component of which will contribute to the radial pressure. In a sense it is a correction to allow for the difference between natural Galactic coordinates and observer's coordinates. The velocity ellipsoid will be oriented along the coordinate system (if there is one) in which the Hamilton–Jacobi equation (which provides the equations of motion) is separable, though unfortunately there is no reason for there to be a simple relationship between this (Stäckel) coordinate system and the coordinate system in which observations are naturally available. We may calculate the term in specific Stäckel potentials that have been devised to approximate our best guess of the form of the Galactic potential, but such calculations are inevitably model-dependent. We therefore consider two representative cases only. If we approximate the Galactic mass distribution by an infinite, cylindrically symmetric, constant-surface-density sheet, and calculate the corresponding potential, then we can show that the velocity ellipsoid will be diagonal in cylindrical polar coordinates, and point always parallel to the Galactic major axis, and at the Galactic minor axis, so that $\sigma_{Rz} \equiv 0$. This is the assumption most commonly adopted.

An alternative idealization is to assume that the potential is dominated by a centrally concentrated or spherical mass distribution, in which case the local velocity ellipsoid points at the Galactic center, and so is diagonal in spherical polar coordinates. This assumption was made by Oort (1965), who found that for $z \ll R$, $\sigma_{Rz}^2 \sim (\sigma_{RR}^2 - \sigma_{zz}^2)z/R$, from which it follows that

$$\frac{R}{\sigma_{RR}^2} \frac{\partial \sigma_{Rz}^2}{\partial z} \approx 1 - \frac{\sigma_{zz}^2}{\sigma_{RR}^2}. \qquad (11.4)$$

An exact derivation for a disk in which $\sigma_{RR}^2 = \alpha^2 \sigma_{zz}^2$ at $z = 0$, when the spherical polar coordinate system in which the ellipsoid is diagonal corresponds with the cylindrical polar system in use here, and in which radial velocities are measured (Kuijken and Gilmore 1989a), provides

$$\sigma_{Rz}^2 = \left[\frac{Rz(\alpha^2 - 1)}{\alpha^2 z^2 + R^2} \right] \sigma_{zz}^2,$$

so that

$$\frac{R}{\sigma_{RR}^2} \frac{\partial \sigma_{Rz}^2}{\partial z} \approx (\alpha^2 - 1) \frac{\sigma_{zz}^2}{\sigma_{RR}^2}. \tag{11.5}$$

For the velocity dispersions quoted above, both Eq. (11.4) and Eq. (11.5) provide the same answer for the old disk,

$$\frac{R}{\sigma_{RR}^2} \frac{\partial \sigma_{Rz}^2}{\partial z} \simeq 0.75, \tag{11.6}$$

whereas for the halo field stars the numerical value from Eq. (11.4) is 0.53. Strong evidence that a value ~ 0.5 is appropriate for this term comes from numerical orbit integrations in potentials derived from recent studies of the Galactic K_z force law (see Chapter 8, and Kuijken and Gilmore 1989a), which show that the velocity ellipsoid tends to point somewhat above the direction to the Galactic center. Thus a numerical value of 0.47 will slightly overestimate the amplitude of the term involving σ_{Rz}. Neglect of this term, which is common, is nevertheless clearly unjustified.

For an exponential disk of radial scale length h_R, and a sample of stars observed in the solar neighborhood, Eq. (11.1) therefore becomes

$$V_c^2 - \langle V_\theta \rangle^2 = \sigma_{RR}^2 \left(2\frac{R_0}{h_R} - 1.4 \right), \tag{11.7}$$

where R_0 is the distance of the Sun from the Galactic center (7.8 ± 0.8 kpc, Feast 1987). Alternatively, for a halo with a power-law density distribution with exponent γ, $\nu(r) \propto r^{-\gamma}$, the several terms in Eq. (11.1), with the values for halo stars derived above, lead to

$$V_c^2 - \langle V_\theta \rangle^2 = \sigma_{rr}^2(\gamma - 0.9), \tag{11.8}$$

where we now use the spherical-polar radial coordinate r for convenience in discussing a power-law density distribution, remembering that near the Sun r is effectively indistinguishable from the cylindrical-polar planar radial coordinate R. Thus a stellar tracer population which belongs to an exponential distribution with scale length ~ 3.5 kpc (a plausible value for the Galactic old disk) will follow an asymmetric-drift relation similar to that of a tracer population which is part of an isothermal distribution describing an r^{-4} spheroidal density distribution. It is also interesting to note that the largest allowed radial velocity dispersion, corresponding to zero net rotation, for such a stellar system with the observed anisotropic velocity dispersions, is $\sim 1.2\, V_c/\sqrt{4}$, or ~ 135 km sec^{-1}. (This comes from a crude attempt to share out the stellar energy among the velocity dispersions.) For a tracer population with any smaller radial velocity dispersion, Eq. (11.1) describes the interplay between the pressure (velocity anisotropy) support and the angular-momentum (rotation velocity) support to the spatial distribution. A higher-velocity-dispersion system than this has larger total energy and will form a more extended system. We might also assume that it would have formed from less-dissipated material, which is, of course, the clue to the *physical* significance of Eq. (11.1) and the reason for this extensive discussion of it. The implications of Eq. (11.1) for Galaxy-formation models are discussed further in Chapter 13. We now compare Eq. (11.1), in the form (11.8), with available observations.

The relevant observational data are shown in Fig. 11.1, where the data points shown have been either collated or calculated from data available in the identified references. The data from the large surveys have been binned by metallicity. Most data for tracer samples with a Galactic rotation velocity greater than about 50 km sec^{-1} are consistent with a single density distribution. A marginally significant exception is the metal-rich globular cluster system, whose radial velocity dispersion is rather low. This datum is, however, somewhat more uncertain than most of the other data shown, because of distance and reddening uncertainties, and the small sample of clusters. The other tracer populations that lie to the left of the majority of the data are the c-type RR Lyrae stars and long-period Mira variable stars with periods from 150 to 200 days. We discuss these stars further below.

An important unanswered question is how best to present the full information content of the data on which Fig. 11.1 is based. A valid statistical description of the distribution function clearly should not be based on the velocity dispersion calculated directly, since this is sensitive to a few extreme points. More robust estimators of the shape of the velocity ellipsoid have not been used as yet. In addition to the validity of the description of the data, there is also the difficulty in allowing for selection effects in the stellar sample selected for study. Proper-motion samples will not find metal-poor stars on circular orbits near the Sun, if they exist with the same low local normalization as the high-velocity subdwarfs. Local surveys will also miss stars on high-angular-momentum, high-energy orbits, since these will always lie beyond the solar circle.

The mean density law consistent with the majority of the data corresponds to an isothermal spheroidal distribution described by a a power law with index ~ -4, or an exponential disk with scale length ~ 3 kpc. The tendency for the tracer populations with the lowest mean rotational velocities to have larger radial velocity dispersions than is consistent with this density profile is of considerable significance, if real. (Numerical calculations show that systematic distance uncertainties move the data roughly parallel to the body of the data with smaller σ_{RR}, so are unlikely to be relevant. With the exception of the data for the metal-poor RR Lyrae stars, however, there are very few stars in the bins with the highest values of σ_{RR}.) The stars with the highest radial velocity dispersions are also the most metal-poor (see below) and hence (presumably) those which formed first in

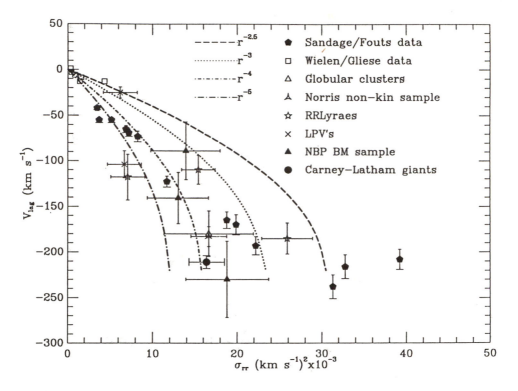

FIGURE 11.1

The relation between the radial velocity dispersion σ_{RR} and the asymmetric drift V_{rot} of samples of old stars in the Galaxy. The data for field stars are binned by metallicity. The key identifies data from the following sources: from the proper-motion sample of Sandage and Fouts (1987) for stars with [Fe/H]$\gtrsim -2$ only, since photometric metallicities and hence distances are very uncertain for the more metal-poor stars; the Wielen (1974) analysis of the Gliese nearby-star catalogue; the analyses by Norris (1986) of the globular-cluster system and of kinematically unbiased low-abundance stars; the analysis by Strugnell *et al.* (1986) of field RR Lyrae stars; long-period variables (binned by period range) from the analysis by Osvalds and Risley (1961); spectroscopically selected low-abundance field stars, from the analysis by Norris *et al.* (1985); and the spectroscopically selected local metal-poor giants, from the study by Carney and Latham (1986). The model lines correspond to different solutions of Eq. (11.8). The following density laws were used: $r^{-2.5}$ (long dashes); r^{-3} (dots); r^{-4} (short dash-dot); and r^{-5} (long dash-dot). The tendency for the data to cross the model lines at low V_{rot} (*i.e.*, very negative V_{lag}) shows that some star formation took place during dissipational collapse of the Galaxy.

the Galaxy. If the most metal-poor stars really do form a more extended spatial distribution than more metal-rich stars (*i.e.*, if there is an abundance gradient in the halo) as suggested by Fig. 11.1, then we may conclude that these stars formed from less-dissipated gas, and hence formed earlier in the evolution of the proto-galaxy than did more metal-rich stars. Their significance is discussed further in later sections.

Although the asymmetric-drift arguments above provide strong evidence that star formation continued *during* a period of dissipational collapse, there is no information in this purely dynamical relationship regarding the *rate* of this collapse. To extend dynamical analyses to include the timescales of Galactic formation we need another clock, which is provided by the chemical abundances.

11.3   KINEMATICS AND CHEMISTRY

Stellar chemical abundance is a clock which measures age in units defined by the formation rate and lifetimes of those (mostly massive) stars which contribute to the enrichment of the interstellar medium. The mean azimuthal streaming motion of a population of stars reflects the amount by which the proto-Galaxy dissipated and collapsed, and the amount of angular-momentum transport that occurred prior to the formation of these stars. Thus the existence of a correlation between kinematics, such as V_{rot}, and [Fe/H] allows us to relate the chemical evolutionary timescale to the dynamical and cooling timescales. Similarly, the amplitude of the peculiar velocity of a star is determined in part by the amount of cooling of the proto-galactic gas before star formation, and hence is a function of time, as also is metallicity. Since this peculiar velocity is now manifest in the star's orbital eccentricity, the relationship between the eccentricity of stellar orbits, projected onto the plane of the Galaxy, and chemical abundance is also commonly discussed. Although there are complexities in their interpretation, correlations (or the lack thereof) between the kinematics of a tracer population of stars and the stars' chemical abundances remain one of the most powerful observationally feasible tests of the early star formation and dynamical history of the Galaxy.

Rotation Velocity *vs.* Metallicity

A correlation between stellar chemical abundance and Galactic rotational velocity was first convincingly established and discussed in detail by Eggen *et al.* (1962; ELS). However, an extensive debate has arisen in the past few years as a result of several new surveys of the kinematics of metal-poor stars, with conflicting claims about the reality of this correlation, particularly for stars more metal-poor than ~ -1 dex. The relevant recent data are collected here. The agreement between the various data sets is acceptable for now for stars more metal-poor than ~ -0.8 dex. The mean value of the rotation velocity decreases from ~ 160 km sec^{-1} at [Fe/H] $= -0.8$ to near zero at [Fe/H] $= -1.5$, and shows

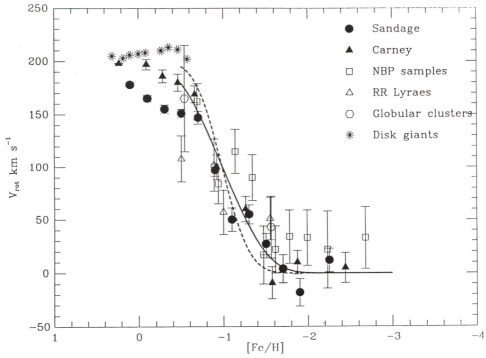

FIGURE 11.2
The relation between rotation velocity relative to the Galactic standard of rest V_{rot} and metallicity, for those samples of field stars with good abundance data discussed in the text. The lines show alternative models, with the solid line being a model involving a smooth correlation between V_{rot} and [Fe/H] over the range $-0.5 \lesssim$ [Fe/H] $\lesssim -1.5$. The dashed line shows a discontinuous relationship between V_{rot} and [Fe/H], with the discontinuity at [Fe/H] $= -1.0$. Both models have been convolved with a Gaussian of dispersion 0.25 dex in metallicity to represent measuring errors.

no significant correlation at lower abundances. Above a metallicity of ~ -1.5 dex there is a transition to a mean rotational velocity of $\gtrsim 160$ km sec^{-1} at ~ -0.8 dex, with these latter values evidently being very poorly decided by these data. (Although this may seem too narrow an abundance interval to be important, it in fact contains almost the entire stellar mass of the halo. Other data for higher-abundance stars are discussed separately below.) Is there a systematic abundance–kinematics correlation in this interval? The answer to this is confused by the question of the precision of the data, and the adequacy of a single mean value to describe data which are not necessarily well fitted by a Gaussian distribution.

The two curves shown in Fig. 11.2 are, in turn, a linear correlation between ($V_{rot} = 0$ km sec^{-1}, [Fe/H] $= -1.5$) and ($V_{rot} = 200$ km sec^{-1}, [Fe/H] $= -0.5$) (solid line) and a discontinuity between $V_{rot} = 0$ km sec^{-1} and $V_{rot} = 200$ km sec^{-1} at [Fe/H] $= -1$ (dashed line). Both curves have been convolved with a Gaussian of dispersion 0.25 dex in [Fe/H] to allow for the effect of measuring

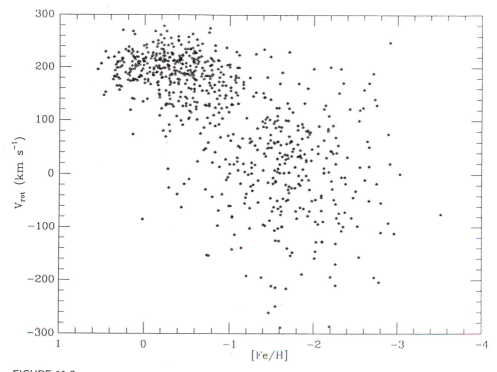

FIGURE 11.3
The relation between rotation velocity relative to the Galactic standard of rest V_{rot} and metallicity, for the sample of proper-motion stars studied by Laird *et al.* (1988). Comparison of this figure with that for the binned data illustrates the difficulty in deducing the reality of a smooth correlation between kinematics and metallicity from data which are averaged into bins, particularly for metallicities near -1 dex.

errors. The greatest uncertainty in distinguishing between the models is now the systematic differences between the results from different data sets. The spectroscopically selected samples discussed by Norris *et al.* (1985) have a systematically higher rotation velocity than do the kinematically selected samples, at almost all metallicities, for reasons which remain to be clarified. Nevertheless, at face value the smoothed correlation is a better description of the binned data than is the discontinuous model.

Sandage and Fouts (1987) interpreted their data as supporting the continuation of a smooth correlation between kinematics and abundance for *all* metallicities. It is clear from the figures above that there is no statistically significant difference (apart from a possible zero-point offset in rotation velocity, which is not relevant to the discussion of a correlation) between their sample and that of Norris *et al.* Rather, the differences between the conclusions of the two groups result largely from different ways of binning data that are clearly (Fig. 11.3) not adequately described by a mean and a standard deviation. We must discuss the distribution function of abundance and kinematic data to derive reliable conclusions. Fortunately, thanks to the very considerable efforts of Sandage and Fouts, of Norris, and of Carney and Latham, a sufficiently large and reliable data base is available for the first time to allow such a discussion.

Orbital Eccentricity *vs.* Metallicity

There is an alternative presentation of the relationship between stellar orbital properties and chemical abundance which is widely discussed; it involves the eccentricity of the stellar orbit in the Galaxy, projected onto the plane, and the star's metallicity. This presentation really depends on the ratio of the components of the star's velocity radially along and tangentially to the line toward the Galactic center, and the stellar metallicity. The derivation of an orbital eccentricity requires the additional assumption of a Galactic potential, adding further uncertainty. Nevertheless, the existence of a correlation between metallicity and the shape of the stellar orbit projected onto the plane of the Galaxy has been cited as the principal evidence to support a rapid-collapse model of the Galaxy (Sandage 1987). Similarly, observational evidence that a tight correlation is violated by a significant fraction ($\sim$ 20 percent) of field metal-poor stars has been cited as evidence that a slow collapse is a preferred model (Yoshii and Saio 1979; Norris *et al.* 1985). This correlation therefore requires explicit discussion.

An interpretation of the existence of stars with very radially biased orbits will be deferred to Chapter 13, after the necessary concepts from Galaxy-formation models have been presented. The analyses of Norris *et al.* (see Norris 1987a for a good discussion), of Morrison *et al.* (1990), and of the Carney–Latham survey of proper-motion stars provide clear evidence that any correlation between orbital eccentricity and stellar metallicity is at best weak.

Correlations *vs.* Discontinuities

A question of some interest is whether or not the appearance of Fig. 11.3 is consistent with a continuous trend (indicative perhaps of significant star formation *during* the period when the proto-galaxy was collapsing and spinning up) or represents a superposition of relatively discrete subsystems (indicative perhaps of the later merger of subsystems which retained a recognizable identity during the early stages of Galactic formation).

Some suggestive rather than conclusive evidence in favor of the picture of discrete substructure in phase space comes from the existence of several apparently intermediate groups of tracers which are identifiably discrete using astrophysical criteria. These include the metal-rich RR Lyrae stars ($\Delta S \lesssim 3$, $V_{rot} \sim 110$ km sec^{-1}; Strugnell *et al.* 1986); c-type RR Lyrae stars ($V_{rot} \sim 100$ km sec^{-1}; Strugnell *et al.* 1986); long-period variables with $150^d \lesssim$ Period $\lesssim 200^d$ ($V_{rot} \sim 115$ km sec^{-1}; Osvalds and Risley 1961); the metal-rich (G-type) globular clusters ($[Fe/H] \gtrsim -1, V_{rot} \sim 100\text{--}200$ km sec^{-1}; Armandroff 1989); and the Arcturus moving group ($V_{rot} \sim 110$ km sec^{-1}; Eggen 1987). The field Type II Cepheids

TABLE 11.1
Velocity dispersion and asymmetric drift of Mira variables

Period range (days)	$V_{rot} - V_{\odot}$ (km sec^{-1})	Total dispersion	Vertical dispersion	Number of stars
< 140	-33 ± 13	81	38 ± 28	22
145–200	-111 ± 22	180	60 ± 67	46
200–250	-61 ± 9	101	62 ± 15	71
250–300	-33 ± 10	88	27 ± 23	77
300–350	-32 ± 6	69	36 ± 10	83
350–410	-23 ± 8	58	26 ± 10	54
> 410	-15 ± 8	50	—	35

(Harris 1981) are another closely related tracer sample, but with less well-known kinematic properties at present. It is remarkable that these different samples of objects all have almost exactly the same rotation velocity. Within the more considerable uncertainties they also have similar abundances, $-0.5 \lesssim [Fe/H] \lesssim -1$.

A careful study of the long-period variables has been provided by Feast (1989; see also Feast et al. 1972, and Plaut 1965), which illustrates the state of the best information and the complexity of the situation regarding the existence of populations of stars with truly intermediate kinematics. The relevant data are collected in Table 11.1. Mira variables exist in many globular clusters, and those in disk globular clusters cover the period range from 191 days (NGC 6712) to 265 days (NGC 6553), and possibly longer. The kinematics of the field Miras over this period range change significantly. Does this mean that the disk-globular-cluster system includes a range of velocity dispersions and a range of asymmetric-drift values? A range in kinematics for specific objects selected from a population with a dispersion in kinematics is, of course, to be expected, but the question is the existence or otherwise of a significant correlation. As in the discussions of Figures 11.1 and 11.3, the situation is complicated when we discuss the distribution function of any property near the region of overlap with an adjacent distribution function. Discrimination between "gradients" as in Fig. 11.1 and complex distribution functions as in Fig. 11.3 is not easy without a very large data set.

Progress in understanding the amount of structure in the phase-space distribution of old stars is likely in two ways. Potentially the most important is further study of the Arcturus moving group. If this really is a physically meaningful and coeval feature in phase space, it contains more new information about the dynamical evolution of the Galaxy than any other currently identified tracer population. The existence of such fine structure in phase space would require considerably more careful dynamical analyses of kinematic data (a "bowl-of-spaghetti" or "can-of-worms" model) but would also explain a variety of marginally significant observational phenomena which are otherwise inexplicable, such as the retrograde globular clusters (Rodgers and Paltoglou 1984) and field stars (Norris and Ryan 1989), and metal-poor moving groups (Eggen 1987; Sommer-Larsen and Christensen 1987). The study which may be more likely to provide robust data would be that of a substantially enlarged sample of the kinematics of field Miras as a function of their periods.

11.4   CHEMICAL ABUNDANCES AND AGES

Calibration of the chemical-abundance enrichment rate onto a timescale which is calibrated independently of the collapse rate, *i.e.*, in years, is necessary to provide direct evidence for the timescale of Galactic evolution. In practice only stars near the main-sequence turnoff have surface gravities which change sufficiently rapidly and monotonically that reliable comparison with evolutionary tracks is possible, although some useful information on a combination of age and chemical abundance can be derived from the color of field giant stars (*e.g.*, Sandage 1987). For single stars near the main-sequence turnoff, the comparison of $uvby\beta$ photometry with theoretical isochrones is by far the most reliable and precise age-dating technique available. If independent abundance estimates are available, then any photometric measure of the temperature of the hottest turnoff stars will measure the age of the *youngest* star in a tracer population. This method has been used to estimate ages for globular clusters, where it also seems that all the member stars are coeval. A similar technique can be applied to field stars (see Gilmore and Wyse 1987), and is illustrated in Fig. 11.4, which shows the color–metallicity data for the high-proper-motion stars studied by Laird *et al.* (1988), as well as the turnoff points of all those globular clusters with recent CCD photometry, and a representative isochrone for old metal-rich stars (VandenBerg and Bell 1985).

The important conclusion from Fig. 11.4 is that essentially all stars with [Fe/H] $\lesssim -0.8$ are, to a good approximation, the same age as the globular-cluster system. More-detailed study has suggested that the *oldest* subdwarfs are ~ 3 Gyr older than the globular clusters (Bell 1988), though such precision in relative age-dating is very sensitive to a careful comparison of observational data and models. Schuster and Nissen (1989) also suggest that there is a real age spread of ~ 3 Gyr in the subdwarf system, and even possibly an age–abundance relation in metal-poor stars. The latter conclusion is still of low statistical weight, depending on only a few points, but if real it will radically alter our perception of Galactic evolution, and deserves considerable attention.

Stars more metal-rich than ~ -0.8 dex have a bluer turnoff than more metal-poor stars, implying that *at least some* of these stars are younger. The *distribution* of ages is, however, unmeasurable from a turnoff color. Some information on the age distribution for stars with [Fe/H] $\gtrsim -0.8$ is provided by studies of open clusters. These form a system with a very large scatter in the age–metallicity plane; clusters exist near the Sun with solar abundance and an age of 12 Gyr (NGC 6791, Janes 1988) and with [Fe/H] ~ -0.5, but an age of only a few Gyr (*e.g.*, Melotte 66). A similar scatter is evident in the age–metallicity relationship for F-dwarfs near the Sun (Twarog 1980; Carlberg *et al.* 1985; Knude *et*

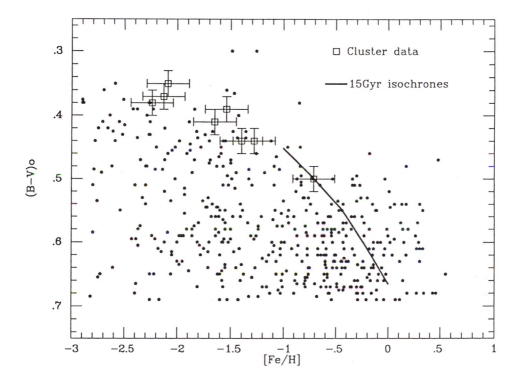

FIGURE 11.4

The *B–V* vs. [Fe/H] relation for all stars observed by Laird *et al.* (1988; points), and for those globular clusters with turnoff colors from recent CCD photometry (Stetson and Harris 1988; boxes). The photometric data are corrected for interstellar reddening. The solid line is a 15 Gyr isochrone calculated with oxygen-enhanced element ratios, and scaled in *B–V* to match the turnoff color of 47 Tuc. The blue edge of the stars with [Fe/H] ≲ −0.7 is adequately defined by the isochrone and by the globular-cluster data, showing that effectively all stars more metal-poor than ∼ −0.7 dex are as old as the globular clusters. At higher abundances the trend for the data to move to the blue of the isochrone shows that at least some stars are younger than the globular clusters.

al. 1987), though unfortunately both qualitative and quantitative differences are found from author to author, even when analyzing the same data, somewhat confusing the real situation. Thus any attempt to deduce a representative age for a stellar population from the turnoff color of the *bluest* field stars with metallicity $\gtrsim -0.8$ dex is unreliable.

Chemical-Element Ratios and Ages

In attempting to deduce the rate of star formation and dynamical evolution in a proto-galaxy, we would like to have available a clock whose rate can be calibrated independently of dynamics, and which runs sufficiently rapidly to resolve the dynamical evolutionary timescales. Such a clock is provided by stellar evolution of high-mass stars, and the fossil record of the clock is observable in the chemical-abundance enrichment patterns in long-lived low-mass stars. Fortunately, there exists a subset of common elements (most importantly oxygen) whose creation sites are restricted to very massive stars, and another subset (most importantly iron) which is also created in lower-mass stars. Since the evolutionary timescales for high- and low-mass stars span the timescale range of interest in galaxy formation, the differential enrichment of oxygen and iron provides an ideal clock to calibrate the rate of star formation in the proto-Galaxy.

Oxygen-to-iron element ratios have now been measured for enough stars to define the systematic trends in the data. The observations are reviewed by Wheeler *et al.* (1989), and will not be discussed in detail here (though see also Chapter 13). The important result for our purposes here is that a significant change of slope occurs in the relationship between the element ratios [O/Fe] and [Fe/H] close to metallicities where there also occurs a change in the stellar kinematics, that is, at [Fe/H] ~ -1. The [O/Fe] ratio is observed to be approximately constant, independent of [Fe/H] for the most metal-poor stars, $-3 \lesssim$ [Fe/H] $\lesssim -1$, while [O/Fe] declines for the more metal-rich stars, [O/Fe] $\sim -1/2$ [Fe/H]. Present data are consistent with a model in which all scatter is due to observational error, so that the amount of cosmic scatter is small.

Assuming that [Fe/H] is a monotonically increasing function of time, we can explain this behavior if the oxygen and iron in the more metal-poor stars has been produced in stars of the same lifetime; whereas for the more metal-rich stars, although the oxygen and iron continue to be produced together, an additional, longer-timescale source now dominates the iron production. Such behavior is in good agreement with supernova nucleosynthesis calculations, which show that oxygen is produced only in Type II supernovae by massive stars ($\mathcal{M} \gtrsim 20\,\mathcal{M}_\odot$), while iron has a contribution from both massive and low-mass stars ($\mathcal{M} \gtrsim 3\,\mathcal{M}_\odot$, Type I supernovae), thereby having an enhanced production once the much more numerous lower-mass stars contribute to its nucleosynthetic yield (Tinsley 1979; Matteucci and Greggio 1986; Wyse and Gilmore 1988). This results in the ratio [O/Fe] decreasing systematically with increasing metallicity [Fe/H].

Clearly, the approximate constancy of the [O/Fe] ratio independently of the [Fe/H] ratio at low metallicities requires that essentially *all* the stars with [Fe/H]

$\lesssim -1$ be formed on a timescale less than that on which a significant number of low-mass (Type I) supernovae exploded. This timescale is rather difficult to estimate precisely, due to uncertainties in the mechanism of Type I supernovae and the fraction of all stars formed which are in binaries of the type that may be expected to be precursors (see Iben 1986). However, the lowest-mass, and hence most numerous, progenitors of the CO white-dwarf binaries thought to become Type I supernovae have main-sequence masses and lifetimes of $\sim 5 \, \mathcal{M}_\odot$ and $\sim 2.5 \times 10^8$ yr. Thus a reasonable estimate for the earliest time after which we expect dominance of iron from Type I supernovae is $\lesssim 10^9$ yr. This general argument appears to be the strongest direct evidence for a rapid formation timescale for the Extreme Population II stars in the Galaxy, and is in agreement with the (currently more contentious) kinematic evidence discussed below.

11.5 THE THICK DISK

The luminosity profile of the Milky Way, as given by star counts, provides evidence for a Galactic thick disk, as was discussed in Chapter 2. Indeed, recent photometric and spectroscopic stellar surveys have emphasized the importance of the Intermediate Population II stars, as introduced at the 1957 Vatican Conference (Oort 1958; O'Connell 1958). The modern characterization of this population assigns to this stellar component a vertical scale height of ~ 1–1.5 kpc, a vertical velocity dispersion of ~ 45 km sec^{-1}, a typical stellar chemical abundance of $\sim 1/4$ of the solar metallicity, and a mean asymmetric drift of ~ 30–50 km sec^{-1}. The detailed values of the descriptive parameters remain poorly known, however, primarily because the offset in the mean values characterizing the thick-disk distribution function over age, metallicity, and kinematics from those mean values characterizing the oldest thin-disk stars is much less than the dispersions in these quantities. Reliable evaluation of the parameters of the distribution function is important, since it allows a discrimination between the several currently viable models of the formation of the thick disk. Such analyses are described in Chapter 13.

Here we summarize the data on the chemical-abundance distribution and the age range of thick-disk stars, and discuss the still-limited information concerning the relationships among the stellar populations near the Sun.

Metallicity of the Thick Disk

The metallicity distribution of stars *in situ* above the thin-disk plane has been the subject of several modern spectroscopic and photometric surveys. A population with a vertical velocity dispersion of ~ 45 km sec^{-1} will dominate samples of stars presently at z heights ~ 1 to a few kpc if it comprises of order 1 percent of the stars in the Galactic plane. Hence distances of ~ 2 kpc from the plane are the most suitable environment to study the properties of the thick disk. Hartkopf

and Yoss (1982; HY) obtained (DDO) metallicity estimates for a spectroscopically selected sample of K giants at the Galactic poles; their data for stars with
distances between 1 and 2 kpc are consistent with a Gaussian in log-metallicity,
with mean $\langle[\text{Fe/H}]\rangle = -0.6$ and $\sigma_{[\text{Fe/H}]} = 0.3$ dex (Gilmore and Wyse 1985).
This conclusion was confirmed by Yoss $et\ al.$ (1987) for an augmented sample,
though it should be noted that Norris and Green (1989) have reobserved many
of the most metal-rich but distant HY stars, and conclude that both the distance and the metallicity were overestimated by HY. Norris and Green therefore
suggest a revised, smaller dispersion for the thick-disk metallicity distribution.
The spectroscopically selected giant samples of Ratnatunga and Freeman (1985,
1989) and of Friel (1987, 1988) also contained relatively few metal-poor stars, in
contradiction to the predictions of models which assume that the halo has metallicity similar to those of the metal-poor globular clusters. In particular, Friel's
observations were in accord with the existence of a thick disk, with scale height
~ 1 kpc and local normalization ~ 3 percent, the stars of which have metallicity
similar to that of the metal-rich globular cluster 47 Tuc, ~ -0.7 dex.

Is the Thick Disk Chemically Discrete?

The kinematically defined sample of stars studied by Eggen (1979), when re-
stricted to stars with vertical velocities $40 \lesssim |W| \lesssim 60$ km sec^{-1}, has a metal-
licity distribution that is bimodal, with well-defined peaks at ~ -0.7 and -1.5
dex, in (remarkable) agreement with the globular-cluster data of Zinn (1985).
The metallicity distributions seen across other restricted velocity ranges are consistent with, but do not require, three discrete metallicity distributions for the
metal-poor halo, the thick disk, and the thin disk (Gilmore and Wyse 1985).
Higher-precision metallicities for large samples of stars are now becoming available; so it is interesting to check their consistency with the results of the analysis
of Gilmore and Wyse. The largest set of appropriate data for metal-rich stars is
that of Trefzger $et\ al.$ (1989a), which is summarized in Fig. 11.5. They studied
a magnitude-limited sample of stars toward the south Galactic pole, deriving
metallicities and gravities from $VBLUW$ Walraven photometry. The abundance
distribution in their data is consistent with a bimodal distribution, with one peak
near $[\text{Fe/H}] = -0.1$, and a second peak, which becomes increasingly dominant
at increasing distances from the plane, near $[\text{Fe/H}] = -0.6$. There is marginal
evidence for an abundance gradient in each of these two distributions. These
values are in good agreement with those derived earlier by Gilmore and Wyse.
Similar values are required by the very large data set of Carney $et\ al.$ (1989), in
their detailed study of the properties of the thick disk. Thus available data are
consistent with the suggestion that the thick-disk abundance distribution is de-
finably discrete from that of the thin old disk. The relationship of the abundance
distribution of the thick disk to that of the halo is discussed below.

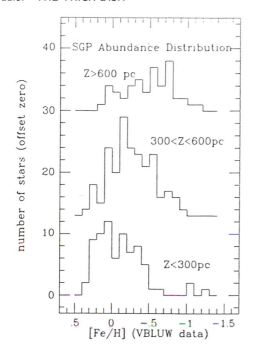

FIGURE 11.5

The abundance distribution in three distance intervals toward the south Galactic pole measured by Trefzger *et al.* (1989a). The data are consistent with two discrete distributions, with the old disk having modal abundance [Fe/H] $\sim$ −0.1, and the thick disk having modal abundance [Fe/H] $\sim$ −0.6.

Is the Thick Disk Kinematically Discrete?

Here we address the kinematics of the thick disk, in particular what we can infer about the evolutionary status of these stars from the similarities and differences of their kinematics from those of other populations in the Galaxy.

(1) Is it discrete from the subdwarf system?

The velocity ellipsoid for the nonkinematically selected Extreme Population II stars is reliably calculated to be $\sigma_{RR} : \sigma_{\theta\theta} : \sigma_{zz} = 131 : 102 : 89$. The vertical velocity dispersion of the thick disk is $\sim$ 45 km sec^{-1} (Ratnatunga and Freeman 1985, 1989; Fig. 12a of Gilmore and Wyse 1987; Sandage and Fouts 1987; Yoss *et al.* 1987; Carney *et al.* 1989; and Kuijken and Gilmore 1989b). The lag behind solar rotation (the asymmetric drift) of the thick disk is apparently 30–50 km sec^{-1} (Ratnatunga and Freeman 1985, 1989; Freeman 1987; Norris 1987c; Sandage and Fouts 1987; Morrison *et al.* 1990), whereas that of the subdwarf population is 180–220 km sec^{-1} (Norris 1986, Fig. 11.1). Thus the kinematics of the thick disk are dominated by rotational support, but those of the subdwarf system are dominated by pressure support from the anisotropic velocity-dispersion tensor. However, do the distributions overlap? This question is equivalent to the question of a correlation between metallicity and rotation velocity discussed above.

The number of stars with abundances and kinematics such that they might plausibly be assigned to either the low-velocity tail of the Extreme Population II, or to the high-velocity tail of the thick disk (*i.e.*, those stars with [Fe/H] ~ -1) is very small; Fig. 11.3 shows the *distribution* over rotation velocity and metallicity for the sample of Laird *et al.* (1988); the binned data are shown in Fig. 11.2. It is clear that there is a relative deficiency of stars with [Fe/H] ~ -1 *and* $-100 \lesssim V$ km sec$^{-1} \lesssim -200$. However, this overlap region *is* populated; in particular, Norris *et al.* (1985) drew attention to metal-poor stars with thick-disk kinematics, whereas Morrison *et al.* (1990), from their spectroscopically selected sample of G/K giants in a field at Galactic latitude 30°, suggest that stars with "disk kinematics" (large rotation velocity) and stars with "halo kinematics" exist in approximately equal numbers with metallicities ~ -1 dex. Thus Morrison *et al.* characterize the binned rotation velocity *versus* metallicity plot of Fig. 11.2 as two vertically offset, horizontal distributions which overlap in metallicity, rather than either the step function or the smooth correlation considered in Fig. 11.3. The existence of a significant "metal-weak thick disk" obviously would have major ramifications for our understanding of the thick disk as a discrete entity. However, the situation remains confused, primarily because very few stars might belong to such a metal-poor tail of the thick disk; so it is difficult to discover their properties and relationships to other groups of stars. It is, however, clear that the *bulk* of the thick disk is not kinematically related to the *bulk* of the halo distribution.

(2) Is it discrete from the thin disk?

The relationship of the thick disk to the high-velocity tail of the old disk is equally problematic, and has been discussed extensively by Sandage (1987) and Norris (1987a). The main point at issue is whether there is a continuous relationship of vertical velocity dispersion with metallicity extending all the way to the ~ 45 km sec^{-1} vertical velocity dispersion of the thick disk, or whether the old-disk velocity dispersion becomes asymptotically constant at the value of ~ 22 km sec^{-1} appropriate for spectroscopically selected samples of old dwarfs near the Sun (Fuchs and Wielen 1987; Sandage 1987). The decomposition of *local* samples of proper-motion stars into components whose kinematics are similar (within a small multiplicative factor) to those of the old disk is fraught with difficulty. Preliminary attempts have been made (Reid and Lewis, private communication) to illustrate this uncertainty, and to show that reliable results must await careful analysis of the several *in situ* surveys which are nearing completion.

The difficulty in deciding if there is a continuous kinematic continuity from the thick disk to the old disk is illustrated in Fig. 11.6. This shows the dependence of the vertical velocity distribution of samples of local giants studied by Norris, together with the data for the F-star sample of the Copenhagen group (Strömgren 1987). The important points to note here are the unsatisfactory disagreement between the two data sets at low abundances, and the level of smoothness or otherwise of the trend seen.

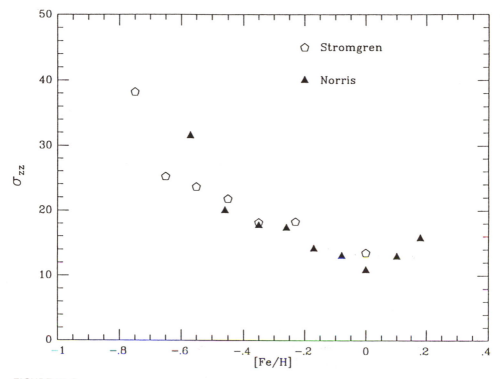

FIGURE 11.6

The relationship between [Fe/H] and vertical velocity dispersion for stars near the Sun, from data by Norris (1987c) and Strömgren (1987). These data provide the basis of the argument that there exists a continuous kinematic relationship between the thick disk and the old disk. Note particularly the degree of consistency between the data sets and the smoothness or otherwise of the trends at low metallicities, where the thick disk first contributes significantly to the data.

It is apparent from Fig. 11.6 that available *local* data cannot decide if the old disk and the thick disk are kinematically discrete. Resolution of this uncertainty, with its important implications for the formation history of the Galaxy, must await completion of the several extant *in situ* surveys of the stellar distribution several kpc from the Galactic plane. The available data marginally favor a model in which the thick disk is a kinematically discrete component of the Galaxy, but the issue remains to be decided by observational test.

The best data set which is available to show the relationship between the thick disk, the old thin disk, and the halo is probably that of Carney *et al.* (1989). In Fig. 11.7 the dependence of the distribution of metallicity on vertical velocity is reproduced. Decide for yourself whether this distribution is adequately described by three discrete metallicity distributions, each with an identifiable vertical velocity dispersion, or whether the data require a continuous distribution function.

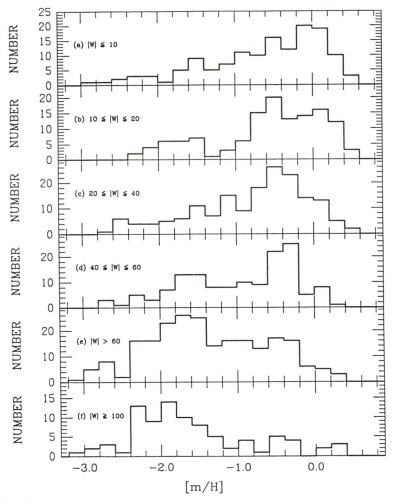

FIGURE 11.7
The dependence of the metallicity distribution on vertical (W) velocity for the sample of stars studied by Carney *et al.* (1989). The question to be answered by this diagram is whether or not the data are better described by three discrete distributions, centered near [Fe/H] = 0.0, -0.5, and -1.6, and with differing W velocity dispersions, or by a continuous relationship between some or all of the three distributions suggested above.

(3) Age of the thick disk

Finding ages for samples of thick-disk stars is an extremely difficult observational problem, partly because of the usual difficulty in assigning a reliable age to anything in astronomy, but here especially because, as we have discussed, there is no obvious *a priori* way to define a sample of purely "thick-disk" stars. Any sample selected by abundance or kinematics will inevitably include some old-disk and/or Extreme Population II stars in addition to the thick disk. Thus calculation of the age of the *youngest* or the *oldest* star in a sample, although tractable, is not an obviously clever way to answer the question of interest. Some information may be derived from Fig. 11.4. It is evident from this figure that the age of the oldest stars with [Fe/H] $\lesssim -0.7$ is comparable to that of the metal-rich globular-cluster system, ~ 14 Gyr (Hesser *et al.* 1987). This abundance range is expected to be dominated by thick-disk stars for $-1 \lesssim$ [Fe/H] $\lesssim -0.7$, suggesting that the most metal-poor thick-disk stars are among the oldest in the Galaxy (*modulo* the age calculations of Bell 1988 and Schuster and Nissen 1989 for field subdwarfs of 18 Gyr). More metal-rich thick-disk stars may or may not be the same age; it is impossible to tell from comparison with diagrams like Fig. 11.4, since some old-disk stars will contaminate the sample. This point is worth emphasizing, as it removes the rigorous basis for the conclusion of Norris and Green (1989) that the thick disk is several Gyr younger than the disk globulars; such a conclusion cannot be derived reliably from photometric data alone. This point is discussed in more detail by Sandage (1988); the interplay between age and metallicity in determining the photometric properties of stellar populations means that we can derive ages from turnoff colors only within the framework of an assumed age–metallicity relationship. The large intrinsic scatter seen in the age–metallicity relation for local F dwarfs (Carlberg *et al.* 1985) and for open clusters (Geisler 1987) will inevitably complicate the calculation of the ages of groups of stars from the color of the main-sequence turnoff.

The morphology of the horizontal branch also in principle contains important information on the age of the thick-disk stars, *modulo* the metallicity and the well-known "second parameter" problem. Rose (1985) found a population of red-horizontal-branch (RHB) candidates with a thick-disk configuration and kinematics, which would imply that the thick disk is old if these are truly RHB stars. The thick-disk nature of the kinematics of these stars was confirmed by the high-precision data of Stetson and Aikman (1987), who derived a vertical velocity dispersion of ~ 45 km sec^{-1}. A complication is that Rose found so many candidate RHB stars. In view of their intrinsic rarity, Rose's result would imply a very large local normalization of the thick disk near the Sun. However, Norris (1987b) and Norris and Green (1989) have argued that these stars are *not bona fide* RHB stars, but rather core-helium-burning "clump" stars similar to those seen in open clusters. They argue that their interpretation would mean a higher metallicity for these stars than that of 47 Tuc, and hence lead to a younger age. They infer that the thick disk is at least 3–6 Gyr younger than the disk globulars, and thus has an age of 8–11 Gyr. However, in addition to the complications with the number of these stars, Janes (1988) has derived an age of 12.5 Gyr for the metal-rich open cluster NGC 6791, which has a well-developed clump, showing explicitly that there is no universal age–metallicity relationship that can be applied.

The *youngest* dated population of RR Lyrae stars is that in the SMC cluster NGC 121, which has an age of ~ 12 Gyr (see Olszewski *et al.* 1987 for a detailed discussion of the age estimates for clump and horizontal-branch stars). The kinematics of the metal-rich RR Lyrae stars ($0 \lesssim \Delta S \lesssim 2$) are those of the thick disk

(Sandage 1981; Strugnell *et al.* 1986), suggesting that 12 Gyr is a lower limit on the age of at least some of the thick disk. Further studies of these metal-rich RR Lyrae stars would be of considerable interest.

Similarly, although the relationship of the globular clusters to field stars is not at all obvious, *if* the disk globular-cluster system studied so well by Zinn (1985) and collaborators (see Armandroff 1989 for the most recent analysis) is indeed part of the thick disk, then the antiquity of the thick disk is reliably established. The rotation velocity of the thick-disk globulars found by Armandroff (1989) is closer to the field thick-disk stars than that derived earlier by Zinn (1985), which could be construed to support the identification of these two populations. However, samples of old-disk open clusters *also* have similar kinematics to the field thick disk, and hence to the disk globular clusters, which under the same logic would suggest an intimate connection between the open and globular clusters. Understanding of the kinematics of the extant sample of open clusters must take account of the destruction processes which operate preferentially to destroy clusters which are confined to the plane, and hence lead to an artificially high scale height and to artificially high velocities for easily observed (*i.e.*, high-latitude) open clusters. The same problem exists for globular clusters, but will be worse for the more loosely bound open clusters.

All these arguments, however, leave open the possibility of a large age *range* in the thick disk. There is no reliable information yet available on this point; obviously if we believe *all* the ages inferred above, then an age spread of several Gyr results. However, this age scatter probably simply reflects the level of uncertainty associated with each of the available age estimates.

REFERENCES

Armandroff, T. 1989. *Astron. J.*, **97**, 375.

Arnett, W. D., Hansen, C. J., Truran, J. W., and Tsuruta, S., eds. 1986. *Cosmogonical Processes*. Utrecht: VNU Science Press.

Baade, W. 1944. *Astrophys. J.*, **100**, 137.

Bell, R. A. 1988. In Philip, p. 163.

Binney, J. 1982. In Martinet and Mayor, p. 1.

Binney, J., and May, A. 1986. *Mon. Not. Roy. Astron. Soc.*, **218**, 743.

Blaauw, A., and Schmidt, M., eds. 1965. *Galactic Structure*. Chicago, IL: Univ. of Chicago Press.

Carlberg, R. G., Dawson, P., Hsu, T., and VandenBerg, D. A. 1985. *Astrophys. J.*, **294**, 674.

Carney, B., and Latham, D. W. 1986. *Astron. J.*, **92**, 60.

Carney, B., Latham, D. W., and Laird, J. B. 1989. *Astron. J.*, **97**, 423.

Contopoulos, G., ed. 1974. *Highlights of Astronomy*. Vol. 3. Dordrecht: Reidel.

Delhaye, J. 1965. In Blaauw and Schmidt, Chapter 4.

Dejonghe, H., and de Zeeuw, P. T. 1988. *Astrophys. J.*, **329**, 720.

Eggen, O. J. 1979. *Astrophys. J.*, **229**, 158.

————. 1987. In Gilmore and Carswell, p. 211.

Eggen, O. J., Lynden-Bell, D., and Sandage, A. 1962. *Astrophys. J.*, **136**, 748.

Elson, E., and Walterbos, R. 1988. *Astrophys. J.*, **333**, 594.

Feast, M. 1987. In Gilmore and Carswell, p. 1.

———. 1989. In Schmidt, p. 205.

Feast, M. F., Woolley, R., and Yilmaz, N. 1972. *Mon. Not. Roy. Astron. Soc.*, **158**, 23.

Freeman, K. C. 1987. *Ann. Rev. Astr. Astrophys.*, **25**, 603.

Friel, E. D. 1987. *Astron. J.*, **93**, 1388.

———. 1988. *Astron. J.*, **95**, 1727.

Fuchs, B., and Wielen, R. 1987. In Gilmore and Carswell, p. 375.

Geisler, G. 1987. *Astron. J.*, **94**, 84.

Gilmore, G., and Carswell, B., eds. 1987. *The Galaxy*. Dordrecht: Reidel.

Gilmore, G., and Wyse, R. F. G. 1985. *Astron. J.*, **90**, 2015.

———. 1987. In Gilmore and Carswell, p. 247.

Harris, W. 1981. *Astron. J.*, **86**, 719.

Hartkopf, W. I., and Yoss, K. M. 1982. *Astron. J.*, **87**, 1679.

Hesser, J. E., Harris, W. E., VandenBerg, D. A., Allwright, J. W. B., Shott, P., and Stetson, P. B. 1987. *Publ. Astron. Soc. Pacific*, **99**, 739.

Iben, I. 1986. In Arnett *et al.*, p. 155.

Illingworth, G., and Schechter, P. L. 1982. *Astrophys. J.*, **256**, 481.

Janes, K. 1988. In Philip, p. 59.

Knude, J., Schnedler Nielsen, H., and Winther, M. 1987. *Astr. Astrophys.*, **179**, 115.

Kormendy, J., and Illingworth, G. 1982. *Astrophys. J.*, **256**, 460.

Kuijken, K., and Gilmore, G. 1989a. *Mon. Not. Roy. Astron. Soc.*, **239**, 571 (Paper I).

———. 1989b. *Mon. Not. Roy. Astron. Soc.*, **239**, 605 (Paper II).

Laird, J. B., Carney. B. W., and Latham, D. W. 1988. *Astron. J.*, **95**, 1843.

Levison, H. F., and Richstone, D. O. 1986. *Astrophys. J.*, **308**, 627.

Martinet, L., and Mayor, M. 1982. *Morphology and Dynamics of Galaxies*. Geneva: Geneva Observatory.

Matteucci, F., and Greggio, L. 1986. *Astron. Astrophys.*, **154**, 279.

McElroy, D. 1983. *Astrophys. J.*, **270**, 485.

Morrison, H. L., Flynn, C., and Freeman, K. C. 1990. Submitted to *Astron. J.*

Mould, J. 1986. In Norman *et al.*, p. 9.

Nolthenius, R., and Ford, H. C. 1987. *Astrophys. J.*, **317**, 62.

Norman, C. A., Renzini, A., and Tosi, M., eds. 1986. *Stellar Populations*. Cambridge, Eng.: Cambridge Univ. Press.

Norris, J. 1986. *Astrophys. J. Supp.*, **61**, 667.

———. 1987a. In Gilmore and Carswell, p. 297.

———. 1987b. *Astron. J.*, **93**, 616.

———. 1987c. *Astrophys. J. (Letters)*, **314**, L39.

Norris, J., Bessell, M. S., and Pickles, A. J. 1985. *Astrophys. J. Supp.*, **58**, 463.

Norris, J., and Green, E. M. 1989. *Astrophys. J.*, **337**, 272.

Norris, J., and Ryan, S. G. 1990. *Astrophys. J. (Letters)*, **336**, L17.

O'Connell, D. J. K., ed. 1958. *Stellar Populations*. Amsterdam: North Holland Press.

Olszewski, E. W., Schommer, R. A., and Aaronson, M. 1987. *Astron. J.*, **93**, 565.

Oort, J. H. 1958. In O'Connell, p. 15.

———. 1965. In Blaauw and Schmidt, Chapter 21.

Osvalds, V., and Risley, A. M. 1961. *Publ. McCormick Obs.*, **11**, part 21.

Philip, A. G. D., ed. 1988. *Calibration of Stellar Ages*. Schenectady: L. Davis Press.

Philip, A. G. D., and Lu, P., eds. 1989. *The Gravitational Force Perpendicular to the Galactic Plane*. Schenectady: L. Davis Press.

Plaut, L. 1965. In Blaauw and Schmidt, Chapter 13.

Pritchet, C. J., and Bergh, S. van den. 1987. *Astrophys. J.*, **316**, 517.

Ratnatunga, K. U., and Freeman, K. C. 1985. *Astrophys. J.*, **291**, 260.

––––––. 1989. *Astrophys. J.*, **339**, 126.

Rodgers, A. W., and Paltoglou, G. 1984. *Astrophys. J. (Letters)*, **283**, L5.

Rose, J. 1985. *Astron. J.*, **90**, 803.

Sandage, A. 1981. *Astron. J.*, **86**, 1643.

––––––. 1987. In Gilmore and Carswell, p. 321.

––––––. 1988. In Philip, p. 43.

Sandage, A., and Fouts, G. 1987. *Astron. J.*, **92**, 74.

Schmidt, E. G., ed. 1989. *The Use of Pulsating Stars in Fundamental Problems of Astronomy*. Cambridge, Eng.: Cambridge Univ. Press.

Schuster, W., and Nissen, P. 1989. *Astron. Astrophys.*, **222**, 69.

Shaw, M., and Gilmore G. 1989. *Mon. Not. Roy. Astron. Soc.*, **237**, 903.

Sommer-Larsen, J. 1987. *Mon. Not. Roy. Astron. Soc.*, **227**, 21P.

Sommer-Larson, J., and Christensen, P. R. 1987. *Mon. Not. Roy. Astron. Soc.*, **225**, 499.

––––––. 1989. *Mon. Not. Roy. Astron. Soc.*, **239**, 441.

Stetson, P. B., and Harris, W. E. 1988. *Astron. J.*, **96**, 909.

Stetson, P. B., and Aikman, G. C. L. 1987. *Astron. J.*, **93**, 1439.

Strömgren, B. 1987. In Gilmore and Carswell, p. 229.

Strugnell, P., Reid, I. N., and Murray, C. A. 1986. *Mon. Not. Roy. Astron. Soc.*, **220**, 413.

Tinsley, B. M. 1979. *Astrophys. J.*, **229**, 1046.

Trefzger, C. F., Pel, J. W., and Blaauw, A. 1989a. Private communication. For further information see Trefzger *et al.* 1989b.

Trefzger, C. F., Pel, J. W., Mayor, M., Grenon, M., and Blaauw, A. 1989b. In Philip and Lu, p. 29.

Twarog, B. A. 1980. *Astrophys. J.*, **242**, 242.

VandenBerg, D., and Bell, R. A. 1985. *Astrophys. J. Supp.*, **58**, 711.

Visvanathan, N., and Griersmith, D. 1977. *Astron. Astrophys.*, **59**, 317.

Wheeler, J. C., Sneden, C., and Truran, J. W. 1989. *Ann. Rev. Astron. Astrophys.*, **27**, 279.

White, S. D. M. 1985. *Astrophys. J.*, **294**, L99.

––––––. 1989. *Mon. Not. Roy. Astron. Soc.*, **237**, 51p.

Wielen, R. 1974. In Contopoulos, p. 395.

Wyse, R. F. G., and Gilmore, G. 1988. *Astron. J.*, **95**, 1404.

Yoshii, Y., and Saio, H. 1979. *Publ. Astron. Soc. Japan*, **31**, 339.

Yoss, K. M., Neese, C. L., and Hartkopf, W. I. 1987. *Astron. J.*, **94**, 1600.

Zinn, R. 1985. *Astrophys. J.*, **293**, 424.

CHAPTER 12

THE DISTRIBUTION OF PROPERTIES OF GALAXIES

Pieter C. van der Kruit

In this chapter the topic will be the distribution and statistics of various parameters that can be measured for spiral galaxies. The aim of this will be twofold. First, this information should lead us to a better understanding of the origin of structure in the early Universe and the subsequent formation and evolution of galaxies. Second, such information will enable us to put our Galaxy in context with other systems, and through local studies we might then be able to better understand both its structure and that of spiral galaxies in general. The emphasis will be on properties that can be derived from observations as described in my two previous chapters (5 and 10), namely, photometry and kinematic studies. I will discuss in general only integral or global properties. A detailed discussion of Local Group galaxies will be given in Chapter 15.

The classic publication in this area is Holmberg (1958). That comprehensive paper is based on a photometric study of 300 extragalactic nebulae and is the culmination of an investigation that took about ten years, starting with photographic studies of galaxies in the late 1940's. The basic data, painstakingly collected, consisted of a homogeneous set of integrated magnitudes, colors, and diameters (and hence surface brightnesses) of the systems. Holmberg especially discussed the dependence of color on the morphological type, which was found to become systematically bluer from elliptical toward irregular systems. Also, he was the first to investigate from a subsample in detail the statistical effects of internal dust absorption on the photometry, and his results have guided those of more recent workers. De Vaucouleurs (1959) also gave a review of integral properties at about the same time, and his discussion included the then-available scarce information on mass distributions.

12.1 THE DISTRIBUTION OF DISK SURFACE BRIGHTNESS AND SCALE LENGTH

In a classic study, Freeman (1970) used published surface photometry of disk galaxies to demonstrate an unexpected effect. For the majority of disks it turned

out that the extrapolated, face-on, central surface brightness μ_0 fell in the surprisingly small range of 21.67 ± 0.30 B-mag arcsec^{-2}. Exceptions were some
S0's on the bright side and dwarf galaxies on the faint side. Immediately it
was suspected that this narrow range constituted a selection effect, such as the
one pointed out earlier by Arp (1965). The effect arises because, for galaxies
to be selected as suitable candidates for surface photometry, their central surface brightness had to fall within a limited range. Fainter surface brightnesses
generally result in smaller angular diameters of the observable extent, and such
galaxies are difficult to see in the first place against the sky background. Brighter
central surface brightnesses occur in systems with small luminosity scale lengths,
and such galaxies appear starlike.

The problem with Freeman's result is that its significance could not be evaluated well, because the sample was not a statistically complete one; indeed, such
a sample did not exist at that time. If a sample is complete in terms of known
and well-defined selection criteria, then it is in principle possible to correct for
these effects. Before discussing that, I will first see how a sample can be judged
to be complete. There are various methods for this available, but the most useful
one is the so-called V/V_m-test, which was used first for studies of the distribution and evolution of quasars by Schmidt (1968). In this method, we calculate
for each object two volumes. The first volume V is the one that corresponds to
a sphere with radius the distance to the object. Then the object is shifted to a
larger distance until it drops out of the sample as a result of the selection criteria.
Note that these can be any number (such as in Schmidt's study radio and optical
brightness), but we should of course use the smallest distance at which it drops
out of the sample. This distance is used to calculate the corresponding volume
V_m. For the sample we then calculate the mean of V/V_m of all the objects. For
a uniform distribution in space (this is a fundamental assumption, as should
always be kept in mind) this mean value $\langle V/V_m \rangle$ should be equal to 0.5. If the
objects are nonuniformly distributed, the calculation of the volumes proceeds as
a weighted integral of the radius squared.

The test was originally designed to investigate space distributions of objects,
and for quasars the cosmological evolution from samples that are presumed complete. The use of the test for statistical completeness of a sample is the reverse
application. For a uniform distribution in space, we could be more demanding
than requiring $\langle V/V_m \rangle$ to be 0.5, since spatial uniformity of the objects predicts
that the distribution of V/V_m is also uniform between 0 and 1. This property
is seldom used, usually because the sample is still too small. The uncertainty in
$\langle V/V_m \rangle$ can be calculated as follows. For a uniform distribution between 0 and 1,
the r.m.s. value is $(12)^{-1/2}$, and for a large number of objects n, the probability
distribution of $\langle V/V_m \rangle$ becomes a Gaussian with a standard deviation equal to
$n^{-1/2}$ times that of the uniform distribution. The uncertainty in $\langle V/V_m \rangle$ then is
$(12n)^{-1/2}$.

Statistically complete samples of galaxies can be drawn from catalogs such
as the Uppsala General Catalogue (UGC; Nilson 1973) in the north or the
ESO/Uppsala Survey (EUC; Lauberts 1982) in the south. These catalogs have

been selected from the Palomar Sky Survey prints and the ESO southern "quick blue survey" in the south, respectively, and are designed to be complete to a diameter of 1 arcmin at a uniform isophote. Since these selections have been done by eye, the precise surface brightness of this isophote needs to be calibrated separately by using calibrated photometry of galaxies. For the UGC, for example, van der Kruit (1987) found that the isophote at which Nilson did the selection (and for which he gives diameters) is at 25.9 B-mag arcsec^{-2}. Application of the $V/V_{\rm m}$ test on these catalogs shows that these are not entirely complete to diameters of 1 arcmin, but that complete samples down to a limiting diameter of 2 arcmin can be drawn. For this diameter the values of $\langle V/V_{\rm m} \rangle$ are 0.497 for the UGC and 0.485 for the EUC.

Disney (1976) was the first to investigate the effects of sample selection in a quantitative manner to see whether Freeman's result could follow from observational bias. He concluded that it could, and for realistic values of the sky background, he could even produce Freeman's number. Furthermore, his analysis also predicted the equivalent for elliptical galaxies ("Fish's law") and the difference of about 6 magnitudes between the mean central surface brightness for ellipticals with an $r^{1/4}$ luminosity distribution[1] and exponential disks. This concept was extended in more detail by Disney and Phillipps (1983), and I will discuss this now in some detail, because it can also be used to extract information on distribution functions from complete samples.

Suppose that a galaxy has a luminosity law

$$L(R) = L(0) \exp\left[-(R/h)^{1/b}\right], \tag{12.1}$$

where $b = 1$ corresponds to the exponential disk and $b = 4$ to the $r^{1/4}$ law. The integrated luminosity then becomes

$$L_{\rm tot} = (2b)! \, \pi L(0) h^2. \tag{12.2}$$

If galaxies are selected to have a diameter at the isophote $\mu_{\rm lim}$ larger than $r_{\rm lim}$ on the sky, we can calculate for each galaxy the maximum distance d out to which it will be included in the sample as

$$d = \frac{(0.4 \ln 10)^b}{[\pi(2b)!]^{1/2}} \frac{(\mu_{\rm lim} - \mu_0)^b}{r_{\rm lim}} \, {\rm dex}\,[0.2(\mu_0 - M + 5)]. \tag{12.3}$$

Here again μ and M are surface brightness and luminosity expressed in magnitudes. This equation is evaluated for a face-on system, so that for an inclined disk, for example, μ_0 is the observed central surface brightness (uncorrected for inclination). This distance d has a maximum as a function of μ_0 that occurs for each value of $r_{\rm lim}$ or M at

$$\mu_0 = \mu_{\rm lim} - \frac{b}{0.2 \ln 10}. \tag{12.4}$$

---

[1]Here again r will denote a spherical and R a planar distance.

Similarly, Disney and Phillipps have derived an equation for a selection based on integrated magnitude $m_{\lim}$ within the limiting isophote $\mu_{\lim}$. In practice we may have to do here with a maximum observed surface brightness μ_M, for example, from overexposure on photographic plates. If we write $s = 0.4 \ln 10\ (\mu_M - \mu_0)$ and $p = 0.4 \ln 10\ (\mu_{\lim} - \mu_0)$, the limiting distance d becomes

$$d = \left[\left(\sum_{n=0}^{n=2b} \frac{s^n}{n!} \right) \exp(-s) - \left(\sum_{n=0}^{n=2b-1} \frac{p^n}{n!} \right) \exp(-p) \right]^{1/2} \mathrm{dex}\,[0.2(m_{\lim} - M + 5)].$$

(12.5)

Again, d has a maximum as a function of μ_0 that now also depends on μ_M. The peaks are now, however, much broader than before.

So for each galaxy we can calculate a volume $(4/3)\pi d^3$ within which it enters the catalog or correspondingly a visibility d^3. In an unbiased sample and assuming uniform space densities, the number of galaxies with a particular value of μ_0 will be present proportional to this visibility. Disney's original argument then was the following. Samples such as were studied by Freeman have been selected according to angular diameter and possibly integrated magnitude. If we select galaxies exceeding a certain diameter limit at an isophote $\mu_{\lim}$ of about 24 B-mag arcsec^{-2} (which is a likely value in practice), it then follows that d^3 has a peak at 21.8 and 15.3 B-mag arcsec^{-2} for $b = 1$ and $b = 4$, respectively. If the selection is done on integrated apparent magnitude, a similar result obtains; if $\mu_M = 19$ B-mag arcsec^{-2}, the peak occurs at 18.5 and 12 B-mag arcsec^{-2} for $b = 1$ and $b = 4$, respectively. The values for the diameter selection now are very similar to those found by Freeman (21.6 ± 0.3) and Fish (14.8 ± 0.9), and consequently Disney has argued that this observational bias has actually produced these "laws."

The difference between the two values is also predictable in the following sense. An elliptical galaxy with an $r^{1/4}$ luminosity law and a disk with an exponential profile with both the same integrated luminosity and the same effective radius (which encloses half the total light) will have central surface brightnesses that differ by about 6 mag. This is a result of only the different central light concentrations of the two luminosity laws. So what these two empirical laws may be saying is that most galaxies have the same length scale as a function of luminosity regardless of whether they are elliptical or disk.

Before we discuss "Freeman's law" further, two points should be made.

1. It has been known from the start that dwarfs with fainter values of μ_0 occur, and their existence is therefore not an issue (what is an issue is whether or not these occur in such numbers that they make a major contribution to the cosmic luminosity density). The early ones known were Local Group members recognized as conglomerations of faint stars, and therefore selected in a completely independent way.

2. A number of studies have followed Freeman's, finding essentially the same result (although the dispersion tended to be larger). However, all of these

were again based on samples that were not complete under well-defined selection criteria, and consequently the resulting distributions could not be corrected for observational bias. The very extensive study by Grosbøl (1985) is also not complete in any statistical sense, because the selection was done on the basis of inclusion in the Reference Catalogue.

An attempt to provide information based on statistically complete samples was reported in van der Kruit (1987). Here galaxies were selected from background fields on deep IIIa-J Schmidt plates (providing J-magnitudes that are close to B), and after scanning, the selection was made quantitative, in that all disk galaxies (with some inclination and morphological-type restrictions) which have a major-axis diameter at the isophote of 26.5 J-mag arcsec^{-2} in excess of 2 arcmin were included. As it turned out—in agreement with the remarks above—this consisted of precisely all galaxies that would be selected with a similar angular size limit in the UGC. This dataset, consisting of 51 galaxies, confirmed Freeman's result for nondwarf galaxies (Sc or earlier), namely $\mu_0 = 21.52 \pm 0.39$ mag arcsec^{-2} (roughly B-band). Dwarfs which are of a morphological type later than Sc, and which turn out to be small in physical size as well when redshifts were available, are fainter and have $\mu_0 = 22.61 \pm 0.47$ mag arcsec^{-2}. Furthermore, selection as discussed above, with the values for the limiting surface brightness μ_{lim} appropriate to the present sample, predicted that a peak should have occurred at a considerably fainter surface brightness than was estimated above, since these deep plates provided a different value for μ_{lim}. So nondwarf galaxies do have a relatively narrow distribution of μ_0, and this is not the result of selection effects. It is interesting to note that in terms of space densities dwarfs dominate by a large factor, but the dwarfs provide only about one-quarter of the cosmic luminosity density.

A major question is, of course, what happens at different wavelengths where absorption effects and contributions from young populations are entirely different. S. Westerhof and van der Kruit (unpublished) have repeated the analysis just described on IIIa-F plates of the same fields, as far as these existed. This gives a photometric band that is between standard V and R, such that $J - F = 1.25(B - V)$, where the J band used above is very close to B. The face-on distributions of the central surface brightnesses of the systems in common are given in Fig. 12.1, after correction for sample selection by weighting each galaxy by the inverse of the volume that it is sampling, calculated from the equations above. The means and the dispersions are as follows: for 33 nondwarfs, $J = 21.54 \pm 0.39$, $F = 20.63 \pm 0.49$; for 14 dwarfs, $J = 22.52 \pm 0.32$, $F = 21.99 \pm 0.44$. The ratio of the J and F scale lengths is 1.07 ± 0.13; so there is no significant change in h with wavelength, and thus again no systematic evidence for radial color gradients in disks. The distributions just given are marginally narrower in J, but the effect is not significant. If absorption by dust and contributions of young populations are significant, we would actually have expected the F distributions to be narrower. In agreement with the bluer colors of the dwarfs, the means separate indeed in going from J to F.

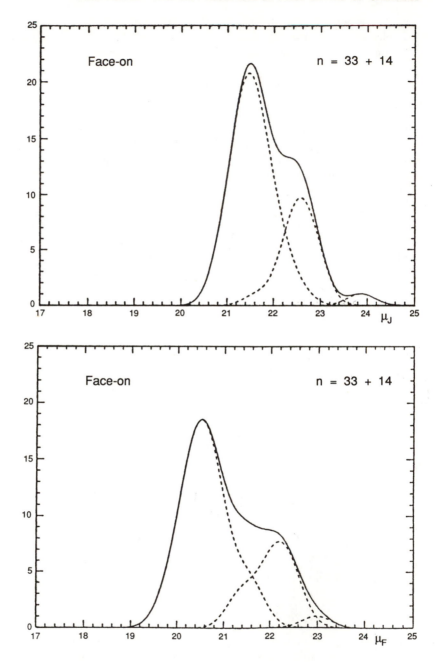

FIGURE 12.1 (Facing page)
Generalized histograms of the face-on, extrapolated central surface brightness of a complete sample of disk galaxies (van der Kruit 1987). The distributions have been corrected for sample selection effects. In both panels the brighter of the two distributions, indicated with dashed lines, contains the systems with morphological type Sc or earlier ("nondwarfs") and the fainter the remaining galaxies ("dwarfs"). The single faintest system is NGC 4392, which should probably be deleted from the sample. The top panel shows data from IIIa-J plates and corresponds roughly to the B-band, whereas the lower panel comes from IIIa-F plates and is between the V- and R-bands.

Elmegreen and Elmegreen (1984) have performed photographic surface photometry of a sample of 34 spirals, both in B and in the I band (0.83 μm). They do not list central surface brightnesses, but they do note that the B and I scale lengths are similar within each galaxy. The scatter is, however, large; from their published data the ratio of B to I scale lengths is 1.16 ± 0.47. If we leave out the four most deviating galaxies, the ratio still is 1.08 ± 0.29. Although there is no systematic trend, the two scale lengths can sometimes be very different in one galaxy. I will comment on observations in the near infrared below.

In the same way we can derive the distribution of scale lengths, again correcting for the known selection effects. These scale lengths exist from smaller than 1 kpc up to 7 kpc, and the frequency distribution is, roughly speaking, exponential with an e-folding of about 1 kpc. This shows the overwhelming number density of dwarf systems (in a physical sense), and the rareness in space of large spirals like our own. On the other hand, it also follows from the data that about 10 percent of all disk stars in the Universe occur in disks with scale lengths larger than 4 kpc, and this is not an improbably low number.

It is also possible to calculate the bivariate distribution function of μ_0 and h from such samples, although doing so requires large numbers. This was done in a preliminary way for the sample above in van der Kruit (1987), but has been improved significantly by R. de Jong and van der Kruit (unpublished) on a larger sample. This sample was drawn from the study of Grosbøl (1985), which was based on galaxies in the Second Reference Catalogue and therefore was not complete. However, since diameters and integrated magnitudes are known from Grosbøl's photometry, we can use the V/V_m test to extract a sample that is complete. The photometry was performed on glass copies of the red Palomar Sky Survey, and magnitudes are approximately R. The largest sample that could be extracted had a limiting diameter of 70 arcsec at the 23.5 R-mag arcsec^{-2} isophote and a limiting integrated magnitude of 11.6, and constituted 299 galaxies. Only 14 galaxies of type Scd or later occurred, and therefore the main study concentrated on the sample without these (this means nondwarfs in the terminology above).

The distribution of central face-on surface brightness is now broader than for the sample above, namely, 20.06 ± 1.19 R-mag arcsec^{-2} after correction for selection effects using the methods described earlier (see Fig. 12.2, top); so each contribution from a galaxy has been weighted according to the volume it samples. We may suspect that the zero-point calibration of the photometry has an effect here, since Grosbøl used a single calibration curve from density to brightness (in mag arcsec^{-2}, including zero-point) for all plates, although his 14 calibrating plates gave consistent results. Grosbøl quotes an uncertainty of about 0.3 mag in μ_0. Comparison of his scale lengths with literature values show a scatter of about 20 percent. Since this slope is generally measured at a mean radius of 2–3 scale lengths, this introduces a further error of about 0.5 mag in μ_0. The effects together reduce the scatter in μ_0 considerably. The distribution of scale lengths again declines exponentially, with an e-folding again of about 1 kpc (Fig. 12.2, bottom). Remember that no (morphological) dwarfs are included in the sample.

It is also possible to calculate the bivariate distribution function of μ_0 and h with these data; it is shown in Fig. 12.3, top, again after weighting each galaxy with the volume it samples. We can see a decrease in the density toward longer scale lengths for fainter central surface brightness; lines of equal density in the figure roughly follow those of constant integrated disk luminosity (or mass, if $\mathcal{M}/L$ is constant). This is remarkable, but there may be effects in the analysis of the luminosity profiles that could produce this in principle. Usually the fit is done in the outer parts, where the plates are not overexposed and the bulge contribution is probably small. Now, if h is underestimated, we would be overestimating the surface brightness (underestimating μ_0, when expressed in magnitudes), and this coupling is in the same sense as observed in Fig. 12.3, top. Grosbøl notes in his paper that bulge light may have introduced a bias and that the effect is "stronger for galaxies with a short length scale of their disk." Although this effect appears unable to produce the relation in Fig. 12.3, top, it shows that the result does need confirmation from at least a study with a different procedure to fit the exponential disk profile. At this stage it is prudent to take the result therefore as preliminary, and not to attach too much physical meaning to it yet. Furthermore, such an effect was not seen in the sample of van der Kruit (1987), although it was much smaller. The major conclusion then is that the most common spiral of type Sc or earlier has a central surface brightness of about 20 mag arcsec^{-2} in R (or 21.5 in B), and that the scale-length distribution drops off rapidly with increasing h.

A final exercise with these data is to weight each galaxy further by its total disk luminosity, so that the diagram measures the relative contributions to the cosmic luminosity density. This is shown in Fig. 12.3, bottom. Galaxies that contribute most have a central surface density of about 20 R-mag arcsec^{-2} and a scale length of about 4 kpc. It should thus be no surprise that the Sun is actually situated in a galaxy only a little bit larger than this.

Another approach has been followed by Phillipps et al. (1987), who obtained a complete sample on the Fornax cluster. They digitized a complete Schmidt plate, selected objects automatically with a threshold of 25.5 B-mag arcsec^{-2}, and then rejected the stellar images. This left them with 1550 galaxies that show a peak in the distribution of μ_0 of 21.8 ± 0.9 B-mag arcsec^{-2}. The ones

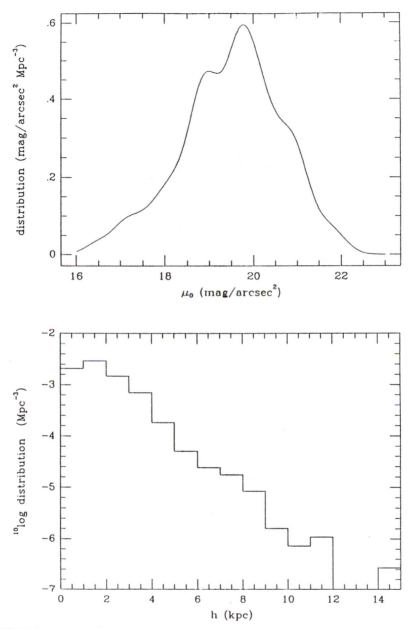

FIGURE 12.2
Results for the distribution of central face-on surface brightness and radial scale length from a statistically complete sample drawn from the study of Grosbøl (1985), consisting of almost 300 galaxies. Only galaxies with types Sc or earlier have been included. (Top) The distribution of the central surface brightness after correction for the sample-selection effects. (Bottom) The distribution of radial scale length for the same data. (From de Jong and van der Kruit, unpublished.)

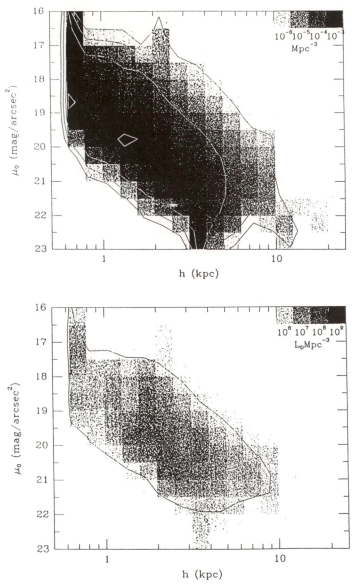

FIGURE 12.3

The bivariate distribution function of central face-on surface brightness and radial scale length in spiral galaxies of type Sc or earlier. Each galaxy has been weighted again by the volume that it samples. (Top) The number density distribution given at the top in Mpc^{-3}. (Bottom) Each galaxy has been weighted further by its luminosity, so that this distribution measures the contribution to the cosmic luminosity distribution. (From de Jong and van der Kruit, unpublished.)

with central surface brightness brighter than 22.5 are uniformly distributed on the sky; after correction for the selection effects, their distribution is somewhat skew, but has a mean and r.m.s. of 21.5 ± 0.6 B-mag arcsec^{-2}. These authors then conclude that "the normal field galaxy population of disk galaxies does have a preferred value." The galaxies with fainter central surface brightness are strongly concentrated toward the Fornax cluster and have scale lengths of 0.2 to 0.6 kpc. These are apparently common in clusters, but not in the field. Their distribution of μ_0 is essentially flat over the range 22.5 to 24.2 B-mag arcsec^{-2}, but their number increases sharply as a function of total luminosity (over the range M_B -14.4 to -11.9) for fainter systems.

12.2 INTERNAL ABSORPTION IN DISK GALAXIES

Although we have not discussed absorption yet in any detail, in studies of surface brightness or integrated magnitudes we must always be careful to correct observations for internal absorption by dust in the galaxies. The first values for such corrections were derived by Holmberg (1958). Absorption may be measured from the systematic change of surface brightness (on top of the usual geometric effect) and color as a function of inclination. However, the effect may be much smaller than is expected intuitively. Take, for example, a disk with a very thin, completely opaque dust layer in its central plane. The absorption then would be half the light, but the color would not be affected; hence there would be no inclination dependency, and we might think that the disk was optically thin.

Holmberg used a sample of 119 galaxies from his own photometry, and calculated the apparent projected surface brightness μ'_{obs} from the total magnitude m and the major axis diameter a, from $\mu'_{\mathrm{obs}} = m + 5 \log a$. The ratio of major to minor axis a/b was used to calculate the inclination, and he then assumed that the absorption effect depended on inclination according to a secant law, just like the Galactic absorption dependence which is inferred from galaxy counts as a function of latitude. (Holmberg's article mentions a cosecant law, but his definition of inclination is the opposite of the usual one today, which is $i = 0°$ for face-on.) Hence he expected that

$$\mu'_{\mathrm{obs}} = \mu'_0 + A_B(\sec \, i \, - 1) = \mu'_0 + A_B \left(\frac{a}{b} - 1 \right). \qquad (12.6)$$

From fits he then found that his data followed the observations (at least for $a/b < 3$), and that A_B is 0.4 mag for Sa–Sb galaxies and 0.28 for Sc. This shows that galaxy disks are not optically thick, and this result has been used extensively since then. Although similar calculations [with somewhat different inclination dependencies, particularly $\log(\sec i)$ rather than $\sec i - 1$] have been published, most recently by Kodaira and Watanabe (1988), this result has not been changed significantly. It should be noted that exceptions exist; for example, Jura (1980) has proposed that disks are optically thick, and that absorption is actually the cause of the central-surface-brightness constancy that I have just discussed.

Holmberg also measured the systematic change of color index with inclination, and again found a secant law with inclination. Holmberg did note that his inclination dependency was that of "the ideal case, when the obscuring matter is located as an absorbing screen entirely in front of the luminous matter." Clearly the actual arrangement in space is that dust and stars are mixed, and the situation more complicated. Holmberg discussed this in some detail, and concluded that his empirical description may still apply to the real world.

The question has received renewed attention with the discovery that galaxies have rather high fluxes of far-infrared radiation as observed with the IRAS satellite. After all, the light absorbed by the dust must be reradiated in the infrared, and many galaxies have IRAS fluxes similar to or even larger than in the optical (de Jong et al. 1984). A recent discussion of the absorption problem in terms of these observations has been given by Disney et al. (1989, but see also references therein), and I will follow that presentation here. The alternative to optical thickness of disks, as suggested by these IRAS fluxes, is that the dust heating comes from extra star formation in optically thick molecular clouds that cover only a small fraction of the area of the disk, leaving the disk in general optically thin.

Disney et al. discuss some alternative simple models. Take a uniform disk of stars with volume emissivity E^* and thickness T. First take a foreground screen of optical depth τ in the B-band. Then the optical surface brightness $L(i)$ is

$$L(i) = E^* T \sec i \, \exp(-\tau \sec i). \qquad (12.7)$$

The Holmberg surface brightness (now not in magnitudes) is always $L'(i) = L(i) \cos i$, and it can be easily seen that this corresponds to Holmberg's empirical law. The face-on extinction is

$$A_B = 1.086 \, \tau. \qquad (12.8)$$

The bolometric surface brightness L_{bol} is, of course, $E^* T \sec i$, and if the far-infrared surface brightness $L_{\text{FIR}} = L_{\text{bol}} - L(i)$, then

$$\frac{L_{\text{FIR}}}{L(i)} = \exp(\tau \sec i) - 1. \qquad (12.9)$$

The ratio of integrated fluxes is, of course, the same.

Next, take a uniform slab with stars and dust perfectly mixed and a mean free path for the optical radiation λ, so that $\tau = T/\lambda$. Then

$$L(i) = E^* \lambda \left[1 - \exp\left(-\frac{T}{\lambda} \sec i \right) \right]. \qquad (12.10)$$

The face-on extinction is

$$A_B = -2.5 \log \left[\frac{1 - \exp(-\tau)}{\tau} \right]. \qquad (12.11)$$

Now, for the optically thick case $\tau \gg 1$, we have

$$L(i) = E^* \lambda = \text{constant}, \tag{12.12}$$

and Holmberg's projected surface brightness varies as $\cos i$. The FIR surface brightness becomes

$$\frac{L_{\text{FIR}}}{L(i)} = (\tau - 1) \sec i. \tag{12.13}$$

For the optically thin case $\tau \ll 1$,

$$L(i) = E^* T \sec i, \tag{12.14}$$

and Holmberg's surface brightness becomes independent of i. Further,

$$\frac{L_{\text{FIR}}}{L(i)} = \frac{\tau}{2} \tag{12.15}$$

and is inclination-independent, as expected intuitively.

An even more realistic case is a "sandwich model," in which the dust has a thickness pT, so that $\tau = pT/\lambda$. The solutions now are

$$L(i) = E^* T \sec i \left(\frac{1-p}{2} \{1 + \exp[-\tau(i)]\} + \frac{p}{\tau(i)} \{1 - \exp[-\tau(i)]\} \right), \tag{12.16}$$

$$A_B = -2.5 \log \left\{ \frac{1-p}{2} [1 + \exp(-\tau)] + \frac{p}{\tau} [1 - \exp(-\tau)] \right\}. \tag{12.17}$$

The FIR-to-optical ratio can be found from this also, using $L_{\text{bol}} = E^* T$ as before, and will not be written out here in full.

Let us consider the opaque case $\tau \gg 1$. Then

$$L(i) = E^* T \sec i \, \frac{1-p}{2}, \tag{12.18}$$

$$A_B = -2.5 \log \left(\frac{1-p}{2} \right), \tag{12.19}$$

$$\frac{L_{\text{FIR}}}{L(i)} = \sec i \, \frac{(1+p)\tau - 2p}{(1-p)\tau + 2p}. \tag{12.20}$$

The surface brightness for the optically thick sandwich behaves just like an optically thin slab in its dependence upon i. Also, the ratio of FIR to optical flux is independent of the inclination in both cases. This latter situation is roughly what is observed, but we may therefore not conclude that disks are indeed optically thin.

In the optically thin case $\tau \ll 1$ we have

$$L(i) = E^* T \sec i \left(1 - \frac{1-p}{2\tau} \sec i \right), \tag{12.21}$$

which for small τ again approaches the thin slab, and Holmberg's surface brightness is again independent of inclination. Also,

$$A_B = -2.5 \log \left(1 - \frac{1-p}{2}\tau \right), \qquad (12.22)$$

$$\frac{L_{\text{FIR}}}{L(i)} = \frac{\tau}{2}. \qquad (12.23)$$

These illustrative models show the strong dependence on the precise model for the distributions. Disney *et al.* consider some more realistic models with exponential distributions, but the results so far suffice to show the main point, namely, that an analysis like Holmberg's is not generally able to distinguish optically thick disks from optically thin ones. This is shown graphically in Fig. 12.4, where some of Holmberg's data are compared to these schematic models. Here the difference $\mu'(i) - \mu_0$ is plotted against the axis ratio, which measures the inclination. The triangles show Holmberg's fit (or the screen model) with $A_B = 0.43$ mag, the stars an opaque slab, and the dashed line an optically thick sandwich with $p = 0.5$. Optically thin slab and sandwich models predict no dependence on i. It is clear from this figure that it is too early to conclude that disks are optically thin, and in agreement with the remarks above, color information essentially adds nothing to this.

It appears that comparison of IRAS fluxes with optical ones as a function of Hubble type might make a clearer case for one of the two possibilities. Certainly these are consistent with optically thick disks, but a more detailed discussion by Disney *et al.*, which I will not repeat here, shows that a nonuniform distribution of the dust and star formation in shielded molecular clouds will make it difficult to find a unique resolution of this issue. All we can say is that a thin, opaque layer in the central plane of a galaxy is possible, in which case the surface brightness is decreased by a factor of 2 or 0.75 mag. Disney *et al.* conclude that in special circumstances even 2 mag is possible; however, the existence of optically thin disks is not disproved.

12.3   PHOTOMETRY IN THE NEAR INFRARED

It has become possible recently to extend mapping of luminosity distributions of galaxies to near-infrared wavelengths. Before the advent of two-dimensional detectors at these wavelengths, we could obtain such information only by a painstaking process of observing many individual positions, although raster scanning on large telescopes could be done with reasonable amounts of observing time and moderate resolutions after the advent of indium-antimonide (InSb) detectors. A special problem at near-infrared wavelengths is the fact that the thermal background surface brightness of the sky relative to that of the galaxies is much higher than in the optical, and thermal emission from the telescope also poses

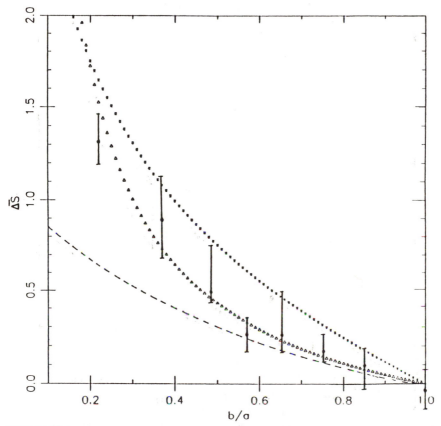

FIGURE 12.4

Effects of absorption on the surface brightness of disks in the analysis of Holmberg (1958). The vertical scale is the difference between observed reduced surface brightness (total luminosity divided by a circular area, with radius the major axis) and that for face-on; the horizontal scale is the axis ratio. Data are for Sa–Sb galaxies from Holmberg. The lines are three models: the triangles are a uniform screen in front of the disk (Holmberg's secant law); stars an optically thick slab model; and the dashed line an optically thick sandwich model with $p = 0.5$. The models are described in the text. (From Disney *et al.* 1989, Royal Astronomical Society.)

special problems. At K (2.2 μm) the photon shot noise from the thermal radiation of the telescope and the sky dominate the noise in the data; fortunately, the number of photons received is so large that this shot noise can be suppressed (with reasonable amounts of observing time) to acceptable levels.

    An important study in this rapidly developing field has been the thesis of Wainscoat (1986), who used the raster-scanning technique on 4-meter-class telescopes (AAT and UKIRT) to map edge-on galaxies at J (1.2 μm), H (1.6 μm), and K (2.2 μm). He also compared the near-infrared photometry to optical work. He found an interesting feature in early-type galaxies (NGC 7814 and NGC 7123), where the luminosity is dominated by that of the bulge. In fact, these galaxies are classified as Sa, because in the optical a strong central dust lane is seen projected onto the bulge, although there is very little evidence indeed of a stellar disk. It turns out that in the near-infrared we can observe a disk in emission just where in the optical the dust lane is seen. This emission clearly is from the stellar disk, and it follows from this observation that the dust and the stars must have similar vertical scale heights. This is completely different from the situation in our Galaxy and other later-type systems, where the stellar disk is always much thicker than the gas-and-dust layer. This means that in early-type galaxies the kinematics and dynamics of the dust and the stars of the disk are closely similar, and no secular evolution of the stellar kinematics occurs. This is most likely related to the low gas density, which would make the typical timescale of the scattering of stellar orbits much longer than, for example, in the solar neighborhood. As Wainscoat points out, this may also open the possibility that dust is present throughout the bulges of early-type galaxies, which would also have an effect on the observed color gradients in these components.

A further interesting observation by Wainscoat concerns the late-type edge-on IC 2531, which has very little bulge. In the K band the observations of a vertical cross-cut through the disk appear to indicate that the vertical distribution of stars continues as an exponential all the way to the plane, and therefore has a much sharper peak than would be predicted by the isothermal distribution with a sech^2 law. Unfortunately, the observations as presented refer to a radial distance along the major axis that is not very far out in the disk, and there possibly is at low z a contribution from bulge light. Furthermore, there is always the possibility that there is recent star formation in the area of the dust lane, where the gas also resides, and therefore significant contributions from red supergiants to the near-infrared luminosity can be expected. These effects need to be investigated further. However, as was discussed in Chapters 7 and 9, the actual distribution of stars is in any case expected to deviate near the plane from the isothermal distribution, because younger populations must have smaller velocity dispersions and therefore be more confined to the plane of the disks.

It is now also possible to use two-dimensional arrays in the near-infrared. The ones that are operational at this time are still limited in extent (roughly 60×60 pixels), but this is likely to change significantly very soon. The use of these devices greatly speeds up the collection of data on extended objects, although for larger galaxies a process of taking adjacent frames with a small overlap (so-called mosaicing) is still necessary to sample the whole image. As an example, Fig. 12.5 shows a preliminary K-band image taken at UKIRT of NGC 891, the edge-on galaxy that was mentioned extensively in the discussions on optical photometry in Chapter 5. Contour levels here are a factor 2 in intensity (about 0.75 mag), and the faintest contour is about 19.35 K-mag arcsec^{-2}. The sky surface brightness at these wavelengths is usually in the range 11 to 12 K-mag arcsec^{-2}, depending especially on the level of thermal emission of the telescope and on the outside temperature. The image in Figure 12.5 required about three hours on the telescope, including frequent blank-sky frames, which are necessary because of the rapid variation of sky background levels. Note the similarity of

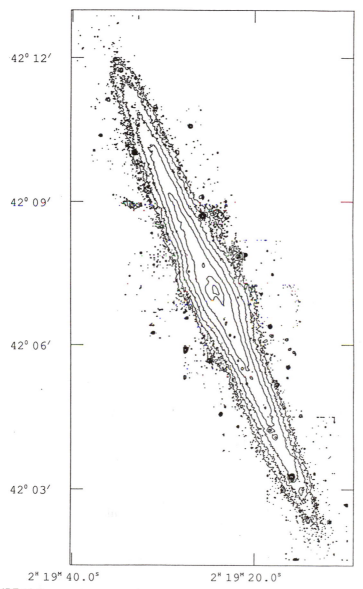

FIGURE 12.5
Preliminary K-band isophote map of NGC 891. The contours are spaced by 0.75 mag (a factor of 2 in intensity), and the faintest one is at about 19.35 K-mag arcsec^{-2}. Note the much-reduced effect of the dust absorption. (From Wainscoat and van der Kruit, unpublished.)

the contours to a map of optical isophotes and the much reduced effect of the dust lane.

With these new detectors it will also be possible to perform more extensive surveys of the luminosity distribution in the near infrared of more face-on systems. This would in particular provide important constraints on the range of variation of central extrapolated surface brightness, independent of younger populations, and would also allow more definite statements on the effects of dust absorption. No such surveys are yet available, but they can be expected in the near future. Using single-point raster-scanning mapping techniques, Giovanardi and Hunt (1988) have performed near-infrared photometry of nine large Sc galaxies in J, H, and K with a range of 3.5 mag in integrated luminosity. They found that the dispersion in the extrapolated face-on central surface brightness is comparable to that in B (namely, 0.6 to 0.9 mag). Also, the scale lengths are not statistically different from those found in the B-band. This study is just a first attempt (the resolution is only 28 arcsec), and the sample is not complete in any statistical sense. Yet the observed small spread in μ_0 is interesting and, if it would hold up for larger and more carefully selected samples, would indicate that the central-surface-brightness constancy is indeed a property of the old-disk population (and therefore presumably of the total disk surface density), and that absorption and mix with younger populations have no appreciable effect on this. The similarity of optical and near-infrared scale lengths shows that there is no appreciable radial absorption effect on optically derived luminosity profiles and no radial effect of the mix with younger populations.

12.4   INTEGRAL LUMINOSITIES, COLORS, AND TOTAL MASSES

In this section I will briefly discuss the integrated luminosities and the colors of spiral galaxies and the Tully–Fisher relation. An extensive review of most of these data has been given by Tinsley (1980), and little has changed fundamentally since then. I will not discuss the galaxy luminosity function here in detail; it has been reviewed recently by Binggeli et al. (1988).

It has been known for a long time that there is a correlation between the integrated color and the morphological type of a galaxy. In effect, these colors follow a more or less straight line in the two-color diagram, ranging roughly from $U - B = -0.3$, $B - V = 0.3$ for Sc and irregular galaxies to $U - B = 0.6$, $B - V = 1.0$ for ellipticals. This correlation has been interpreted as a result of different histories of star formation but similar ages for these various types, starting with the work of Tinsley (1968) and Searle et al. (1973). These methods are all in principle the same. First, assume an initial mass function (IMF, the mean distribution over masses for star formation) and calculate what the change is in luminosity and colors of such a burst of star formation as a function of time. Then add such individual contributions with a weight as a function of age that follows the rate of star formation (SFR) over the life of a galaxy. This function ranges from a single initial burst of star formation for the early-type galaxies to a constant SFR for late types. The basic result of these early studies was that the observed two-color diagram could be reproduced with a single IMF as inferred for the solar neighborhood and ages of about 10^{10} years for all galaxies.

Following and extending Tinsley (1973, 1980), we can illustrate this result analytically in the following way. Let the IMF be a simple power law (masses in $\mathcal{M}_\odot$)

$$\phi(\mathcal{M}) = C\mathcal{M}^{-(1+x)}d\mathcal{M} \quad \text{for} \quad \mathcal{M}_{\mathrm{L}} < \mathcal{M} < \mathcal{M}_{\mathrm{U}}. \tag{12.24}$$

For $\mathcal{M}_{\mathrm{L}} \ll \mathcal{M}_{\mathrm{U}}$ and $x > 0$ the constant C becomes, after normalization of the IMF over all masses, equal to $x\mathcal{M}_{\mathrm{L}}^x$. For the so-called Salpeter function (one of the earliest equations for the IMF) we have $x = 1.35$. Now, approximate the main-sequence mass–luminosity relation with $L = \mathcal{M}^\alpha$ (L and $\mathcal{M}$ in solar units) and the time that a star spends on the main sequence as $t_{\mathrm{MS}} = \mathcal{M}^{-\gamma}$ (the unit of time is then the main-sequence lifetime of a one-solar-mass star, or about 10^{10} years). Reasonable values for α are 4.9 in U, 4.5 in B, and 4.1 in V, and 3 for γ. Further, we assume that each star becomes a giant for a period of time 0.03 with luminosities (in solar units) 35 (U), 60 (B), and 90 (V). The calculation of the luminosity evolution of a single burst of star formation with (in total) ψ_0 stars at time $t = 0$ then proceeds as follows. At time t, the stars with mass $\mathcal{M} > \mathcal{M}_t = t^{-1/\gamma}$ have evolved away from the main sequence, and the total light from main-sequence stars therefore is

$$L_{\mathrm{MS}}(t) = \int_{\mathcal{M}_{\mathrm{L}}}^{\mathcal{M}_t} \psi_0\, \mathcal{M}^\alpha\, \phi(\mathcal{M})\, d\mathcal{M}$$

$$= \frac{x}{\alpha - x}\, \mathcal{M}_{\mathrm{L}}^x\, \psi_0\, \mathcal{M}_t^{\alpha-x}. \tag{12.25}$$

The number of giants at time t (for $t_{\mathrm{g}} \ll t$) is

$$N_{\mathrm{g}}(t) = \psi_0\, \phi(\mathcal{M}_t)\, \left|\frac{d\mathcal{M}}{dt_{\mathrm{MS}}}\right|_{\mathcal{M}=\mathcal{M}_t} t_{\mathrm{g}}$$

$$= \frac{\psi_0\, x}{\gamma}\mathcal{M}_{\mathrm{L}}^x\, \mathcal{M}_t^{\gamma-x}\, t_{\mathrm{g}}. \tag{12.26}$$

The luminosity of the single burst then becomes

$$L_{\mathrm{SB}}(t) = L_{\mathrm{MS}}(t) + N_{\mathrm{g}}(t)L_{\mathrm{g}}. \tag{12.27}$$

With $\mathcal{M}_{\mathrm{L}} = 0.1$ (the low-mass end of the main sequence) and with (adopted; see explanation two paragraphs below) absolute magnitudes for the Sun of $U_\odot = 5.40$, $B_\odot = 5.25$, and $V_\odot = 4.70$, we get results that are very close to more-detailed calculations. For $t < 0.03$ we have to take $t_{\mathrm{g}} = t$. Some representative numbers are shown in Table 12.1. The single-burst model rapidly becomes redder with time, while at the same time the luminosity (see $\mathcal{M}/L$) decreases. Of course, in the calculation of $\mathcal{M}/L$ we assume that no stars form with mass below $\mathcal{M}_{\mathrm{L}}$. For very short times t the actual value of $\mathcal{M}_{\mathrm{U}}$ needs to be taken into account also.

TABLE 12.1
Colors and mass-luminosity ratios for a single-burst model

$t = 0.01$	$U - B = -0.34$	$B - V = 0.12$	$(\mathcal{M}/L)_B = 0.15$
0.03	−0.06	0.45	0.38
0.1	0.18	0.64	1.12
0.3	0.38	0.79	2.79
1	0.56	0.90	6.95
3	0.66	0.96	14.9

We can then proceed to calculate the evolution of a galaxy by taking a SFR $\psi(t)$, from

$$L(t) = \int_0^t \psi(t - t')\, L_{\mathrm{SB}}(t')\, dt'. \tag{12.28}$$

For an initial burst (elliptical and S0 galaxies) we naturally get back the values above. For a constant SFR (and taking $\mathcal{M}_U = 32$), we get, at $t = 1$, $U - B = -0.25$, $B - V = 0.24$, and $(\mathcal{M}/L)_B = 1.03$, and these values correspond to late-type galaxies. Actually, the precise form of the SFR is not important (see also Larson and Tinsley 1978); the final colors and $(\mathcal{M}/L)$ depend only on the ratio of the present rate of star formation (over the last 10^8 years or so) to the mean SFR over the whole life of the galaxy.

The approximation given is only indicative (the absolute magnitudes of the Sun are incorrect, but are chosen to provide good answers), but allows us to estimate effects. Two-color diagrams from observations and from more detailed models are compared in Fig. 12.6. It can be seen there that the observations are reproduced very well by the calculations. In the figure the effects of making different assumptions (age, slope x of the IMF, $\mathcal{M}_U$, and metallicity Z) are also shown. It follows that these have a very minor effect, and the observed two-color diagram cannot be used to discriminate between these various possibilities.

Larson and Tinsley (1978) have also investigated the colors of peculiar galaxies. These turn out to have a much larger spread around the mean line of Fig. 12.6; this spread can easily be explained as the result of bursts of star formation of various strengths and ages in galaxies with various colors at the time of the burst. These bursts have strengths of up to 5 percent (ratio of mass in the burst to that of the stars already formed by that time) and durations of up to 2×10^7 years.

For spiral galaxies there is also an observed relation between the integrated luminosity and the amplitude of the rotation curve. The latter is usually measured from the width ΔV of the integrated H I-line profile at 21 cm (and corrected for inclination). Magnitudes need to be corrected statistically for absorption effects. Galaxies at moderate inclination are most suitable for this: for edge-ons the absorption correction to the magnitude is uncertain, and for face-on galaxies

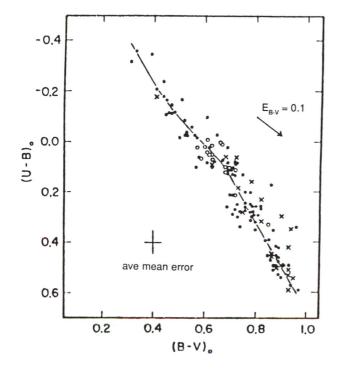

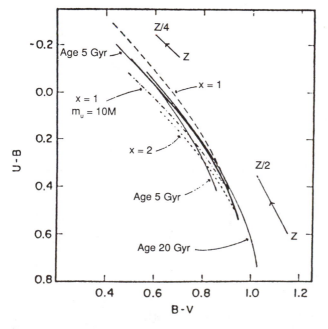

FIGURE 12.6
The two-color diagram of galaxies. (Top) The observed distribution; filled circles are galaxies in the Hubble Atlas, crosses E and S0 galaxies in the Virgo cluster, and open circles Galactic open and globular clusters. (Bottom) Model distributions. The solid line, which fits the data very well, has the IMF and the metallicity of the solar neighborhood and an age of 10 Gyr. It ranges from an initial burst of star formation to a constant SFR. The effects of changing the age, the slope x of the IMF, the upper mass limit $\mathcal{M}_U$, and the metallicity Z have been indicated. (From Tinsley 1980.)

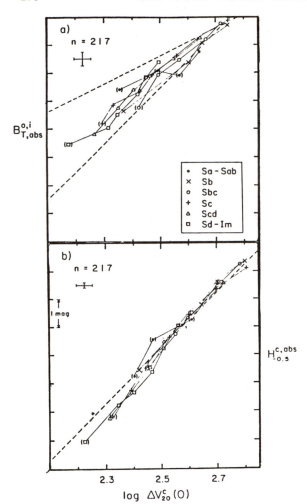

FIGURE 12.7
The Tully–Fisher relation in the optical (B, top) and the near IR (H, bottom). The sample of 217 galaxies has been binned using regressions of the two variables, and morphological types are distinguished by different symbols. The lines have slopes corresponding to α = 2 and 4 in Eq. (12.29), but at H only the latter is shown. The absolute-magnitude scales are arbitrary (that is, independent of the actual Hubble constant), and distances have been calculated using a Virgocentric inflow model. (From Aaronson and Mould 1983.)

the inclination is difficult to estimate. This relation, named after the discoverers (Tully and Fisher 1977), has the approximate form

$$L \propto \Delta V^{\alpha}. \tag{12.29}$$

The value of α has been measured for various samples and at various wavelengths. A very extensive study is that presented by Aaronson and Mould (1983), who base their investigation on a sample of about 300 galaxies with radial velocities less than 3000 km sec^{-1}, for which they have integrated magnitudes in both B and H (1.6 μm). The result is shown in Fig. 12.7. Distances have been derived from the redshifts, using a best-fit model for the pattern of streaming motions from infall toward the Virgo cluster (about 330 km sec^{-1} at the Local Group). This excludes galaxies with redshifts less than 300 km sec^{-1}, and some in a conical shell around the Virgo cluster, and leaves them with 217 systems. Three inferences can be made. The scatter is less in the near IR than in the optical, probably because of unreliable absorption corrections and more severe effects from young stellar populations in the B-band. The slope is wavelength-dependent, being steeper at H. For Sb and Sc galaxies α is about 3.5 in B and about 4.3 in H. A small type-dependence appears in B, but no significant change in the relation is apparent with type at H, again probably because of effects from young stars at B.

For exponential disks the mass is proportional to $\sigma_0 h^2$ and $V_{\max}$ to $(\sigma_0 h)^{1/2}$, so that we expect the mass to be proportional to $V_{\max}^4/\sigma_0$. Now, taking Freeman's constancy of central surface brightness and a constant $\mathcal{M}/L$ between galaxies, we expect for pure disks that $\alpha = 4$, roughly as observed. However, there are dark halos, and this implies two possibilities. If rotation curves are at all radii dominated by the dark matter, there should be a correlation not only between the amounts of dark and visible matter but also between their distributions. If, on the other hand, the inner rotation curves are provided essentially by the disk alone, then the Tully–Fisher relation follows immediately, especially if a "disk-halo conspiracy" exists. I will return to this question later.

Finally, I note that the values given in this book for the disk color, luminosity, and rotation speed of the Galaxy fit in very well with those of external galaxies in terms of the relations discussed here.

REFERENCES

Aaronson, M., and Mould, J. 1983. *Astrophys. J.*, **265**, 1.

Arp, H. C. 1965. *Astrophys. J.*, **142**, 402.

Binggeli, B., Sandage, A., and Tammann, G. A. 1988. *Ann. Rev. Astron. Astrophys.*, **26**, 509.

Disney, M. J. 1976. *Nature*, **263**, 573.

Disney, M. J., and Phillipps, S. 1983. *Mon. Not. Roy. Astron. Soc.*, **205**, 1253.

Disney, M. J., Davies, J., and Phillipps, S. 1989. *Mon. Not. Roy. Astron. Soc.*, **239**, 939.

Elmegreen, D. M., and Elmegreen, B. G. 1984. *Astrophys. J. Supp.*, **54**, 127.

Freeman, K. C. 1970. *Astrophys. J.*, **160**, 811.

Giovanardi, C., and Hunt, L. K. 1988. *Astron. J.*, **95**, 408.

Grosbøl, P. J. 1985. *Astron. Astrophys. Supp.* **60**, 261.

Holmberg, E. 1958. *Medd. Lund Astron. Obs.*, ser. 2, no. 136.

Jong, T. de, Clegg, P. E., Soifer, B. T., Rowan-Robinson, M., Habing, H. J., Houck, J. R., Aumann, H. H., and Raimond, E. 1984. *Astrophys. J.*, **278**, L67.

Jura, M. 1980. *Astrophys. J.*, **238**, 499.

Kodaira, K., and Watanabe, M. 1988. *Astron. J.*, **96**, 1593.

Kruit, P. C. van der. 1987. *Astron. Astrophys.*, **173**, 59.

Larson, R. B., and Tinsley, B. M. 1978. *Astrophys. J.*, **219**, 46.

Lauberts, A. 1982. *The ESO/Uppsala Survey of the ESO(B) Atlas*. Garching bei München: European Southern Observatory.

Nilson, P. 1973. *Uppsala General Catalogue of Galaxies*. Privately printed.

Phillipps, S., Disney, M. J., Kibblewhite, E. J., and Cawson, M. G. M. 1987. *Mon. Not. Roy. Astron. Soc.*, **229**, 505.

Schmidt, M. 1968. *Astrophys. J.*, **151**, 393.

Searle, L., Sargent, W. L. W., and Bagnuolo, W. G. 1973. *Astrophys. J.*, **179**, 427.

Tinsley, B. M. 1968. *Astrophys. J.*, **151**, 547.

———. 1973. *Astrophys. J.*, **186**, 35.

———. 1980. *Fund. Cosm. Phys.*, **5**, 287.

Tully, R. B., and Fisher, J. R. 1977. *Astron. Astrophys.*, **54**, 661.

Vaucouleurs, G. de. 1959. *Handbuch d. Phys.*, **53**, 311.

Wainscoat, R. J. 1986. Ph.D. thesis, Australian Natl. Univ.

CHAPTER 13

FORMATION AND EVOLUTION
OF THE GALAXY

Gerard Gilmore

In principle, an understanding of the formation and early evolution of the Galaxy is a well-defined purely theoretical problem. All we need is a detailed knowledge of the spectrum of perturbations in the early Universe and their subsequent evolution; an understanding of the physics of star formation in a variety of environments, with particular emphasis on a prediction of the distribution of orbital elements of those intermediate-mass massive-star binaries which will evolve to supernovae; a description of the hydrodynamics of a proto-galaxy, particularly including the effects of a high supernova rate, the efficiency of mixing of the chemically enriched ejecta, and the incidence of thermal and gravitational instabilities; the growth and transport of angular momentum and their effect on the growth of a disk; the origin, growth, and effects of magnetic fields; the effects of a time-dependent gravitational potential on the dynamics of any stars formed up to that time; the accretion history of the Galactic disk; and the history of the star-formation rate. In practice, there remain some limitations in our understanding of at least some of these physical processes. Hence it is still useful on occasion to try to deduce the important physics involved in galaxy formation by comparison of available idealized models with observations of old stars which are currently near the Sun, and whose present properties contain some fossil record of the Galaxy's history. The discussion in parts of this chapter follows that of Gilmore *et al.* (1989).

13.1 THE FORMATION OF DISK GALAXIES

Current understanding of the formation and early evolution of disk galaxies allows a description of the important physical processes at various levels of complexity and generality. At one extreme, we simply consider the global evolution of a gas cloud, and assume that mean values of relevant parameters suffice for an adequate description of generic properties. Alternatively, we give up general applicability, instead adopt specific numerical values for those parameters which quantify the important physics, and attempt a detailed confrontation of model predictions with observed stellar populations. The relation of any model prediction to detailed observations at a single radius in a specific galaxy clearly needs to be considered with some care. Mindful of this caveat, we outline here the most important timescales and physical processes which are likely to play a role in determining the observable properties of galaxies like the Milky Way.

Dissipational Disk-Galaxy Formation

The existence of cold, thin, galactic disks has strong implications for galaxy formation. To see this, consider a standard picture whereby galaxies form from growing primordial density perturbations, which expand with the background universe until their self-gravity becomes dominant and they collapse upon themselves. Were there to be no loss of energy in the collapse, and neglecting angular momentum, the transformation of potential energy into thermal (kinetic) energy would lead to an equilibrium system with final radius equal to half its size at maximum expansion, supported by random motions of the constituent particles. Thus an equilibrium, purely gaseous proto-galaxy should have temperature

$$T \equiv T_{\text{virial}} \sim \frac{G\mathcal{M}m_{\text{p}}}{kR}, \qquad (13.1)$$

and a stellar proto-galaxy should equivalently have velocity dispersion

$$\sigma^2 \sim T_{\text{virial}} \frac{k}{m_{\text{p}}}, \qquad (13.2)$$

where k and m_{p} are the Boltzmann constant and mass of the proton, respectively. Numerically, $T_{\text{virial}} \sim 10^6 R_{50}^{-1} \mathcal{M}_{12}$ K for gravitational (half-mass) radius, R, in units of 50 kpc, and mass $\mathcal{M}$ in units of $10^{12} \mathcal{M}_\odot$. Since the disks of spiral galaxies are cold, with $T \ll T_{\text{virial}}$, energy must have been lost. Since this lost energy was in random motions of individual particles, the only possible loss mechanism is through an inelastic collision, leading to the internal excitation of the particles and subsequent energy loss through radiative de-excitation. Clearly, particles with small cross-section per unit mass for collisions, such as stars, will not dissipate their random kinetic energy efficiently, so that dissipation must occur prior to star formation, while the galaxy is still gaseous. The virial temperature of a typical galactic-sized potential well is $T_{\text{galaxy}} \sim 10^6$ K, with corresponding one-dimensional velocity dispersion of ~ 100 km sec^{-1}.

The physical conditions in the Universe at the epoch of galaxy formation ($z \sim$ a few), as deduced from observations of quasar absorption lines (the Gunn–Peterson test for neutral hydrogen), are such that hydrogen is ionized, and correspond to temperatures of the proto-galactic gas of $\sim 10^4$ K, with a sound speed of only ~ 10 km sec^{-1}. Thus collapse of this gas in galactic potential wells will induce supersonic motions, and lead to both thermalization of energy through radiative shocks and subsequent loss of energy by cooling. It is this conversion

of potential energy, first to random kinetic energy as described by the virial theorem, and then to radiation *via* atomic processes, the net result of which is an
increase in binding energy of the system, which is termed dissipation.

The rate at which excited atoms can cool is obviously a fundamental limit on
the amount and rate of dissipational energy loss, and hence on the maximum rate
at which a gas cloud can radiate its pressure support and collapse. A convenient
measure of this timescale is the *cooling time* of a gas cloud, which is the time
for radiative processes to remove the internal energy of the cloud. Defining the
cooling rate per unit volume to be $n^2\Lambda(T)$, where n is the particle number
density, and where the functional form of Λ depends on the relative importances
of free-free, bound-free, and bound-bound transitions, and thus is an implicit
function of the chemical abundance, gives

$$t_{\text{cool}} = \frac{3nkT}{n^2\Lambda(T)} \propto \frac{T}{n\Lambda}. \tag{13.3}$$

It is usually of most interest to compare this timescale with the global *gravitational free-fall collapse time* of a system, which is the time it would take to
collapse upon itself if there were no pressure support. This timescale depends
only on the mean density of the system, and is given by

$$t_{\text{ff}} \sim 2 \times 10^7 n^{-1/2} \quad \text{yr.} \tag{13.4}$$

The term *rapid* is often used to describe evolution that occurs on about a free-fall
time.

An example of the role of atomic processes in allowing dissipation and increase
of binding energy during galaxy formation, which though idealized is still of interest in comparisons with observation, was discussed in three contemporaneous
papers: Rees and Ostriker (1977), Binney (1977), and Silk (1977), which developed an earlier suggestion by Lynden-Bell (1967). These authors investigated
the nonlinear (collapse-phase) evolution of (baryonic) cosmological density perturbations in the density–temperature plane. (These ideas are straightforwardly
adapted to allow for a significant nonbaryonic component of galaxies. See, *e.g.*,
White and Rees 1978; Fall and Efstathiou 1980; and Blumenthal *et al.* 1984.)
In the Rees–Ostriker model, the perturbation is hypothesized to be initially sufficiently lumpy and chaotic that collisions between local irregularities lead to
efficient shock-thermalization of the kinetic energy of collapse, resulting in a hot
($T \sim T_{\text{virial}} \sim 10^6$ K), pressure-supported system. The subsequent evolution will
then depend on the efficiency with which the heated gas can radiate, and can be
calculated readily if we assume for simplicity that the system is also of uniform
density (these simplifying assumptions of course remove the possibility of any
useful discussion of star formation, which depends on *local* cooling and instability, in this model). We may then define a curve in the density–temperature plane
where the gas-cooling time equals the free-fall collapse time of the perturbation
itself. Systems which formed with a *short* cooling time will occupy a locus inside
this curve.

The $t_{\rm cool} = t_{\rm ff}$ locus has an upper boundary corresponding to $\mathcal{M} \sim 10^{12}\,\mathcal{M}_\odot$, $R \sim 100$ kpc; the fact that these limits also correspond to the upper bound of masses and radii characteristic of observed galaxies is very suggestive. Indeed, when we translate observed surface brightnesses and velocity dispersions of galaxies to put them on the density–temperature plane, we find that present-day galaxies of all Hubble types lie within the *cooling time = free-fall time* curve, whereas groups and clusters of galaxies lie outside (Silk 1983; Blumenthal *et al.* 1984). This can be interpreted to imply that the luminous parts of galaxies cooled and collapsed rapidly, at least for those galaxies of high enough central surface brightness to have been studied to date (see Disney 1976, Bothun *et al.* 1987). Indeed, Gunn (1982) finds that the "bulge" of the Milky Way Galaxy individually also lies within the $t_{\rm cool} \lesssim t_{\rm ff}$ boundary of the density–temperature plane, suggesting that it too dissipated on timescales comparable to its free-fall collapse time. Other observational constraints on the duration of the formation of the metal-poor halo of the Milky Way are discussed further below.

The evolution of very massive ($\mathcal{M} \gtrsim 10^{12}\mathcal{M}_\odot$) proto-galaxies is, however, only poorly predicted by this theory. They may have had an early pressure-supported, quasistatic collapse phase, provided, of course, that the cooling time is less than the Hubble time; such density perturbations may plausibly evolve along a constant Jeans-mass[1] track in the density–temperature plane, with density and temperature increasing and little or no star formation, until conditions are such that the cooling time is less than the dynamical time, and again rapid collapse is expected to ensue. An alternative to this last conjecture was suggested by Fall and Rees (1985), who argued instead that conditions within a proto-galaxy, once the global cooling and collapse times became comparable, may lead to thermal instability, with a background plasma with temperature $T \sim T_{\rm virial} \sim 10^6$ K, and embedded dense condensations with $T \sim 10^4$ K. The important mass scale of the condensations is still set by gravitational instability, however. Fall and Rees suggest that this phase of galactic evolution represents an epoch of globular-cluster formation, since the Jeans mass under these conditions is $\sim 10^6\,\mathcal{M}_\odot$. However, the Jeans mass will be continually reduced, presumably to stellar masses, unless further cooling by molecular hydrogen is suppressed in some way. Thus continuing formation of globular-cluster-sized objects requires some additional special conditions. Regardless of the fine tuning required in theories of the formation of globular clusters, the basic idea that thermal instability might cause a sufficiently massive proto-galaxy to "hang up" with $t_{\rm cool} \gtrsim t_{\rm ff}$ in its early stages of evolution suggests that the halos of at least very massive galaxies may have collapsed less rapidly than on a free-fall time, with continuing star formation in thermally unstable condensations perhaps being possible over this long a time.

---

[1]The *Jeans mass* is that minimum mass at which gravity overwhelms pressure so that density perturbations of mass $\mathcal{M} \gtrsim \mathcal{M}_{\rm J} \sim 10^8\,T_4^{3/2}\,n^{-1/2}\,\mathcal{M}_\odot$ are unstable and collapse upon themselves, where the numerical factor is for temperature T in units of 10^4 K and number density, n, in units of particles cm^{-3}.

The preceding discussion is based on an extremely idealized model of a proto-galaxy, in that only the *global* cooling and collapse timescales of a *uniform* gas cloud are considered. No analytic descriptions of more plausible models exist as yet. A first step has been made by White (1989), who considers an idealized spherically symmetric gas cloud which has an imposed initial density gradient. White's models are motivated by cosmologies dominated by cold dark matter (CDM), and assume that 90 percent of the galaxy is nonbaryonic, and that the (remaining 10 percent) gas is initially distributed in proportion to the total density, which has a profile consistent with observed flat rotation curves. The inner, more dense regions then cool on a shorter timescale than the outer regions, so that we can define a time-dependent "cooling radius" for each proto-galaxy within which the gas is sufficiently dense to cool on a Hubble time. What happens to the gas within the cooling radius is as indeterminate in this model as in those discussed above. However, we can imagine earlier and slower star formation in the central regions than the earlier models suggested.

The preceding discussion can say nothing about when or how local Jeans-mass condensations actually form stars; the inherent assumption is that cooling is necessary and sufficient for efficient star formation, though the critical distinction between global and local timescales is rarely made explicit. However, it is clear that the existence of *gaseous disks* requires that the star-formation efficiency be low during the early stages of disk formation. A realistic discussion of galaxy formation must consider the hydrodynamics of the gas in a proto-galaxy. Numerical computation of hydrodynamic models of galaxy formation can contain an explicit formulation of the rate of star formation, along with the other important timescales of gaseous dissipation, viscous transport of angular momentum, and free-fall collapse. Larson's (1976) models still offer the most detailed discussion of the effects of gas processes within the prescription of galaxy formation, despite the computational limitations to the hydrodynamical approach at that time. These models identified the major requirements to produce galaxies that contain *both* a high-central-surface-brightness, nonrotating halo, and an extended cold disk. Initially both the star-formation rate and the viscosity must be high, to form the nonrotating but centrally concentrated halo, but both these quantities must be suppressed to form a thin disk that has lower central surface density and is centrifugally supported. Carlberg (1985) has pioneered the "sticky-particle," modified N-body approach, and several groups have initiated studies of disk-galaxy evolution using the smoothed-particle-hydrodynamics scheme. These developments promise to provide substantially more plausible models of galaxy formation over the next few years. The general conclusion from available studies is that, although we can build models which are somewhat like observations, it is necessary to specify the most sensitive parameters (viscosity and in effect the star-formation rate) in a quite *ad hoc* way. Considerably more sophisticated numerical experiments are required to ensure a plausible treatment of the hydrodynamics of a multiphase interstellar gas in a system with a high supernova rate, even assuming that we understood how to parameterize viscosity and star formation, and knew the initial conditions.

An important general feature of recent models of disk-galaxy formation is exemplified by Gunn's (1982) continual-infall models. These models hinge on the existence of loosely bound material surrounding a density peak whose central regions are collapsing (rapidly) to form a galaxy, and are extensions of the cosmological secondary-infall paradigm of Gunn and Gott (1972), modified to include dissipation by the infalling gas. The free-fall collapse timescales for the outer regions can be of the order of a Hubble time, leading to a picture of disk-galaxy formation whereby the central regions collapse rapidly to form the bulge, followed by accretion of proto-disk material. Insofar as the subsequent evolution of the gas may be modeled through the process of shock heating, cooling, and star formation, the general features of continual-infall models are in good qualitative agreement with observed galactic disks. Thus we might reasonably expect that galactic *disks* formed on a longer timescale, though still without pressure support having played a major role, than did galactic halos, the slower collapse arising simply because their lower-density initial conditions allowed a longer free-fall collapse time. The lack of disk-dominated systems in dense environments such as rich clusters of galaxies is consistent with this picture of unperturbed, continual accretion of disks (Larson *et al.* 1980; Frenk *et al.* 1985).

The angular momenta of galaxies that formed in environments of different density might also be expected to differ. Following Eddington (1928) *"we wonder how such a state of rapid revolution of the stars all in one direction round the center of the Galaxy could ever have arisen."* The answer was provided by Strömberg (1934), who realized that tidal torques from adjacent masses at the time of maximum expansion of a system would generate systematic rotation. The specific-angular-momentum distribution of the material surrounding density peaks has been investigated analytically by Ryden (1988) in the context of CDM-dominated cosmological models, and using N-body techniques by Barnes and Efstathiou (1987) for various assumed cosmological power spectra. A consensus is that the effect of tidal torques is to produce a system with specific angular momentum increasing with radius. Zurek *et al.* (1988) and Frenk *et al.* (1988) have shown that in CDM, or any scenario where chaotic aggregation of smaller systems is part of the formation of galaxy-sized systems, dynamical friction of dense clumps on the smoother background causes transport of both energy and angular momentum from the orbiting clumps to the smooth outer regions. Thus during the buildup of structure initially strongly bound particles lose both energy and angular-momentum, whereas the weakly bound particles gain energy (become more weakly bound) and also gain angular momentum. There is overall alignment of the angular-momentum vector of different shells in binding energy. These authors argue that slowly rotating stellar systems, such as giant elliptical galaxies or halos of disk galaxies, form in direct analogy to the dissipationless dark halos that they model, *i.e.*, with lots of dynamical friction and merging of stellar clumps, the dark halo and outer stellar envelope taking up the angular momentum transported outward. Disks of spiral galaxies would then form without significant angular-momentum transport because the baryons will remain gaseous until the virialization of the dark halo, and shock heating, as described earlier, would homogenize the gas. The predictions of these models could be tested in detail if we knew the angular-momentum distribution of the outer halo of our Galaxy; all we know at present is that the *kinematically selected* subdwarfs in the solar neighborhood have a lower specific angular momentum than the disk stars, by roughly a factor of five, and that the metal-poor globular-cluster system is consistent with zero net rotation to Galactocentric distances of ~ 30 kpc.

13.2   GALACTIC CHEMICAL EVOLUTION

The evolution of the chemical elements in the Galaxy is probably the most-studied and best-understood aspect of Galactic evolution. There are many excellent articles reviewing the basic models of Galactic chemical evolution and the available data sets. Hence we need not discuss the models in detail here. Rather we will outline the physically important general features of basic models of chemical evolution, and relate them to the description of the Galaxy provided in earlier chapters of this book. The original idea was presented by van den Bergh (1962), and an excellent review of basic chemical-evolution models is provided by Tinsley (1980). In addition, the collection and analysis of local stellar data by Pagel and Patchett (1975), the analytical extension of basic models to include the effects of gas flows by Lynden-Bell (1975), and the application of gas-flow models to stellar data by Hartwick (1976) are well worth reading.

To describe the chemical evolution of the Galaxy it is convenient to imagine a box initially containing a mass of gas of primordial composition. As stars form and evolve in this gas, chemical elements are created and dispersed into the interstellar medium. Thus the mean abundance of the gas is a monotonically increasing function of time. The essential features of the physics are retained if we adopt several other simplifying hypotheses. These include perfect mixing of the newly created elements with the remaining gas, so that the abundance is a function only of time, and not of position; instantaneous recycling of gas through stars, so that lifetimes of stars are either infinitely short or infinitely long (this removes any dependence of the metallicity on the history of the star-formation rate); a constant stellar mass function, so that a unit mass of gas forms some fraction of long-lived stars and some fraction of newly created chemicals, with both these values being time-independent; and a fixed total mass, a *closed box*, so that gas neither enters nor leaves the system. At face value all these simplifications seem absurd. However, there is a subset of elements, of which the most important is oxygen, which are synthesized only in high-mass short-lived stars, and whose abundance can be measured in stars of sufficiently low mass that they have not evolved in the age of the Universe (G dwarfs). Thus the instantaneous recycling assumption is well founded in this case. The other assumptions are less obviously justifiable, and a substantial literature has evolved round discussions of their several failings. We discuss the constraints on these assumptions from available data in the Galaxy below. In view of its lack of complexity, this description of Galactic chemical evolution is known as the *simple model* (Searle and Sargent 1972).

A particularly convenient parameter which can be defined is the *true yield, p,* which is the mass of newly synthesized heavy elements ejected into the interstellar medium in a generation of stars per mass of gas locked into long-lived stars. The fraction of the mass formed into stars which remains locked up in long-lived stars is termed α. In principle the yield can be calculated from stellar evolutionary models, though in practice it cannot be to useful precision as yet. If we then define the *gas fraction, f,* to be the fraction of the total mass of the system which is in gas, then the metallicity z of the gas can readily be calculated from mass-conservation arguments to be

$$z = p \ln \left(1/f \right) \qquad (13.5)$$

and the mean yield

$$\langle z \rangle = p \left[1 + \frac{f \ln f}{1 - f} \right] \Longrightarrow p \quad \text{as} \quad f \Rightarrow 0. \qquad (13.6)$$

The *apparent* or *effective yield*, p_{eff}, is defined as the yield a system would have were it to be analyzed as if it were described by the simple model

$$p_{\text{eff}} \equiv z/\ln \left(1/f \right). \qquad (13.7)$$

Thus, if we were to analyze two systems of stars and gas, and deduce two different values of p_{eff}, then we could deduce either that the true yield (*i.e.*, the stellar mass function for high-mass stars) differed between the two systems, or that one of the other assumptions (initially primordial gas or no gas flow in or out of the region of interest during its evolution) was different in the two systems. The assumption of an *ad hoc* nonzero initial metallicity for a subsystem in the Galaxy is a pure *deus ex machina* unless some physically plausible mechanism for generating the initial enrichment is identified. We therefore do not consider models of rapid early enrichment independently of models of the enriching region.

This leaves the effects of gas flows into and out of the (no longer closed) "box" to be considered. Following Hartwick (1976) we define the parameter c to be the mass-loss rate, and assume c is proportional to the star-formation rate dS/dt, with $S(t)$ the cumulative number of stars formed up to time t ($S(0) = 0$). This assumption removes an explicit dependence on the star-formation rate from the analysis, but retains the essential physical features of the effects of gas flows on the stellar abundance distribution. If we define $D(t)$ to be the cumulative amount of mass removed from the system ($D(0) = 0$), conservation of gas mass M_{g}, ($M_{\text{g}}(0) = 1$) means

$$M_{\text{g}} = 1 - \alpha S - D = 1 - (\alpha + c)S, \qquad (13.8)$$

and conservation of heavy-element abundance z means

$$M_{\text{g}} \frac{dz}{dS} = p\alpha. \qquad (13.9)$$

These last two equations give, where the subscript 1 refers to parameters of stars and gas remaining in the original box,

$$\frac{S}{S_1} = \frac{1 - \exp\{-[(\alpha + c)z]/p\alpha\}}{1 - \exp\{-[(\alpha + c)z_1]/p\alpha\}}. \tag{13.10}$$

The average metal abundance of the mass of gas removed from the "box" is

$$\bar{z} = \frac{p\alpha}{\alpha + c} - \frac{z_1}{\exp\{[(\alpha + c)z_1]/p\alpha\} - 1}. \tag{13.11}$$

The general effects of gas flow into or out of the volume in which star formation is proceeding on the metallicity distribution of long-lived stars remaining in the volume may be summarized as follows.

(1) The *effective yield*, p_{eff}, is at most *slightly* increased. Outflow or inflow of unenriched gas both reduce the effective yield, whereas inflow of already enriched gas will either reduce or slightly increase it. Thus, the simple model gives (more or less) the largest value for p_{eff}.

(2) The abundance distribution tends to be narrowed. Outflow leaves the abundance distribution unchanged, but inflow of unenriched or enriched gas produces a narrower distribution.

To illustrate these changes, Fig. 13.1 shows the cumulative abundance distribution of stars in a Galactic model which is based on Eq. (13.10), though applied twice, and for comparison the predictions of the basic simple model. The model with gas flows started with initially primordial gas, evolved with outflow (with outflow parameter $c \approx 15$) to fit the observed abundance distribution of halo stars, used the resulting outflow gas to begin the cycle anew to create the thick disk (for which $c \approx 1$ was required), and again used the outflow as the beginnings of the Galactic thin disk (Gilmore and Wyse 1986). In such modeling we are constrained not only by the shape of the abundance distribution, but also by its normalization, since the model makes explicit predictions of the mass ratios in each component, which must also be fit.

It is important with models involving gas flows into and out of the "box" to understand what is meant by the "box." In terms of the models as formulated above, the physically important thing is the gas available for star formation. Thus inflow and outflow refer to inflow and outflow of gas into the temperature–position plane where star formation can occur. Outflow and inflow of gas need not involve its physical movement at all; heating of gas will remove it from consideration in the models (whereas cooling will be analogous to inflow), provided that the cooling time is longer than the timescale for significant star formation to proceed. Thus it is perfectly sensible to have "outflow" from the zone in which halo stars form, and "inflow" of the same gas into the region in which thick-disk stars form, and then to use the stellar abundance distributions for halo and thick-disk stars in the same coordinate volume near the Sun to compare the models with observation.

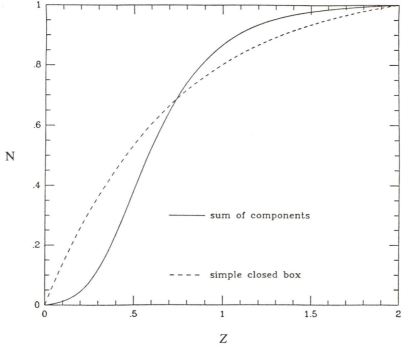

N

Z

FIGURE 13.1

The cumulative abundance distribution of long-lived stars predicted by a model including the effects of gas flows out of the star-forming region. The primary effect of gas flows is to reduce substantially the number of low-abundance stars expected, relative to the predictions of the simple model. This model is from Gilmore and Wyse (1986), and has parameters optimized to fit observations of stars near the Sun.

Modifications to the Simple Model for Halo Stars

The abundance distribution of unevolved halo dwarfs has been calculated to high precision by Laird *et al.* (1988b) from a survey of proper-motion stars. These authors show that the distribution of halo field stars is quite well described by the predictions of the simple model outlined above, and they derive a value for the effective yield of $p_{eff} \approx -1.6$ dex. (There is, of course, some uncertainty due to the difficulty in defining a clean sample of "halo" stars, as discussed in earlier chapters; this should, however, not be a problem here if we restrict our attention to the lower-metallicity part of the distribution.) Numerical calculations from stellar mass functions like that seen near the Sun, together with fits of the simple model to other regions, suggest that the true yield $p \approx 0.5$–2 dex. Thus the simple model requires enough modification of one of its parameters, either the stellar initial mass function (IMF) or the "closed-box" assumption, by a sufficiently large amount to change the yield by an order of magnitude. We discuss each assumption in turn.

(1) Variable stellar initial mass function?

The physics that determines the stellar IMF is presumably sensitive to some-thing as yet undiscovered. Chemical abundance, which controls the cooling rate of the gas, is one of the more likely possibilities; hence there have been many suggestions that the IMF is metallicity-dependent. The usual form assumed for this dependence is for there to be a relative *excess* of high-mass stars formed at low abundances. If so, we would expect more high-mass stars, per unit of mass formed into stars, than occur near the solar neighborhood today. However, high-mass stars are those which create and release, through supernovae, the elements whose evolution we are discussing. Since these elements are predominantly oxy-gen and iron (for present purposes), and oxygen entirely and iron in part are produced in high-mass stars, enhancing the formation of high-mass stars will *increase* the true yield, exacerbating the disagreement with observations. We must therefore appeal to a decrease of high-mass star formation to generate a low effective yield.

Observational evidence that the stellar IMF was different at very low abun-dances can, of course, now be found directly only for low-mass stars ($\mathcal{M} \lesssim 0.8\,\mathcal{M}_\odot$), and in practice over only a fairly narrow mass range. Studies by Eggen (1983) and by Hartwick *et al.* (1984) show there to be no significant differ-ences between the luminosity function of the field subdwarfs and that of solar-abundance stars near the Sun for the range $4 \lesssim M_V \lesssim 14$. The limit on any allowed differences is less than roughly a factor of two. To confuse the mat-ter somewhat, there is increasing evidence (McClure *et al.* 1986) for real dif-ferences between the luminosity functions of different globular clusters for the range $4 \lesssim M_V \lesssim 8$. The suggestion of the latter authors that the cluster luminos-ity functions change systematically with metallicity, however, are not supported by data available more recently. Thus the situation at low masses is that real cluster-to-cluster variations in the luminosity function are seen, but with no sys-tematic dependence on metallicity, and with the conversion of these differences to differences in the cluster initial mass functions being a nontrivial task. The field-star mass function, though poorly known, is not significantly different from the mass function for solar-abundance stars calculated by averaging over large volumes, though again small-scale variations are seen in the field (see Gilmore and Roberts 1988 for further examples and discussion). For our purposes here, however, the only constraint on low-mass stars required is that some exist to provide the record of the chemical evolution. This requirement seems to be met.

Some suggestive evidence that the IMF did not vary a lot at high masses and low metallicities is provided by the explanation of the systematic oxygen over-abundance in low-metallicity halo stars discussed later in this chapter. Briefly, the [O/Fe] ratio as a function of [Fe/H] is approximately constant over the range $-3 \lesssim$ [Fe/H] $\lesssim -1$, which includes almost the entire abundance range covered by halo stars. The value of the [O/Fe] ratio is in good agreement with the value calculated by assuming current nucleosynthesis models and an IMF like that in the Galactic disk at present. This agreement would have to be coincidental, with several fortuitously compensating unrelated effects, if the high-mass IMF really was very different at low metallicities. Thus it seems unlikely that the true yield could have been as low as the observed effective yield during formation of the halo.

(2) Gas flows?

Is it reasonable to expect that something like 90 percent of the gas mass of the proto-Galaxy would be removed from the star-forming region, either by being physically in a different place or by being heated to a sufficiently high temperature that it could not cool during the remaining period of halo formation? Yes, it is.

One of the better-known properties of the Milky Way is that it has a disk. This fact is an important constraint on possible models of Galactic chemical evolution. In earlier sections we discussed the evidence that the halo forms a (basically) nonrotating pressure-supported system, and that the disk forms a rotating, angular-momentum-supported system. Models of galaxy formation provide convincing evidence that the angular momentum in the Galaxy at present was created in tidal torques applied to the proto-Galaxy as it turned round and collapsed out of the background cosmological expansion. Since angular momentum is conserved once created, the present specific angular momentum of some mass in the Galaxy is a measure of how far it collapsed in radius, and hence spun up. The specific angular momentum of the disk is much greater than that of the halo, hence it collapsed more. The unavoidable consequence of this is that the gas now in the disk flowed through the volume of the proto-Galaxy in which the halo formed. Thus, if halo formation removed gas from the star-formation process, that gas will have been swept into the disk, and must be considered in chemical-evolution models.

The preceding discussion, however, merely locates the resting place of the lost gas. It is also necessary to know if gas flows took place *during* star formation. The evidence here comes from the discussion of asymmetric drift in Chapter 11, and from Fig. 11.1. That figure shows the systematic decrease in radial velocity dispersion (equivalently, an increase in binding energy) which took place during early formation in the halo. That is, star formation was accompanied by some dissipative collapse during formation of the halo. Such a dissipative collapse is, of course, simply a flow of gas during the time of halo formation. Gas flows occurred during formation of the halo. The similarity in age of the halo and metal-poor thick-disk stars discussed in earlier sections strongly suggests that this gas flow continued until the chemical abundance reached ~ -0.8 dex, by which time the ejected gas had been swept up and was part of the thick disk. Thus the existence of a disk in the Galaxy is strong evidence that gas flows occurred during the early evolution of the Galaxy, and ought to be considered in models of chemical evolution. Other possible violations of the assumptions underlying the simple model, though not rigorously excluded, have no direct supporting evidence.

13.3   THE SIMPLE MODEL AND OBSERVATIONS

The numerical parameters needed to calculate simple models of chemical evolution are the true yield p, and the gas fraction f. Since violations of the several assumptions underlying the simple model tend to reduce the effective yield to below the true yield, one method to estimate the true yield is to find the largest apparent yield in a system which follows the simple model adequately. There are two such calculations, for the solar neighborhood, and for the central Galactic bulge. Both have their difficulties, but provide values of $\sim 0.6 Z_\odot$ and $\sim 2 Z_\odot$, where $Z_\odot$ is the solar metallicity. Theoretical calculations tend to be closer to the lower value, which is that more commonly adopted.

Strong evidence that the true yield cannot have varied significantly, that is, that the high-mass stellar initial mass function has not changed measurably during the lifetime of the Galactic disk, is provided by the calculation of the Galactic $K_z(z)$ force law discussed in Chapter 8. The important relevant conclusion from that analysis is that the difference between the dynamically deduced and the observationally identified mass associated with the Galactic disk is $-2 \pm 12 \, \mathcal{M}_\odot \, \mathrm{pc}^{-2}$, not significantly different from zero. This measurement excludes the possibility that there exists a significant population of old, cool stellar remnants (white dwarfs, neutron stars, or black holes) left as evidence of an enhanced production of high-mass stars early in the evolution of the disk. The true yield cannot have been substantially higher at early epochs than it is today.

Attempts to adjust the predictions of the simple model to lower the true yield are also strongly constrained by the fact that there is no evidence for any significant unidentified mass associated with the Galactic disk. The fraction of gas cycled through stars in each "generation" in the simple model and ejected as enriched material clearly depends on the low-mass IMF as well as the high-mass IMF. If a very large fraction of the mass is locked into low-mass stars, then proportionately less is available for recycling. The fraction of mass removed from the system by being locked into long-lived stars in each generation is described by the α parameter in Eq. (13.10), and is about 0.8 for a mass function like that near the Sun today. This number could be increased substantially if a large population of very low-mass stars were formed. Such stars would have to be of too low mass to burn hydrogen (i.e., $\mathcal{M} \lesssim 0.08 \, \mathcal{M}_\odot$), or they would have been detected in direct studies of low-luminosity stars (see Reid and Gilmore 1982 and Gilmore et al. 1985 for this argument). The same constraint on low-mass low-luminosity stars as that on old high-mass low-luminosity stellar remnants discussed above is appropriate. There cannot be enough very low-mass stars in the Galactic disk to affect the predictions of models of Galactic chemical evolution.

Explanations of an apparently variable effective yield (that is, a deficit of metal-poor stars—the G-dwarf problem) then must rely on the effects on chemical evolution of the finite time required for the gas currently in the thin disk to have got there. The time-scale for accretion of a galactic disk is mentioned above under the alias of the "cosmological secondary-infall paradigm."

The gas fraction in the halo and the thick disk are both zero, since both systems appear to be exclusively old. The thin-disk gas fraction is derived as part of the calculation of the Galactic $K_z(z)$ law, in Chapter 8. The observationally determined value is $13 \pm 3 \, \mathcal{M}_\odot \, pc^{-2}/46 \pm 9 \, \mathcal{M}_\odot \, pc^{-2}$, or $\sim 0.3 \pm 0.1$.

Abundance Distributions in Old Stars

The most detailed and careful study of halo stars and comparison with the halo globular clusters is that by Laird *et al.* (1988a). As noted above, the field halo stars follow a distribution which is in remarkable agreement with the predictions of the simple-model Eq. (13.10). The peak of the distribution is at $\langle[Fe/H]\rangle = -1.6$. The halo globular-cluster system has the peak of its distribution at the identical metallicity, though it has a somewhat narrower distribution in abundance. In effect, the halo globular-cluster system lacks a few (~ 4) very metal-poor clusters. The metal-rich tails of the halo cluster and the halo field-star distributions cannot be discussed reliably, because of the complications of disentangling the abundance distributions from that of the thick disk. The identity of the maximum values of the abundance distributions, and the fact that the distributions are so similar (though not identical) in shape, is strong evidence that the formation and chemical evolution of the globular-cluster system proceeded intimately related with the formation and chemical evolution of the halo field stars. Should unevolved globular-cluster stars be shown to have the same patterns of element ratios as a function of [Fe/H] as do the field stars, the evidence would be overwhelming. The present differences in their spatial distributions must then be an evolutionary effect.

The thick-disk abundance distribution is discussed in earlier chapters. The data are consistent with a Gaussian distribution, with mean value $\langle[Fe/H]\rangle = -0.6$, and a poorly known dispersion about this mean of about 0.2 dex. The abundance distribution for old-disk stars is also consistent with a Gaussian distribution, with mean metallicity $\langle[Fe/H]\rangle = -0.1$, and a dispersion about this mean of perhaps 0.2 dex. The numerical values are evidently poorly known because of the complexity of disentangling overlapping distribution functions, and the effects of gradients in one or both populations. The shapes of the distribution functions are even less well known.

These data refer to the sample of stars identified and studied in a volume of radius a few hundred parsecs centered on the Sun. Thus if large-scale gradients in the abundance distributions exist, these values will not be representative of the total abundance distribution of these populations in the Galaxy. Nevertheless, we can discuss the chemical evolution of a region near the Sun regardless of larger-scale gradients. Extension of this analysis to the chemical evolution of the Galaxy after the first few billion years requires that we know the abundance distribution of disk stars near the Sun. Since there are age–metallicity and age–velocity dispersion relations for disk stars, some care is required in deriving the appropriate abundance distribution of old-disk stars. Simply counting those stars very near the Sun is inadequate, for such a sample will overrepresent young,

TABLE 13.1
Weights for conversion of solar neighborhood vertical
velocities to oscillation periods

Vertical velocity (km sec^{-1})	Vertical period (Myr)	Relative weight
10	77	1.00
30	105	1.36
50	136	1.75
100	183	2.53

low-velocity-dispersion (and metal-rich) stars. We need to know the abundance distribution of stars in a column through the disk. This can be derived either by direct counting of stars *in situ* or by weighting the contribution of each star in a local sample by the fraction of the time it spends near the Sun. Thus high-velocity (and old and metal-poor) disk stars are given higher weights than low-velocity (and young and metal-rich) disk stars, since the high-velocity stars spend less time near the Sun, and are therefore found in a local sample less frequently than in a column through the disk.

The correct weighting to use is derived from the $K_z(z)$-law analysis discussed earlier. All published analyses of the local abundance distribution have used a weighting scheme based on an approximate analysis of the relationship between velocity in the Galactic plane and orbital period which is in error by up to a factor of two. The correct weights are summarized in Table 13.1. With these weights applied to data for G stars near the Sun, the resulting abundance distribution is in quite tolerable agreement with the model distribution shown in Fig. 13.1. A much more detailed and reliable analysis of the chemical evolution of the Galactic thin disk will be possible once the several current surveys of large samples of stars several hundred parsecs from the Galactic plane are complete. Extrapolation from local data on a few bright stars will then no longer be necessary.

13.4   AGE *VS.* METALLICITY RELATIONS

If we can derive reliable ages for stars, then knowing both the stellar age and the chemical abundance will allow us to calculate the age–metallicity evolution of the Galactic disk empirically. Since the chemical abundance of stars being formed at any time depends on the true yield, the amount of enrichment of the interstellar medium up to that time, and the amount of inflow and outflow of gas, a substantial amount of information concerning the evolution of the Galactic disk is potentially available. Many analyses of this type have been carried out. The most detailed and extensive are those by Twarog (1980), Carlberg *et al.* (1985), and Lacey and Fall (1985), where detailed discussions can be found.

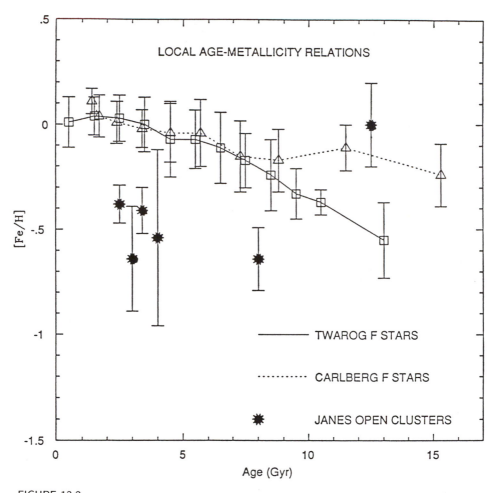

FIGURE 13.2
Age–metallicity relations derived by Twarog (1980; solid line and square symbols) and by Carlberg *et al.* (1985; dotted line and triangles) from analyses of the same sample of F stars near the Sun. The open stars show some extreme data for open clusters near the Sun. These data illustrate the significant uncertainties remaining in analyses of the local age–metallicity relation, and also that there is a real scatter in the relation at any age.

The essential features of the observational results are summarized in Fig. 13.2, which shows some of the available age–metallicity relations. The agreement for stars from ∼ 2 Gyr to ∼ 10 Gyr old is quite satisfactory. For younger stars it is still unclear if the mean trend continues rising, or if it has flattened off at roughly solar abundance. The most recent data sets currently in preparation favor a continuing rise. For stars older than ∼ 10 Gyr and more metal-poor than ∼ −0.25 dex, the agreement between different analyses is poor, because of a combination of systematic differences in age and abundance calibration methods, as well as isolation of slightly different subsets of stars from small samples.

The information in this diagram regarding the accretion history of gas by the disk and the constancy of the IMF is well discussed in the references above. For our purposes here we will illustrate the potential of such analyses by considering just one aspect of the results in Fig. 13.2, the scatter. The scatter in the observed abundance distribution of stars at a given age appears to be real "cosmic" scatter, and not simply observational error. The amplitude of the scatter can be estimated from the open clusters shown in the figure, though interpretation of the data for such objects is complicated by the possibility of a radial gradient in their abundance distribution. Since there is an age–velocity dispersion relation for disk-population objects, older objects will have larger velocities, and hence larger radial excursions in the Galaxy, than will younger objects. Thus radial and temporal abundance gradients must be disentangled with care. This is especially true for open clusters, since weakly bound clusters are tidally destroyed if they stay in the plane. Thus only clusters with anomalously high velocities will survive into old age.

An alternative measure of the scatter and the trend in the age–metallicity relation comes from the data of Strömgren (1987). In his sample of F/G stars near the Sun, 3 percent of stars with $+0.15 \gtrsim [\text{Fe/H}] \gtrsim -0.15$ are older than 10 Gyr, but 80 percent of those with $-0.6 \gtrsim [\text{Fe/H}] \gtrsim -0.8$ are older than 10 Gyr. Thus, although there is a clear trend in the mean, a significant number of the oldest stars in the disk are more metal-rich than the youngest stars near the Sun. The individual age–metallicity data from the analysis by Carlberg *et al.* make this clear, and are shown in Fig. 13.3. This abundance dispersion has been shown recently to be a real feature on the shortest timescales resolvable. Boesgaard (1989) has derived abundances of unprecedented precision for six clusters and groups, which are presented in Fig. 13.3. There is clearly a real scatter in the enrichment of the interstellar medium.

Since we do not, by assumption, expect scatter in the age–metallicity relationship in the simple model, ought we to be surprised by this result? As always when analyzing the Galaxy, it is valuable to compare our data with our visual impression of the sky. Spiral galaxies contain spiral arms, and spiral arms contain giant-molecular-cloud (GMC) complexes, in which the massive stars which create the chemical enrichment reside. These GMC complexes are sufficiently large that they survive more than one supernova explosion, and hence will be self-enriched during their existence. It is even possible that the supernova explosions act as a trigger to allow new star formation to proceed. Thus each GMC complex will, to some extent, act like a small simple model, though most of the local enrichment will be in the products of high-mass supernovae. After enough supernovae (or whatever other disruptive processes we prefer) have occurred to disrupt the cloud complex, the remaining enriched gas will mix back into the remaining interstellar medium, and be diluted. Since most, if not all, star formation occurs in GMC complexes, the evolution (self-enrichment, then dispersal) of many such complexes during the evolution of the Galactic disk will lead to a saw-tooth progression of metallicity with age, and a large range of abundances

(a)

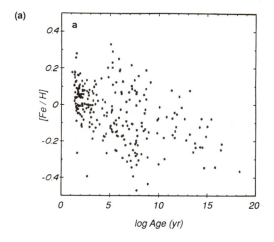

FIGURE 13.3a
Age and metallicity
data for F stars near
the Sun from the
analysis by Carlberg *et
al.* (1985). The scatter
in the data is
significantly greater
than the observational
errors, indicating real
inhomogeneity in the
interstellar medium.

(b)

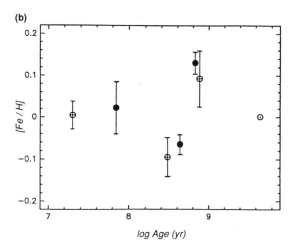

FIGURE 13.3b
The age–metallicity relation for nearby clusters and
groups, illustrating the real scatter in the abundance
distribution of the interstellar medium at any time.
These data are from Boesgaard (1989), and in order
of increasing age are for: α Per, Pleiades, UMa group,
Coma, Hyades, Praesepe, and the Sun.

of newly formed stars at any time. Another effect that will also lead to inhomogeneities in the interstellar medium is the accretion of small gas clouds by the
disk. Although an obvious effect, it is rather hard to quantify.

13.5 EVOLUTION OF CHEMICAL-ELEMENT RATIOS

Element ratios have been measured for several samples of nearby stars, both
giants and dwarfs, selected in various ways. These data are collated and discussed
by Gilmore *et al.* (1989) and by Wheeler *et al.* (1989), and are presented here
for the oxygen-to-iron ratio, which adequately illustrates the important general
features of the results. The important feature of the data for our purposes here is
that a significant change of slope occurs in the relationships between such element
ratios as [O/Fe] or [α/Fe] (where the "alpha" elements are those synthesized by
successive capture of alpha particles, such as Mg, Si, Ca, and Ti) and [Fe/H], close
to metallicities where there also occurs a change in the stellar kinematics, that
is, at [Fe/H] ~ -1 (for [O/Fe], [α/Fe]) and ~ -0.4 ([α/Fe]), although this latter
break is less well established. The [O/Fe] ratio is observed to be approximately
constant, independent of [Fe/H] for the most metal-poor stars, $-2.5 \lesssim$ [Fe/H]
$\lesssim -1$, whereas [O/Fe] declines for the more metal-rich stars, [O/Fe] $\sim -1/2$
[Fe/H]; present data sets indicate an extremely small intrinsic dispersion, all
scatter being consistent with observational error.

The appearance of Fig. 13.4 can be explained if the oxygen and iron in the
more metal-poor stars have been produced in stars of the same lifetime, whereas
for the more metal-rich stars, although the oxygen and iron continue to be produced together, an additional, longer-timescale source now dominates the iron
production. This results in the ratio [O/Fe] decreasing analogously to that of the
ratio of a primary element to a secondary element. This occurs because oxygen
is produced in massive stars only, whereas iron has a contribution from both
massive and low-mass stars (Type II and Type I supernovae, respectively), and
has an enhanced production once the much more numerous, lower-mass stars
contribute to its nucleosynthetic yield (Tinsley 1979; Wyse and Gilmore 1988).
Detailed models to quantify these statements are still problematic, because of
difficulties in calculating the abundance pattern of elements released in a supernova explosion. The basic problem is that presupernova model stars do not
explode, whereas presupernova real stars do. In the most careful available models, the canonical 25 $\mathcal{M}_\odot$ Type II precursor produces a solar enrichment pattern,
assuming that the entire outer layers of the star are ejected (Woosley and Weaver
1986), whereas the observations show a constant oxygen enhancement with respect to solar for the metal-poor stars, [O/Fe] ~ 0.5. Here we will, for simplicity,
consider only the classic Type II supernovae, due to exploding massive stars,
and Type I supernovae, due to either the coalescence of a white-dwarf/white-
dwarf binary or accretion of matter onto a white dwarf from a companion during
Roche-lobe overflow.

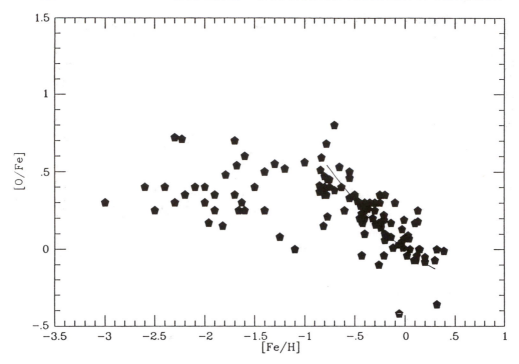

FIGURE 13.4

A compilation of oxygen-to-iron element-ratio measurements from the literature, plotted as a function of the iron-to-hydrogen ratio. This figure is adapted from Wyse and Gilmore (1988). The smooth curve through the data for [Fe/H]$\gtrsim$ −1 is the prediction of a simple model of chemical evolution with constant supernova rates in the ratio 1.5 : 1.0 for Type I : Type II, resulting in twice as much oxygen as iron being produced per unit time. The existence of a change in the slope of the [O/Fe] $vs.$ [Fe/H] relationship shows that the Galaxy was too young to have a significant number of Type I supernovae when the metallicity was below [Fe/H] $\approx$ −1.

At metallicities above ∼ −1 dex the [O/Fe] ratio declines toward the solar value. The slope of this relation is ∼ −0.5, implying that the amount of iron synthesized in association with oxygen is small compared to the amount synthesized independently of oxygen. This statement is valid even though the relevant [Fe/H] range encompasses the entire thick disk and most of the thin disk. We may utilize this constraint to calculate the relative rates of Type I and of Type II supernovae during the evolution of the disks, which in turn constrains the star-formation rate during the formation of most of the stellar mass of the Galaxy. If the IMF is constant, the star-formation rates in the thick- and thin-disk formation stages must be low enough to prevent imprinting a feature through their associated massive stars. Assuming also that single stars more massive than $8\,\mathcal{M}_\odot$ explode as Type II supernovae, and that binary systems with a minimum primary mass of $5\,\mathcal{M}_\odot$ eventually evolve into Type I supernovae (the results are not sensitive to these values), integration of the Miller and Scalo (1979) IMF predicts that the number of potential progenitors of Type I supernovae is a factor ~ 1.5 higher than the number of progenitors of Type II supernovae. The time for an individual star to become a supernova depends sensitively on the initial orbital properties and could easily be several Hubble times. The rate of Type I supernovae is therefore impossible to predict analytically. Observations of the present relative rate of Type I and Type II supernovae offer the most reliable constraint. Tammann (1982) finds that the Type I : Type II ratio is $\sim 1.5 : 1$ in a range of spiral types, Sab–Sd. Such galaxies have derived average star-formation rates that vary little over the past $\sim 6 \times 10^9$ years (Gallagher *et al.* 1984), despite differences in gas content and metallicity. Thus the present relative rates of Type I and Type II supernovae in disk galaxies may be assumed to be roughly constant over a large fraction of a Hubble time, and most potential candidates for Type I supernovae actually explode in less than a Hubble time.

Current models agree that each Type I event ejects $\sim 0.6 \, \mathcal{M}_\odot$ of iron (*e.g.*, Woosley and Weaver 1986), although there is less consensus about the yield from Type II supernovae, in part because of the uncertainties in how much of the progenitor envelope is actually ejected unchanged. The bare-core calculations of Arnett (1978) produced only $\sim 0.25 \, \mathcal{M}_\odot$ of iron for stars of main-sequence mass $20 \lesssim (\mathcal{M}/\mathcal{M}_\odot) \lesssim 30$, assuming that 60 percent of the "Si" yield is ejected as iron, and adopting his transformation between core and main-sequence masses. Woosley and Weaver (1986) find about a factor 2 more iron than does Arnett for their standard $25 \, \mathcal{M}_\odot$ model. Here we will assume that each Type I event yields a factor 2 more iron than every Type II event, or $0.6 \, \mathcal{M}_\odot$ and $0.3 \, \mathcal{M}_\odot$, respectively. The observation that current rates, averaged over a range of spiral galaxies with fairly constant star-formation rates, are $\sim 1.5 : 1$ for Type I : Type II implies that roughly 3/4 of the iron production averaged over the lifetime of the Galaxy is from Type I events. Thus adopting an oxygen yield of $0.6 \, \mathcal{M}_\odot$ for each typical Type II event (Woosley and Weaver 1986) and the iron yields above implies that $1.2 \, \mathcal{M}_\odot$ of iron and $0.6 \, \mathcal{M}_\odot$ of oxygen, or twice as much iron as oxygen, are returned to the interstellar medium per unit time. The return of fixed amounts of iron and oxygen, rather than fixed increments, of course results in the prediction of a nonlinear slope to the [O/Fe] : [Fe/H] relation, in good agreement with the observations shown in Fig. 13.4. This agreement suggests that the star-formation rate has indeed been fairly constant over the lifetime of the disk, and supports Tammann's *relative* supernova rates. It must be noted, however, that the uncertainties in the supernova yields and in the observed supernova rates are so large that no unique model exists. The relative supernova rates of van den Bergh *et al.* (1987), Type I : Type II $\sim 1 : 1.5$, adding together the Type Ia and Type Ib, would lead to approximately equal amounts of iron from each type of supernova, and equal amounts of iron and oxygen returned to the ISM per unit time, assuming naively that both Type Ia and Type Ib have the same yield of iron. The enhanced rate of Type II supernovae leads to a predicted relation between [O/Fe] and [Fe/H] that has a less-steep initial decay than that obtained using Tammann's relative rates, but that is also consistent with the observations because of the large observational scatter.

Scatter in Chemical-Element Ratios

One of the most important constraints on the homogeneity of the star-formation in the halo, and hence chemical evolution, is readily derivable from the existence of a break in element ratios as a function of [Fe/H]. The inferred lack of cosmic scatter implies little spatial variation in the star-formation history of the proto-Galaxy, which is obviously of considerable importance in determining the levels of independent substructures in the gas cloud which became the Galaxy. Models such as those of Searle (1977; Searle and Zinn 1978) whereby the halo is a result of the merging of many independent "fragments" each with its own chemical-enrichment history, seek to explain the entire range of metallicity seen in halo stars as due to the statistical effects of chemical inhomogeneities, rather than as a trend with age. The fragments are assumed to be disrupted by some process after some time, producing the field stars of the halo (see Fall and Rees 1977). Such models also predict a large spread in the element ratios observed from field star to field star if, for any one of the fragments, each of which is presumed to contain many globular-cluster-sized objects, star formation continued long enough that Type I supernovae became a significant source of iron.

The behavior of element ratios in a fragment model depends on the star-formation rate and the stellar initial mass function (IMF). The arguments above show that Type I supernovae will contribute to the iron-production rate in any system at some fixed time after the onset of star formation. Assuming the IMF to be constant, we see that fragments of higher star-formation rate will have attained a higher [Fe/H] prior to that fixed time, so that the break in the [O/Fe]–[Fe/H] relation will occur at higher [Fe/H]. Similarly, fragments of substantially lower star-formation rate would lead to a break at much lower [Fe/H]. Thus samples of stars which originated in many different fragments which were subsequently disrupted would not show a well-defined break at one given metallicity, contrary to the observations. The small scatter seen in the element-ratio relations, together with the excellent description of the halo field-star distribution by the simple model with a single value for the effective yield, then requires that all fragments had evolved to nearly the same gas fraction at the time when they were disrupted, which in all cases was $\lesssim 10^9$ yr after significant star formation began. Such an effect is implicit in the models of Hartwick (1976) and Gilmore and Wyse (1986), whereby the metallicity distribution of the extreme halo is a consequence of gas being lost from the star-forming process at a rate proportional to, and greater than, the star-formation rate. The required rate of gas removal is a factor ten to fifteen higher than the star-formation rate, so that it is possible for each fragment to evolve to completion, by losing all its gas, in a time shorter than that required by the break in the [O/Fe]–[Fe/H] relation, or $\lesssim 10^9$ yr.

Fragment models require also that star-formation began over all of that part of the proto-Galaxy which became the extreme halo in the solar neighborhood at nearly the same time. This requirement arises since, if some fragments lagged behind and began their chemical evolution substantially later than most, their oxygen-rich ejecta would have enriched the gas which was to form the younger disk populations, contrary to the observations. A further constraint on the spatial variations of the star-formation process in the halo comes from the fact that metals may be transported at no more than the local sound speed, whereas the free-fall velocity may well be supersonic (Fall 1987). This would lead to large local chemical inhomogeneities if the star-formation rate varied rapidly from place to place, and if the local sound speed were low. The sound speed will be low if the material remains near 10^4 K. However, if the gas has been heated to the galactic virial temperature, then the sound speed equals the free-fall velocity (by definition), so that efficient mixing is not constrained by the distance metal-enhanced ejecta can travel in a free-fall time. In general it seems more plausible to identify any fragments with short-lived condensations which become the sites of star formation in a gaseous background, rather than as discrete structures merging to form the proto-Galaxy. In general the existence of structure in the element-ratio relations is strong evidence that there was a continual increase with time of the heavy-element abundance of the Galaxy, at least for stars of lower metal abundance than that at which the break is apparent, and is difficult to reconcile with stochastic chemical-evolution models of the halo.

We may quantify the maximum allowed variations in the star-formation rate by making the conservative assumption that the (2-sigma) scatter about the mean trend in [O/Fe] against [Fe/H], of a few tenths of a dex or about a factor of 2, reflects real cosmic scatter rather than measurement uncertainties, even though all the observed scatter is consistent with measurement uncertainties. The number of Type I supernovae at a given time is determined by the star-formation rate some 1 to a few Gyr previously, but that of Type II supernovae is determined by the star-formation rate at that time. The scatter would then be caused by short-lived increases in star-formation rate, leading to enhanced rates of Type II supernovae. Scatter of a factor of 2 implies that the Type II formation rate be increased by a factor of $\lesssim 4$ for a 2-sigma effect. Hence the mean star-formation rate in the Galaxy cannot have increased by as much as a factor of 4 since the Galaxy attained a metallicity of -1 dex.

The average star-formation rate during the formation of the low-mass, metal-poor halo was $\sim 10 \, \mathcal{M}_\odot \, \mathrm{yr}^{-1}$, of order a typical present-day disk star-formation rate. Thus disk galaxies, by far the most common nondwarf galaxies in the Universe, do not go through a sustained, highly luminous burst of star formation in their earliest stages, and are not good candidates for detection in primeval galaxy searches.

The apparent lack of cosmic scatter in the [O/Fe] *vs.* [Fe/H] data shown in Fig. 13.4 leads to a paradox when compared to the [Fe/H] *vs.* age data shown in Figures 13.3a and 13.3b, which show a substantial scatter. The upper limit on the element-ratio scatter from Fig. 13.4 is a few tenths of a dex, and the real scatter in Figures 13.3a and 13.3b is similarly a few tenths dex. The reconciliation of the paradox is simply that Figures 13.3a and 13.3b show the iron abundance as a function of time, but Figure 13.4 shows the oxygen-to-iron ratio as a function of iron abundance. Thus, even if the intrinsic dispersion in the age–iron relationship is entirely caused by an excess of Type II supernovae, as in the discussion above, with a consequent excess production of oxygen at that time, the slope of the oxygen-to-iron production ratio of ~ 2 will project out the scatter in the oxygen–age relation when the data are plotted as in Fig. 13.4.

Kinematics and Chemical-Element Ratios

If the preceding discussion contained the whole story of the history of star formation in the Galaxy, then it would have to be mere coincidence that the change in the predominant production mechanism of iron occurred close to a metallicity, or epoch, at which the stellar kinematics change from those of a pressure-supported system to those of an angular-momentum-supported system. Rather, the coincidence of the value of [Fe/H] at which the Galaxy changed from a pressure-supported system to an angular-momentum-supported system with the value of [Fe/H] at which the interstellar medium became diluted by the products of long-lived stars provides a diagnostic of the relative star-formation and dissipation rates in the proto-Galaxy.

Because the elemental yields are a fairly slow function of progenitor mass, and hence lifetime, we do not expect discontinuities in element ratios to occur in a situation where the star formation proceeds at a reasonably constant rate. A discontinuity in kinematic properties implies that the ratio of the dissipation rate to the star-formation rate changes rapidly. A possible explanation is that at metallicities [Fe/H] $\gtrsim -1.5$, the efficiency with which a gas cloud cools from $\sim 10^6$ K (a typical galactic virial temperature) increases markedly, because of a transition of the dominant cooling mechanism from free-free radiation, independent of metallicity, to line radiation, proportional to the number density of metals. Thus a rapid increase in the dissipation rate, and hence collapse to a disk-like angular-momentum-supported structure, is not implausible at a metallicity of ~ -1 dex. It is not crucial for these arguments that the breaks in kinematics and element ratios occur at *exactly* the same metallicity.

When sufficient data and reliable massive-star evolutionary models all the way through the supernova explosion, with corresponding elemental yields, become available, it will be possible to quantify these arguments, subject to the assumption of a constant stellar IMF, and provide a real timescale in years for the duration of that period of proto-Galactic evolution which was dominated by collapse (possibly nondissipational) on a dynamical timescale, and that period when angular-momentum transport (in dissipational collapse) became an important physical process, and when angular-momentum support thus became the dominant dynamical process.

13.6   SPECIFIC MODELS OF MILKY WAY GALAXY FORMATION

The most widely referenced model of the formation of our Galaxy is that of Eggen *et al.* (1962, henceforth ELS), which was developed primarily to understand their observations of the kinematics and chemical abundances of stars near the Sun. This model requires that the stellar halo formed during a period of rapid collapse of the entire proto-Galaxy, subsequent to which the remaining gas quickly dissipated into a metal-enriched cold disk, in which star formation has continued until the present. The ELS model was designed to provide conditions under which the oldest stars populated radially anisotropic orbits, but stars that formed later had increasingly circular orbits, in accord with their data, which implied that the most metal-poor stars, assumed to be the oldest stars, were on more eccentric, lower-angular-momentum orbits than the more metal-rich stars. This model is based on two crucial assumptions. First, that a pressure-supported, primarily gaseous galaxy (where $T = T_{\rm virial}$) is stable against star formation; here the *global* cooling time is the shortest timescale of interest, and thermal instabilities of the type invoked by Fall and Rees (1985) and discussed briefly above must be suppressed. Second, that stellar orbits cannot be modified to become more radial after formation of the stars.

If the first assumption were valid, the observed high-velocity stars must have formed from gas clouds which were not in equilibrium in a pressure-supported system. If the second assumption above were valid, these clouds formed stars while on radial orbits at large distances from the Galactic center. Thus in this picture, these clouds must have turned around from the background universal expansion, and been collapsing toward the center of the potential well. Hence the oldest stars of the Galactic halo must have formed as the proto-Galaxy coalesced. To estimate the *rate* of the collapse, ELS analyzed the evolution of the radial anisotropy of a stellar (or gas cloud) orbit as the Galactic potential changed, and showed it to be approximately conserved during a slow collapse but to become more radially anisotropic in a fast collapse. They argued against a slow collapse on the grounds that such a collapse requires tangentially biased velocities (remembering that pressure support has been excluded by assumption), and this tangential bias will be unaffected by the resulting slow changes of the gravitational potential. The observed radial anisotropy of the stellar orbits then implies an initially radially biased velocity ellipsoid, and the calculations of ELS show that such a velocity ellipsoid will have become more radially anisotropic during collapse. Hence ELS deduced that the gas clouds were in free-fall radial orbits, and that the consequent collapse must have been rapid, meaning comparable to an orbital or a dynamical timescale, which is a few times 10^8 years. It should be noted that Isobe (1974) came to the opposite conclusion from his analysis of the ELS data, and favored a slow collapse, but Yoshii and Saio (1979) augmented the ELS sample and also concluded that the halo collapsed over many dynamical times. We return to the difficulty of inferring a timescale below.

Clearly if either of ELS's assumptions were violated, there need be no correlation between the *time* of a star's formation, which they infer from a star's metallicity, and its *present* orbital properties. If their first assumption were violated, then stars formed everywhere at all times would inevitably be on highly radial orbits, because the star has too small a surface area to be pressure-supported by the gas. As we mentioned above, assumptions about star formation in pressure-supported systems must be treated as *ad hoc* until we understand better the physics of star formation, so that conclusions based on such assumptions are at best uncertain. If their second assumption were violated, then the stars which are now the high-velocity stars near the Sun could have originated from more-circular orbits interior to the Sun, and have present orbital properties which depend only on dynamical processes subsequent to their formation. The realization that a forming galaxy undergoes changes in its gravitational potential which are of the same order as the potential itself (violent relaxation) means that stellar orbits can be modified considerably. Recent *N*-body models (see, *e.g.*, May and van Albada 1984, and McGlynn 1984, for excellent descriptions of representative experiments) for systems in which dissipation does not play a major role show that the final state, of the collapsed system depends on both the degree of homogeneity and the temperature of the initial state. As seen in the cosmological *N*-body simulations discussed above, clumps cause angular-momentum and energy transport. Violent relaxation never goes to completion, so that final and initial orbital binding energies and angular momenta are correlated, with the interior regions becoming more centrally concentrated and the outer regions being puffed up. The typical final steady-state velocity distribution is highly anisotropic exterior to (roughly) the half-light radius, and more isotropic interior to that radius (if violent relaxation were 100 percent efficient, all systems would reach the same final state, with isotropic velocity distribution). In the Galaxy, the halo half-light radius is ~ 3 kpc, well interior to the Sun's orbit. Thus the expected velocity distribution of old stars near the Sun after virialization of the halo is anisotropic, as observed by ELS, even though the dynamical evolution of the system is not as they envisage, and a correlation between kinematics and age is no longer an inevitable conclusion. We might, for example, imagine a situation where later (rapid) collapse of either the disk or the dark halo, or the merger of a few large substructures, could lead to rapid dynamical evolution of a central halo component which had previously formed on a longer timescale. Models of this type have yet to be studied in detail.

13.7   STELLAR KINEMATICS AND GALAXY FORMATION[2]

The kinematic properties of stars in the Galaxy are related, through the gravitational potential ψ, to their spatial distribution. The scale length of the spatial distribution is determined by the total energy (kinetic and potential) of the stellar orbits, as well as by the gradient of the potential (*i.e.*, the force on the star). The shape of the spatial distribution depends on the relative populations of the orbits supported by the potential, and on the relative amounts of angular-momentum (rotational) and pressure (stellar velocity anisotropy) balance to the

---

[2]The broad discussions of this section draw heavily on material already presented in Chapters 2 and 11. It has therefore seemed appropriate to repeat some of that material (usually in condensed form) here.

potential gradients. The total orbital energy and angular momentum of the gas which will become a star depends on the maximum distance from the center of the Galaxy which it reached before falling out of the background expansion of the Universe, the angular momentum of its orbit at that time, the depth of the potential well through which it fell, the fraction of the total orbital energy which was lost (dissipated) before the gas formed into a star, and the subsequent dynamical evolution of the stellar orbit. That is, the present kinematic properties of old stars in the solar neighborhood are determined in part by initial conditions in the proto-Galaxy at the time of the first star formation, and in part by physical processes during galaxy formation. Hence, local kinematic studies can help us understand the detailed physics of galaxy formation by relating local kinematics to large-scale structural properties of the Galaxy.

Stellar chemical abundance is determined by the fraction of the available interstellar medium (ISM) at the time and place of the star's formation that had been processed through the nuclear-burning regions of massive stars. It provides a valuable chronometer for the early evolution of the Galaxy. The chemical abundance of the ISM at any time depends on the local history of formation and evolution of stars sufficiently massive to have created new chemical elements, and the mixing of local gas with more-distant material. This more-distant gas may or may not itself be enriched, so that the time-dependence of the chemical abundance of newly forming stars depends on both the local and the global star-formation rates, the rate of infall of primordial gas, and the efficacy of mixing in the ISM. Thus, although the chemical abundance of newly formed stars is a valuable timepiece, this chronometer need not be a smooth or even a single-valued function of chronological time. Clearly, however, the distribution function of stellar kinematics, chemistry, and age contains a wealth of information on the distribution of proto-Galactic gas, the dissipational and star-formation history of that gas, the subsequent dynamical history of the resulting stars, and the Galactic gravitational potential.

Correlations Between Kinematics and Chemistry

Stellar chemical abundance is a clock which measures age in units defined by the formation rate and lifetimes of those (mostly massive) stars which contribute to the enrichment of the ISM. The mean azimuthal streaming motion of a population of stars reflects the amount by which the proto-Galaxy dissipated and collapsed, and the amount of angular-momentum transport that occurred prior to the formation of these stars. Thus the existence of a correlation between kinematics, such as V_{rot}, and [Fe/H] allows us to relate the chemical evolutionary timescale to the dynamical, cooling, and viscous transport timescales.

The existence or otherwise of an abundance gradient in the halo is often cited as an important diagnostic of the timescale of galaxy formation, in the sense that the lack of a gradient is taken to mean a slow and/or chaotic collapse (see Searle and Zinn 1978; Carney et $al.$ 1990), and the presence of a gradient, manifested by a smooth correlation between kinematics and metallicity, to mean a rapid collapse (Sandage 1987). In general this is not correct. The presence of an abundance gradient means that dissipation was an important process during formation of the stellar component of the halo, but does not necessarily define timescales. In a dissipationless collapse there is no arrow of time in the kinematics, and so no correlation between kinematics and metallicity. In a dissipative collapse, however, a star-forming gas cloud will be continually transferred onto lower-energy orbits with time, and will be successively enriched with time because of the continuing star formation, creating a correlation between chemical abundance and orbital energy, which is an abundance gradient. Only low-metallicity stars will be found in the outer regions of galaxies that formed dissipatively, and high-metallicity stars will be found only in the inner regions.

However, since the cooling time of a proto-galaxy may be expected to be less than the free-fall collapse time, dissipation need not slow a collapse significantly. In a rapid (free-fall) dissipationless collapse, no abundance gradient will arise, although one present in the initial conditions may survive, since violent relaxation in practice never goes to completion. A slow dissipationless formation of the halo could occur if the field stars originated in many independent "fragments" that were captured and disrupted over a Hubble time (see Searle and Zinn 1978), each fragment having its own star-formation history. This last model would also not lead to an abundance gradient, due to the assumed randomness of the "fragments." However, consideration of the element ratios of halo stars, as discussed above, poses stringent constraints on such a model for formation of the field stars. It may remain viable for the formation of the outer parts of the globular-cluster system, as originally motivated.

Correlations (or the lack thereof) between the angular momentum of a tracer population of stars and the stars' chemical abundances, however, remain one of the most powerful observationally feasible tests of the early star-formation and dynamical history of the Galaxy.

The amplitude of the peculiar velocity of a star also may be expected to be correlated with that star's metallicity. Since this peculiar velocity is now manifest in the star's orbital eccentricity, the relationship between the eccentricity of stellar orbits, projected onto the plane of the Galaxy, and chemical abundance is also commonly discussed. The general arguments above may be applied in support of its use. In addition to the uncertainties arising from the potential effects of violent relaxation, an extra note of caution is required in its interpretation, however. Unless we were confident that the levels of substructure in the proto-Galaxy, and the consequent interactions (more precisely, the hydrodynamics of the dissipational and heating processes during Galactic collapse, and the mixing of the hot stellar ejecta into the infalling gas) were clearly understood, then we should not be entirely confident of the evolution of the *random* motions of newly forming stars.

One important conclusion from the preceding discussion which is not widely appreciated is that the absence of a clear correlation between rotation velocity and metallicity at very low abundances does *not* mean that a rapid-collapse model of the Galaxy is invalid. Thus the conclusions reached by Sandage and Fouts (1987) are not incompatible with the most precise modern data, which show that there is no correlation of rotation velocity with stellar metallicity, even though Sandage and Fouts's description of their data is not consistent with the absence of such a correlation. We emphasize that the presence or absence of a correlation between rotation velocity and stellar metallicity does *not* distinguish between fast- and slow-collapse models of the Galaxy. We may learn something about a complex combination of the importance of dissipation and the efficiency of violent relaxation, but timescales enter the interpretation of the relationship between rotation velocity and stellar metallicity only by assumption.

Orbital Eccentricity *vs.* Metallicity

There is an alternative presentation of the relationship between stellar orbital properties and chemical abundance which is widely discussed. It involves the eccentricity of the stellar orbit in the Galaxy (projected onto the plane) and the star's metallicity. This presentation really depends on the ratio between the components of the star's velocity (radially along and tangentially to the line toward the Galactic center) and the stellar metallicity. The derivation of an orbital eccentricity requires the additional assumption of a Galactic potential, adding further uncertainty. Nevertheless, the existence of a correlation between metallicity and the shape of the stellar orbit (projected onto the plane of the Galaxy) has been cited as the principal evidence leading ELS to support a rapid-collapse model of the Galaxy (Sandage 1987). Similarly, observational evidence that a tight correlation is violated by a significant fraction ($\sim$ 20 percent) of field metal-poor stars has been cited as evidence that a slow collapse is a preferred model (Yoshii and Saio 1979; Norris *et al.* 1985). This correlation is therefore worthy of explicit discussion.

The observational status of such a correlation is a little confused at present. The large kinematically selected samples studied recently by Sandage and Fouts (1987), Laird *et al.* (1988), and Norris and Ryan (1989) are naturally biased against stars with nearly circular orbits, since such stars will, because of the trade-off between the various terms in the asymmetric-drift equation, tend to have smaller peculiar space velocities than will stars on lower-angular-momentum orbits with the same total energy. Thus careful modeling of the available data will be required to test that any correlation between orbital planar eccentricity and stellar chemical abundance is not due to a selection bias. Low-abundance stars are intrinsically sufficiently rare, and those with both sufficiently large peculiar velocities to enter a kinematically selected sample and the correct angular momentum to reach the solar neighborhood occupy such a tiny fraction of the whole of phase space that we would not expect to find them readily in any case. Low-metallicity stars on nearly circular orbits may well exist in very large numbers nearer the center of the Galaxy, irrespective of the rate of star formation during collapse of the Galaxy.

We emphasize that what is really at issue is the existence or otherwise of a significant number of stars with low metallicities, high angular momenta, and small peculiar velocities. These stars are further discussed elsewhere under the guise of the "metal-poor thick disk." In the ELS rapid-collapse picture-stars acquire large systematic rotation velocities (random velocities are ascribed to initial conditions) because of the spin-up of the proto-Galaxy as it collapses with conservation of angular momentum. Thus mean rotational velocity is a direct measure of the collapse factor of the star-forming gas (if viscosity is unimportant), and is therefore a clock for the collapse, in the same way that stellar abundance is a clock for star formation. Since the natural timescale for an increase in rotational velocity because of Galactic collapse is a free-fall time, ELS deduced from their observed orbital-shape–abundance correlation that star formation also occurred on a free-fall timescale. The basic rapid-collapse model of ELS assumes that star formation, chemical enrichment, and collapse proceed at the same time, so that, in their model, we would not expect to find a substantial number of stars which are metal-poor (*i.e.*, among the first stars formed according to the chemical clock) but on roughly circular orbits (*i.e.*, among the later stars formed according to the collapse clock). The ELS model would allow a substantial number of metal-poor stars with both high angular momentum and high orbital energy, as they emphasized. Such stars would, however, still be on eccentric orbits (see ELS's discussion of their Figure 9 for this point).

Since there exist some stars with low metallicity on nearly circular orbits (Norris *et al.* 1985; Norris 1987), does this rule out the rapid-collapse model, and does it support a slower-collapse alternative? To answer these questions, we must rediscuss the important assumptions behind the argument. For our purposes here, these assumptions are: that the collapse was synchronized, so that metal enrichment follows the collapse factor locally (equivalently, the collapse clock and the enrichment clock are in phase everywhere, with a similar timescale for the two processes; this is necessary, since no correlation would be seen if the two timescales were very different); that the Galactic gravitational potential changed slowly at all times; and that the concept of a "gas cloud" retaining kinematic parameters during the collapse-and-enrichment process up to formation of a low-mass star is reasonable.

Since the correlation of orbital eccentricity and metallicity discussed by ELS is apparently violated by many stars, a variety of possibilities arises. Searle and Zinn (1978) developed a model in which the synchronization of the collapse-and-enrichment clocks was put out of phase, by allowing the existence of substantial substructure, which survived as gravitationally bound entities for longer than a Galactic collapse time. In effect their model is a sum of a large number of small ELS-like collapses, but with a different *rate* of enrichment (equivalently, star formation) in each element. The most useful constraints on models of this type come from the cooling-time arguments discussed earlier in this chapter, and from study of the relative enrichment rates of different chemical elements, which provides a higher-temporal-resolution clock. These constraints are also discussed further above. An alternative possibility has been suggested by Norris (1987), that the timescale for the Galactic collapse was much longer than a dynamical timescale, though it is not clear how this model can be consistent with the data either; the Larson (1976) model that Norris uses as an example predicts smooth, steep, vertical metallicity gradients. Merely changing the absolute timescale (in years) of both the collapse and the star-formation rates will, of course, have no effect on the predictions of the ELS model if both rates are changed proportionately, since neither has been formulated in years. Although hydrodynamical support of the gas against collapse might well act to damp out random motions in the gas clouds during a slow collapse, this possibility on its own is not consistent either with the observations or with the cooling timescale arguments. Clearly, changing the relative rates of the collapse and enrichment will change the *slope* of any resulting correlation, but it will not change the scatter. It is increased scatter which must be explained. Such scatter can be explained, under the assumptions that chemical abundance is a monotonic clock and that violent relaxation has not erased the fossil record, by inhomogeneity in either the chemical enrichment rate or the gaseous-collapse factor (Searle and Zinn 1978), or by a wide diversity in the distribution of initial (precollapse) angular momenta among gas clouds, but not merely by changing the rapidity of the proto-Galactic collapse. The absence of a tight correlation between orbital eccentricity and metallicity does not support a slow-collapse model of the Galaxy, and in fact argues against a slow, dissipative collapse.

Perhaps the most important feature of the eccentricity–metallicity relation is that stars on highly radial orbits exist at all. A star can now be on a radial orbit either simply because it has dropped out of the quasistatic collapse of a pressure-supported gaseous Galaxy, or because it formed from gas which was on a highly radial orbit (and which would be undergoing rapid collapse, because of the cooling-time arguments), or because it formed on some other orbit which was subsequently made more radial by violent relaxation. The first alternative, which was not considered feasible in principle by ELS, is in fact ruled out by the lack of an abundance gradient in the metal-poor halo, as discussed above. The remaining two mechanisms involve rapid collapse of something, either the halo itself, or the dominant mass of the Galaxy (which is not the halo). Thus the combination of kinematic and chemical data implies that the existence of a significant number of subdwarf stars on radial orbits requires a rapid collapse of the halo and/or the dominant contribution to the mass of the central few kiloparsecs of the Galaxy.

Similarly, in a rapid-collapse model we also expect any *tight* correlation which might have existed between stellar kinematic parameters and time-independent internal properties of the star (*e.g.*, chemical abundance) to be smeared out by the violent relaxation. That is, contrary to the expectation of ELS, who (it must be emphasized) were working before the discovery of the concept of violent relaxation, we do *not* expect a strong correlation of orbital eccentricity and metallicity in a rapid-collapse model. The very existence of a large fraction of halo stars on highly radial orbits is strong evidence for rapid collapse of most of the mass of the Galaxy, but does not, of course, provide any evidence about whether the formation of most halo stars was completed before, or took place during the collapse of the dominant contribution to the mass of the Galaxy.

13.8   THE TIMESCALE OF GALAXY FORMATION

In view of the convoluted nature of the preceding discussion, we provide here a brief summary of the observational constraints on the timescale on which the Extreme Population II stars in the Galaxy formed. The most important observational constraint is provided by the approximate constancy of the oxygen-to-iron element-abundance ratio as a function of iron abundance for metal-poor stars. These data show the stars in the Galaxy with metallicities less than ~ -1 dex to have formed on a timescale of $\lesssim 1$ Gyr. Analysis of the asymmetric-drift data (Chapter 11) shows that the most metal-poor stars formed during a period of dissipational collapse of unconstrained duration. Study of the correlations between stellar kinematics and stellar metallicities (Chapter 11), together with analysis of the oxygen-to-iron element ratios in metal-poor stars, shows that the stars which formed during the dissipational collapse all formed within 1 Gyr of the onset of significant star formation. It is not yet clear if the apparent absence of evidence for dissipation during formation of the more metal-rich halo stars (Chapter 11) indicates that the dissipational collapse had effectively ceased, or at least become dissipationless, during their formation. It may well be simply that a subsequent period of violent relaxation of the Galactic potential well has disturbed the fossil record of the state of collapse of the proto-Galaxy during the formation of the more metal-rich Extreme Population II stars. Since a characteristic timescale for dynamical evolution is ~ 0.5 Gyr for halo stars, and two to three times that for the infalling disk material, we may deduce that the period of star formation leading to the present Extreme Population II stars occurred during a dissipational collapse which lasted for no more than a few dynamical times. Hence the Galaxy formed its first generations of stars during a period of "rapid" collapse. Further constraints on the homogeneity of the proto-Galaxy during this collapse are discussed on pp. 299–304.

13.9   FORMATION OF THE THICK DISK

The luminosity profile of the Milky Way, as calculated from star counts, the vertical distribution of chemical abundances, and the vertical distribution of stellar kinematics all provide incontrovertible evidence for a Galactic thick disk, as was discussed in Chapter 2. The descriptive parameters relevant to this stellar component are: a vertical scale height of ~ 1–1.5 kpc; a normalization near the Sun of a few (1–5) percent of the number of thin-disk stars; a vertical velocity dispersion of ~ 45 km sec^{-1}; a typical stellar chemical abundance of $\sim 1/4$ of the solar metallicity; and a mean asymmetric drift of ~ 30–50 km sec^{-1}. We do not have detailed values of the descriptive parameters, however, primarily because the offset in the mean values characterizing the thick-disk distribution function over age, metallicity, and kinematics from those mean values characterizing the oldest thin-disk stars is much less than the dispersions in these quantities. Does it matter what the thick disk is like? Reliable evaluation of the parameters of the distribution function is worth while, since it may allow a discrimination between the several currently viable models of the formation of the thick disk. Possible formation mechanisms for the thick disk include:

1. A slow, pressure-supported collapse phase following rapid formation of the Extreme Population II system, similar to the sequence of events, though not their relative duration, in Larson's (1976) hydrodynamical models of disk-galaxy formation (Gilmore 1984).

2. Violent dynamical heating of the early thin disk by satellite accretion (see Gilmore *et al.* 1985; Hernquist and Quinn 1989), or by violent relaxation of the Galactic potential (Jones and Wyse 1983).

3. Accretion of the thick-disk material directly, by, *e.g.*, satellite accretion with preferential population of suitable orbits (Statler 1989).

4. An extended period of enhanced kinematic diffusion of stars formed in the thin disk to high-energy orbits (Norris 1987).

5. A rapid increase in the dissipation and star-formation rates because of more-rapid cooling once the metallicity is above ~ -1 dex (see Wyse and Gilmore 1988).

An alternative model, that the thick disk is the response of the halo stars to the potential of a massive disk (Gilmore and Reid 1983), can now be excluded, since the surface mass density of the disk has recently been calculated, and is too low to dominate the Galactic potential well (Kuijken and Gilmore 1989a,b; and Chapter 8).

Discrimination between these several types of models is possible from appropriate age, metallicity, and kinematic data. The first type of model in the list will lead to an intermediate-age system, with an abundance gradient. The second will have a small internal age range, but is unlikely to have an extant abundance gradient (*modulo* the details of the dynamical evolution). The third has a wide variety of allowed combinations of age and abundance, though it is difficult to reconcile with available understanding of Galactic chemical evolution. The fourth will have a range of ages and a similar chemical abundance to the oldest thin disk, but a kinematic discontinuity between the old disk and the thick disk. The last model predicts that the thick disk will be kinematically distinct from the metal-poor halo, and have metallicity $\gtrsim -1$ dex. Because being able to settle on the correct model may allow us to deduce the evolutionary history of the Galaxy, much work is now going into establishing a reliable description of the kinematic, abundance, and age structure of the thick disk (see Sandage 1987, Norris 1987, and Chapter 11). Available data marginally favor a model in which the thick disk is a kinematically discrete component of the Galaxy, which would favor a formation process involving one or a few short-lived events, but the issue remains to be decided by observational test.

REFERENCES

Arnett, W. D. 1978. *Astrophys. J.*, **219**, 1008.

Barnes, J., and Efstathiou, G. 1987.. *Astrophys. J.*, **319**, 575.

Bergh, S. van den. 1962. *Astron. J.*, **67**, 48ʋ

Bergh, S. van den, McClure, R. D., and Evans, R. 1987. *Astrophys. J.*, **323**, 44.

Binney, J. 1977. *Astrophys. J.*, **215**, 483.

Boesgaard, A. 1989. *Astrophys. J.*, **336**, 798.

Bothun, G. D., Impey, C. D., Malin, D. F., and Mould, J. R. 1987. *Astron. J.*, **94**, 23.

Blumenthal, G. R., Faber, S. M., Primack, J. R., and Rees, M. J. 1984. *Nature*, **311**, 517.

Brück, H. A., Coyne, G. V., and Longair, M. S., eds. 1982. *Astrophysical Cosmology*. Vatican City: Pontifica Academia Scientiarum.

Carlberg, R. G. 1985. In van Woerden *et al.*, p. 615.

Carlberg, R .G., Dawson, P., Hsu, T., and VandenBerg, D. A. 1985. *Astrophys. J.*, **294**, 674.

Carney, B., Aguilar, L., Latham, D. W., and Laird, J. B. 1990. *Astron. J.*, **99**, 201.

Disney, M. 1976. *Nature*, **263**, 573.

Eddington, A. S. 1928. *Mon. Not. Roy. Astron. Soc.*, **88**, 331.

Eggen, O. J. 1983. *Astron. J.*, **88**, 813.

Eggen, O. J., Lynden-Bell, D., and Sandage, A. (ELS). 1962. *Astrophys. J.*, **136**, 748.

Ellis, R., Frenk, C., and Peacock, J., eds. 1989. *The Epoch of Galaxy Formation*. Dordrecht: Reidel.

Fall, S. M. 1987. In Kron and Renzini, p. 15.

Fall, S. M., and Efstathiou, G. 1980. *Mon. Not. Roy. Astron. Soc.*, **193**, 189..

Fall, S. M., and Rees, M. J. 1977. *Mon. Not. Roy. Astron. Soc.*, **181**, 37P.

———. 1985. *Astrophys. J.*, **298**, 18.

Frenk, C. S., White, S. D. M., Efstathiou, G., and Davis, M. 1985. *Nature*, **317**, 595.

Frenk, C. S., White, S. D. M., Davis, M., and Efstathiou, G. 1988. *Astrophys. J.*, **327**, 507.

Gallagher, J. S., Hunter, D. A., and Tutukov, A. V. 1984. *Astrophys. J.*, **284**, 544.

Gilmore, G. 1984. *Mon. Not. Roy. Astron. Soc.*, **207**, 223.

Gilmore, G., and Carswell, B., eds. 1987. *The Galaxy*. Dordrecht: Reidel.

Gilmore, G., and Reid, I. N., 1983. *Mon. Not. Roy. Astron. Soc.*, **202**, 1025.

Gilmore, G., Reid, I. N., and Hewett, P. C. 1985. *Mon. Not. Roy. Astron. Soc.*, **213**, 257.

Gilmore, G., and Roberts, M. S. 1988. *Comments on Astrophys.*, **12**, 123.

Gilmore, G., and Wyse, R. F. G. 1986. *Nature*, **322**, 806.

Gilmore, G., Wyse, R. F. G., and Kuijken, K. 1989. *Ann. Rev. Astron. Astrophys.*, **27**, 555.

Gunn, J. E. 1982. In Brück *et al.*, p. 233.

Gunn, J. E., and Gott, J. R. 1972. *Astrophys. J.*, **176**, 1.

Hartwick, F. D. A. 1976. *Astrophys. J.*, **209**, 418.

Hartwick, F. D. A., Cowley, A., and Mould, J. 1984. *Astrophys. J.*, **286**, 269.

Hernquist, L., and Quinn, P. J. 1990. *Astrophys. J.*, in press.

Isobe, S. 1974. *Astron. Astrophys.*, **36**, 333.

Jones, B. J. T., and Wyse, R. F. G. 1983. *Astron. Astrophys.*, **120**, 165.

Kron, R., and Renzini, A., eds. 1987. *Towards Understanding Galaxies at High Redshift.* Dordrecht: Reidel.

Kuijken, K., and Gilmore, G. 1989a. *Mon. Not. Roy. Astron. Soc.*, **239**, 571 (Paper I).

———. 1989b. *Mon. Not. Roy. Astron. Soc.*, **239**, 605 (Paper II).

Lacey, C., and Fall, M. 1985. *Astrophys. J.*, **290**, 154.

Laird, J. B., Carney. B. W., and Latham, D. W. 1988. *Astron. J.*, **95**, 1843.

Laird, J. B., Rupen, M. P., Carney. B. W., and Latham, D. W. 1988. *Astron. J.*, **96**, 1908.

Larson, R. B. 1976. *Mon. Not. Roy. Astron. Soc.*, **176**, 31.

Larson, R. B., Tinsley, B. M., and Caldwell, C. N. 1980. *Astrophys. J.*, **237**, 692.

Lynden-Bell, D. 1967. In van Woerden, p. 257.

———. 1975. *Vistas in Astron.*, **19**, 299.

May, A., and van Albada, T. J. 1984. *Mon. Not. Roy. Astron. Soc.*, **209**, 15.

McClure, R. D., VandenBerg, D. A., Smith, G. H., Fahlman, G. G., Richer, H. B., Hesser, J. E., Harris, W. E., Stetson, P. B., and Bell, R. A. 1986. *Astrophys. J.*, **307**, L49.

McGlynn, T. 1984. *Astrophys. J.*, **281**, 13.

Miller, G. E., and Scalo, J. M. 1979. *Astrophys. J. Supp.*, **41**, 513.

Norris, J. 1987. In Gilmore and Carswell, p. 297.

Norris, J., Bessell, M. S., and Pickles, A. J. 1985. *Astrophys. J. Supp.*, **58**, 463.

Norris, J., and Ryan, S. G. 1989. *Astrophys. J. (Letters)*, **336**, L17.

Pagel, B. E. J., and Patchett, B. E. 1975. *Mon. Not. Roy. Astron. Soc.*, **172**, 13.

Rees, M. J., and Ostriker, J. P. 1977. *Mon. Not. Roy. Astron. Soc.*, **179**, 541.

Rees, M. J., and Stoneham, R. J., eds. 1982. *Supernovae: A Survey of Current Research.* Cambridge, Eng.: Cambridge Univ. Press.

Reid, I. N., and Gilmore, G. 1982. *Mon. Not. Roy. Astron. Soc.*, **201**, 73.

Ryden, B. S. 1988. *Astrophys. J.*, **329**, 589.

Sandage, A. 1987. In Gilmore and Carswell, p. 321.

Sandage, A., and Fouts, G. 1987. *Astron. J.*, **92**, 74.

Searle, L. 1977. In Tinsley and Larson, p. 219

Searle, L., and Sargent, W. 1972. *Astrophys. J.*, **173**, 25.

Searle, L., and Zinn R. 1978. *Astrophys. J.*, **225**, 357.

Silk, J. 1977. *Astrophys. J.*, **211**, 638.

———. 1983. *Nature*, **301**, 574.

Statler, T. S. 1989. *Astrophys. J.*, **344**, 217.

Stromberg, G. 1934. *Astrophys. J.*, **79**, 460.

Strömgren, B. 1987. In Gilmore and Carswell, p. 229.

Tammann, G. 1982. In Rees and Stoneham, p. 371.

Tinsley, B. M. 1979. *Astrophys. J.*, **229**, 1046.

———. 1980. *Fund. Cosmic Phys*, **5**, 287.

Tinsley, B. M., and Larson, R. B., eds. 1977. *The Evolution of Galaxies and Stellar Populations.* New Haven, CT: Yale Univ. Press.

Twarog, B. A. 1980. *Astrophys. J.*, **242**, 242.

Wheeler, J. C., Sneden, C., and Truran, J. W. 1989. *Ann. Rev. Astron. Astrophys.*, **27**, 279.

White, S. D. M. 1989. In Ellis *et al.*, p. 15.

White, S. D. M., and Rees, M. J. 1978. *Mon. Not. Roy. Astron. Soc.*, **183**, 341.

Woerden, H. van, ed. 1967. *Radio Astronomy and the Galactic System.* London: Academic Press.

Woerden, H. van, Burton, W. B., and Allen, R. J., eds. 1985. *The Milky Way Galaxy.* I.A.U. Symposium no. 106. Dordrecht: Reidel.

Woosley, S. E., and Weaver, T. A. 1986. *Ann. Rev. Astron. Astrophys.*, **24**, 205.

Wyse, R. F. G., and Gilmore, G. 1988. *Astron. J.*, **95**, 1404.

Yoshii, Y., and Saio, H. 1979. *Publ. Astron. Soc. Japan*, **31**, 339.

Zurek, W. H., Quinn, P. J., and Salmon, J. K. 1988. *Astrophys. J.*, **330**, 519.

CHAPTER 14

CHEMICAL EVOLUTION
AND DISK-GALAXY FORMATION

Pieter C. van der Kruit

In this chapter I will discuss some aspects of the formation of galaxies in view of the various properties and their distributions that I have discussed in other chapters of this book. For that purpose I will need to review first some aspects of the distributions of metallicity in disks and bulges, and some schematic background relating to chemical evolution in galaxies. Those aspects that are particularly related to the solar neighborhood are also covered by other chapters of this book.

14.1 ABUNDANCE GRADIENTS IN GALAXIES AND CHEMICAL EVOLUTION

In previous chapters I have discussed in detail the evidence for abundance gradients in bulges. This comes from color changes of the integrated light as a result of a hotter giant branch, a horizontal branch extended more to the blue, and less line blanketing in old, metal-poor populations. As a result the radial bluing in bulges is interpreted as a progressive decrease in mean stellar metallicity with increasing galactocentric radius. The amplitude of the effect is comparable to the total color variation in the integrated light of Galactic globular clusters, where the known abundance [Fe/H] has been measured independently, and therefore the color variation can be calibrated. Consequently, the observed variation of abundance is of the rough order of one to two dex.

   In disks the abundances are most easily measured in the gas from emission lines from the H_{II} regions. It was advocated first by Searle (1971) that the radial change in the line ratio of [O_{III}] at 4959 and 5007Å and Hβ (often called the "excitation") that he observed in a few nearby galaxies was a result of variations of the oxygen-to-hydrogen ratio and therefore probably the heavy-element abundance in the H_{II} regions. This is not entirely obvious, because the observed line ratios in emission-line spectra depend also on the continuum spectrum (effective temperature) of the ionizing star(s) and therefore on the ionization level of the gas and on the electron temperature. Searle proposed three alternatives to interpret his observations, namely, a gradient in the stellar temperatures, a changing dust content, which would affect the stellar radiation field, and an abundance gradient. He then concluded that the main effect of the changing line ratios had to be attributed to variations in chemical composition of the gas. Since then many studies have been done that confirm both the observed changes in line ratios and the interpretation that these are most likely due to abundance gradients in the disks.

The excitation [O III]/Hβ increases with increasing stellar temperature and decreases with increasing oxygen abundance relative to hydrogen [O/H]. The latter effect has the following reason. The cooling of the H II regions occurs through emission in optical lines; in particular, the nebular lines from O^+ and N^+ in the neutral-He zone, and O^{++} in the He^+ zone, dominate this effect. Now, a lower abundance (of oxygen) reduces the cooling and increases the electron temperature. Then the [O III] (and also [O II]) lines become stronger relative to Hβ.

As a example of observed gradients, I will take the recent observations of the Sab spiral M81 (Garnett and Shields 1987), which is the earliest type in which such a study has been done. These authors use two methods to derive abundances from the observed line ratios. For this interpretation of the spectra we need to know the electron temperature. The first method is the empirical one of Pagel et $al.$ (1980), which is based on a range of H II regions in various galaxies that span the range of observed spectra. Pagel et $al.$ have used bright regions to measure the electron temperatures (which is possible using weak emission lines, such as [O III] at 4363 Å, so that the population of two upper levels of an atom can be inferred), so that the abundance can be inferred; for others they used photoionization models. This calibrates the relation between excitation and metallicity. Furthermore, Garnett and Shields used their own photoionization models to infer abundances. The result is a radial gradient of [O/H] of -0.08 dex kpc^{-1} between 3 and 15 kpc radius. This is a typical value also for late-type galaxies. There is no discernible gradient in [O/N].

Before proceeding to discuss the observations, we need some general results from the theory of chemical evolution. For this, define in a closed region of a galaxy the mass in gas as $\mathcal{M}_g$, in heavy elements $\mathcal{M}_Z$, and in stars $\mathcal{M}_*$. The abundance then is $Z(t) = \mathcal{M}_Z(t)/\mathcal{M}_g(t)$. Next define the yield y (Searle and Sargent 1972) as the mass in new metals ejected by heavy stars, when a unit mass is locked in long-lived stars. Then assuming that short-lived stars evolve instantaneously and that the products of nucleosynthesis are also instantaneously mixed in the interstellar medium, we have

$$\frac{d\mathcal{M}_Z}{dt} = y \frac{d\mathcal{M}_*}{dt} - Z(t) \frac{d\mathcal{M}_*}{dt}, \qquad (14.1)$$

$$\frac{d\mathcal{M}_g}{dt} = -\frac{d\mathcal{M}_*}{dt}. \qquad (14.2)$$

Starting with an initial abundance $Z_0 = 0$, this gives the "simple model," which has as solution

$$Z(t) = y \ln\left(\frac{1}{x}\right) \quad \text{with} \quad x = \frac{\mathcal{M}_g(t)}{\mathcal{M}_{\text{tot}}}. \qquad (14.3)$$

The fraction of stars with abundance $\leq Z$ at time t [when the gas abundance is $Z(t) = Z_1$ and $x(t) = x_1$] is then

$$F(Z) = \frac{1-x}{1-x_1}. \tag{14.4}$$

This solution gives for the solar neighborhood the well-known G-dwarf problem, which states that there are too few stars of low metal abundance. It is also possible to calculate the mean stellar metallicity at any time. For this, $F(Z)$ is first converted by differentiation into a number distribution. The result then is, after dropping the subscript 1,

$$\langle Z \rangle = y \, \frac{1 - x(1 - \ln x)}{1 - x}. \tag{14.5}$$

We see, as gas consumption is completed and $x \to 0$, that $\langle Z \rangle \to y$, the yield itself. This is the mean metallicity of the stars, which should be distinguished from the abundance $Z(t)$ of the gas, which goes to infinity in this instantaneous-recycling approximation.

There are various ways of solving the G-dwarf problem with minor extensions of the simple model. The first is to set $Z_0 \neq 0$ (prompt initial enrichment). This can occur if chemical enrichment of the gas has proceeded in another position of the galaxy prior to settling in the disk, particularly in the "thick disk," as has been suggested by Gilmore and Wyse (1986). A second possible solution is that in the early phases formation of massive stars is more pronounced than that of light stars, so that few stars of low metallicity survive until the present time. This is a possibility in Larson's (1986) hypothesis of bimodal star formation, in which formation of both mass-groups of stars is independent and should then be heavily weighted toward massive stars in the initial phases of disk-star formation. The equations above for the simple model survive then as long as we everywhere replace Z with $Z-Z_0$. For complete gas consumption $\langle Z \rangle \to y + Z_0$. I will refer to this as the extended simple model.

A second possibility is to postulate an inflow of unprocessed material into the disk, such that

$$\frac{d\mathcal{M}_g(t)}{dt} = -\frac{d\mathcal{M}_*}{dt} + f(t). \tag{14.6}$$

The extreme model of this kind has $\mathcal{M}_g = $ constant, and then the solution becomes

$$Z(t) = y \, [1 - \exp(-\mu)] \quad \text{with} \quad \mu = \frac{\mathcal{M}_*(t)}{\mathcal{M}_g}, \tag{14.7}$$

$$F(Z_1) = \frac{\mu}{\mu_1}, \tag{14.8}$$

$$\langle Z \rangle = y - \frac{y}{\mu} + \frac{y}{\mu} \exp(-\mu). \tag{14.9}$$

Eventually we will have $\mathcal{M}_* \gg \mathcal{M}_g$ and $\mu \gg 1$; then $\langle Z \rangle \to y$. The important feature of this model is that $Z(t)$ very quickly increases in the early phases (because there is less gas, it also is more easily enriched) and $Z(t)$ approaches y asymptotically.

For the discussion of the gradients in bulges, we also have to consider what happens if enriched material is lost from the area. In that situation we have

$$\frac{d\mathcal{M}_Z}{dt} = y\,\frac{d\mathcal{M}_*}{dt} - Z(t)\,\frac{d\mathcal{M}_*}{dt} - Z(t)g(t), \qquad (14.10)$$

$$\frac{d\mathcal{M}_g}{dt} = -\frac{d\mathcal{M}_*}{dt} - g(t). \qquad (14.11)$$

The illustrative model here has $g(t)$ proportional to the star-formation rate $g(t) = \alpha\,d\mathcal{M}_*/dt$. This could in practice happen when star formation through supernova explosions heats up the interstellar medium, which subsequently is blown away. It is straightforward to see that the equations of the simple model recover in this situation, but with $d\mathcal{M}_*/dt$ multiplied by a factor $(1+\alpha)$ and an "effective yield" $y' = y/(1+\alpha)$. The solution then has the same form as the simple model, with y replaced by y'.

Now let us look at observations, first the abundance gradients in the bulges. I will discuss below that any gradient is probably conserved through the collapse. The hypothesis then is that in the inner regions of the proto-galaxy the interstellar medium, sitting in the bottom of the potential well, had more difficulty in escaping than in the outer parts. So, although star formation would heat up the gas, the value of α in the above model was much lower in the central areas than in the extremities. After depletion of the gas, the mean stellar metallicity will be equal to the effective yield, as we have just seen. This process thus sets up an abundance gradient as a result of gas loss.

In comparison to the Galaxy, we should note that the observed gradients are currently limited to the brighter parts of the bulges. Although such a gradient may be present for the inner bulge in the Galaxy as well, it has been suggested by Searle and Zinn (1978) that there is no such gradient in the outer parts of the globular-cluster system (roughly beyond the solar radius). They suggested that the loosely bound outer clusters have a broader range of ages, and resulted from continuing infall of fragments after completion of the inner-bulge formation. Likewise, these fragments lost gas, and the mean abundances of the stars in the resulting globular clusters remained low, with no radial gradient in their present space distribution. The discussion of the color gradients above and their possible origin then only applies to the inner parts of halo Population II.

Observations of elliptical galaxies also have revealed a correlation between the metallicity and the luminosity. If the latter corresponds to mass, this relation can also be explained qualitatively by the simple description. After all, in small systems the potential well is less deep and the gas more prone to evaporate from the system. Such systems will therefore have lower effective yields and lower mean stellar abundances after completion of gas consumption by star formation. The correlation is indeed in the sense expected in this picture.

Now let us return to the disks. In their study of M81, Garnett and Shields have also estimated the gas fraction in the disk as a function of radius. The resulting

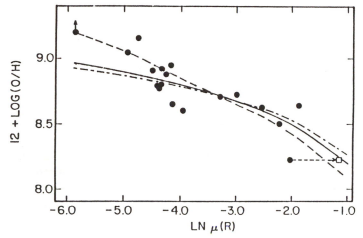

FIGURE 14.1

The correlation between O/H abundance ratio and the gas fraction in the disk of M81. The data points are from observed H II regions. The full line is a fit with the simple model of chemical evolution; the other two are more complicated models. (From Garnett and Shields 1987.)

variation of the O/H ratio from their H II-region spectra with this fraction is shown in Fig. 14.1. Similar figures have been produced also for other galaxies by other workers. The solid line is that for the simple model with a constant yield with radius. In spite of the fact that molecular gas was not included, the figure shows reasonable qualitative agreement. The other curves are more complicated models. Although this should not be taken as confirmation of the simple model, it follows that general considerations of chemical evolution (the extended simple model) can explain the basic trends observed.

There also is a correlation between the mean gaseous abundance (*e.g.*, measured at the effective radius) and the galaxy's integrated magnitude. This ranges by two orders of magnitude between small systems like the LMC and the largest spirals, with the latter having the highest abundances. To some extent this is also a trend in morphological type, since late-type galaxies tend to be less luminous. Also, the gas fraction decreases toward earlier galaxies, and this trend therefore also fits qualitatively with the predictions of the extended simple model.

There is some uncertainty concerning the abundances of nitrogen. In most galaxies there is little radial trend in the N/O ratio, but sometimes there is in the N/S ratio. The latter also seems to correlate with galaxy luminosity. Nitrogen may at least in part be a secondary element, which means that it can be produced in nucleosynthesis only from other ("seed") elements, and therefore its production in a particular generation of stars depends on the abundance of primary elements in the gas out of which these stars form. The simple model then predicts that its abundance is roughly proportional to the square of the abundance of primary elements. The evidence on this point is not yet conclusive.

In his discussion of the K giants used to measure velocity dispersions in the old disk of our Galaxy, Lewis (1986; see also Lewis and Freeman 1989) found that these stars show essentially no radial gradient in their abundance [Fe/H]. This is in contrast to the observed abundance gradient in the gas. This feature, however, also follows from the models given above. In the extended simple model we see that the dependence of $\langle Z \rangle$ on x is rather different from that of $Z(t)$ in the gas (compare Equations [14.3] and [14.5]). For $Z_0 \neq 0$, the mean stellar abundance is very close to this initial abundance, especially if we remember that we should take the relevant value for x at the end of formation of this population, which is then not much less than unity. The resulting small dependence on x might then be the reason for the absence of such a gradient. For the gas we need to take x at the present time, and $Z(t)$ becomes Z_0 plus a few times the yield, and has a strong dependence on x.

14.2   FORMATION OF SPIRAL GALAXIES

I will now discuss a simple scenario for the formation of disk galaxies, with a view toward an understanding of the various observations that I have discussed for external systems. Before doing this I will first describe some general notions that I will then use to describe a possible scenario.

In the first place I recall the basic two-component structure of disk galaxies in their light distributions (van der Kruit and Searle 1982). The possible "thick disks" in our and other galaxies have a small total luminosity compared to that of the disk, and so are a minor component; their existence would not contradict this basic feature of the stellar distributions. In NGC 7814, van der Kruit and Searle did observe a color gradient in the spheroid and noted that the isochromes and the isophotes have similar flattening. Now models have been presented where bulge formation proceeds by a dissipational process, in which stars form out of a relatively slowly contracting gas sphere, which is continually enriched (e.g., Larson 1976). It seems a general property of such models, especially when rotation is included, that the isochromes are considerably flatter than the isophotes, with the inner ones being flatter than those further out. This is so because on this hypothesis a galaxy is a superposition of a chronological sequence of stellar populations in which flattening and metal abundance both increase with time. The observed color gradients suggest that the binding energy of the stars is correlated with metallicity. These observations indicate that the bulge stars formed before the galaxy collapsed, and that the abundance structure existed in the nonequilibrium proto-galaxy. I have discussed above how such a gradient could have originated. Disks clearly form in a dissipational process, so that the basic two-component structure suggests two different epochs of star formation, one occurring before and the other after virialization of the bulge and the collapse of the disk.

Van Albada (1982) has performed N-body simulations of the dissipationless-collapse process of galaxy formation. His code was able to describe the process of violent relaxation (relaxation of stellar orbits resulting from strong variations of the gravitational field) sufficiently well for the results to be indicative of the evolution involved and the behavior of a proto-galaxy under dissipationless collapse. He found two surprising things. In the first place, he found that the resulting distributions in collapsed systems with a variety of irregular initial conditions could be fit excellently to the $R^{1/4}$ law over a region between radii containing 10 and 99 percent of the total mass, provided that the collapse factor is large. The range of surface brightness represented was 12 mag. A second important finding was that the binding energy of the particles before and after collapse was correlated. This means that any structure in the proto-galaxy, such as an abundance gradient, would survive statistically in the process of dissipationless collapse and violent relaxation. This study thus indicates that bulges with their observed luminosity distributions and abundance gradients can indeed be produced without invoking dissipational processes.

There is some dispute about the assumption that bulges and Population II are the first components to collapse, mainly because apparently there are many stars in the inner bulge of the Galaxy (within 0.5 kpc or so) that are considerably younger than the globular clusters of the outer halo (Harmon and Gilmore 1988 and references therein). It should, however, be kept in mind that the inner bulge probably consists of more than just the oldest halo population. At small R (say, smaller than 0.5 kpc) there is no longer a clear-cut distinction between the Population II bulge stars and those of the old disk (including the "thick disk"), since such stars will have similar kinematics (velocity dispersions of order 100 km sec^{-1} and little rotation), and all will be relatively metal-rich. But stars that formed in the inner disk after completion of disk formation over a period of several Gyr (until the gas supply was exhausted) will also have large velocity dispersions due to the expected strong effects of secular evolution of their orbits. It would not be surprising, therefore, if at distances less than 0.5 kpc or so from the plane in the inner bulge, there is a mix of stars from Population II, old-disk population as well as somewhat younger disk generations, that cannot be distinguished by their kinematics or metallicities. The younger stars in the inner bulge should then be identified with those that formed over the first few Gyr after disk formation, and this fits with their observed vertical scale height of about 370 pc (Harmon and Gilmore 1988).

Disk galaxies rotate. Peebles (1969) has studied the possibility that proto-galaxies acquire angular momentum from tidal torques from their neighbors in the early Universe. He found that the amount of angular momentum predicted by this mechanism could be described by a single dimensionless parameter

$$\lambda = J \, |E|^{1/2} \, G^{-1} \, \mathcal{M}^{-5/2}, \tag{14.12}$$

where J is the total angular momentum, E the total energy, and $\mathcal{M}$ the total mass. Peebles estimated that λ would be about 0.08; numerical experiments have shown it to be about 0.07 with a standard deviation of about 0.03. A problem at that time was that the estimated amount of angular momentum that results from this mechanism was far too little to explain the observed rotation.

This problem was addressed by Fall and Efstathiou (1980). They included the effects of a dark halo and assumed that the luminous part of a galaxy settles in the potential well of this dark halo. They considered the case that the halo (of which the bulge may or may not be a part) and the disk have the same distribution of specific angular momentum (angular momentum per unit mass), and that the dissipational collapse of the disk occurred with detailed conservation of angular momentum; in other words, the distribution of specific angular momentum before and after collapse is the same. This makes it possible to calculate from observations the angular-momentum distribution of the proto-disk and therefore of the proto-halo. For this purpose they used observed density distributions (from surface photometry) and rotation curves to calculate the properties of the proto-galaxy. Their specific analytical fit to flat rotation curves will be used below; it is given by

$$V_{\text{rot}}^2(R) = \frac{V_{\text{m}}^2 R^2}{R_{\text{m}}^2 + R^2} \left[1 - \gamma \ln \left(\frac{R^2}{R_{\text{m}}^2 + R^2} \right) \right]. \tag{14.13}$$

This applies up to the radius R_{H} of the halo, after which it will become Keplerian. For actual galaxies R_{m} is 0.1 to 0.5 scale lengths h. From these fits to galaxies, Fall and Efstathiou could also calculate the value of Peebles' λ if the radius R_{H} is assumed. The requirement that tidal torques provide the angular momentum observed then translated into the requirement that the mean collapse factors R_{H}/h are about 20, and mean halo-to-disk mass ratios are of order 10. It is important to stress that they did not presuppose any distribution parameters for the proto-galaxy, and started only from observations with the assumption stated.

Mestel (1963) had made an interesting inference from the then-known rotation curve and mass distribution of the disk of the Galaxy. He calculated the distribution function of specific angular momentum, and noted that it was rather similar to that of a sphere with a uniform density distribution and with uniform (angular) rotation speed. He then proposed that galaxies indeed collapsed with detailed conservation of angular momentum (as Fall and Efstathiou also proposed later) and started from uniform, uniformly rotating spheres. This is sometimes referred to as Mestel's hypothesis. Gunn (1982) used this hypothesis to show that if such a sphere settles in the force field of a dark halo with a flat rotation curve, a roughly exponential disk results. Actually, Freeman (1970) had noted that the self-gravitating exponential disk also has a specific-angular-momentum distribution similar to that of the Mestel sphere, but this was before the flat rotation curves were discovered.

I will now take the following approach in my discussion of a possible scenario of disk-galaxy formation (see also van der Kruit 1987). Assume that proto-galaxies form in the early Universe, and acquire angular momentum from tidal torques from neighbors, and that these proto-galaxies can roughly be described as uniform, uniformly rotating spheres. Then let the dark matter quickly settle (dissipationlessly) in a distribution that resembles an isothermal sphere. This means that the dark halos have at large R a density distribution proportional to R^{-2} and provide a potential field that at larger radii corresponds to a flat rotation curve. Early on, some stars also form, especially in the inner regions, and settle quickly thereafter in the inner bulge. If indeed this star formation is concentrated toward the central areas of the proto-galaxy, it is mostly self-gravitating and will collapse, as van Albada simulated, in a density distribution that closely follows that of the $R^{1/4}$ law. These bulges will have a small amount of rotation, as is observed. The amount of mass that is turned into stars at this stage is controlled by an unknown process, but it is this amount that determines the Hubble type of the galaxy that is forming. It may have to do with details of the angular-momentum distribution in the proto-galaxy. An abundance gradient, which will naturally develop in the proto-galaxy, will survive the collapse, and result in the abundance structure observed in the bulges.

Like Fall and Efstathiou, I will assume that the dark halo has the same angular-momentum distribution as the proto-galaxy. This means that at all positions in the proto-galaxy there is an equal ratio of dark to luminous matter. The distribution of specific angular momentum in the Mestel sphere is given by $\mathcal{M}(h_s)/\mathcal{M}$, which is the fraction of matter with specific angular momentum $\leq h_s$:

$$\frac{\mathcal{M}(h_s)}{\mathcal{M}} = 1 - \left(1 - \frac{h_s}{h_{max}}\right)^{3/2}. \tag{14.14}$$

The proto-galactic sphere has density ρ_0 and therefore radius $R_m = (3\mathcal{M}/4\rho_0)^{1/3}$. The gravitational potential energy is $\Omega = -3G\mathcal{M}^2/5R_m$, and the total angular momentum $J = \frac{2}{5}\mathcal{M}h_{max}$. Now, assume that at the time of the start of the collapse, when the proto-galaxy detaches itself from the expanding Universe and reverses its expansion into contraction, its total energy is essentially gravitational. Then $|E| = |\Omega|$; it cannot be much smaller: if it is a factor 2 smaller, the proto-galaxy is in virial equilibrium. Then we find from Peebles' parameter λ that

$$h_{max} = \frac{5}{2}\left(\frac{5}{3}\right)^{1/2} G^{1/2}\lambda\mathcal{M}^{1/2}R_m^{1/2}. \tag{14.15}$$

Now assume that a mass fraction $1 - \Gamma$ is in the form of dark matter, and that it settles in a roughly isothermal sphere with radius R_H. Then its gravitational energy after collapse is in good approximation

$$\Omega_H = -\frac{G\mathcal{M}_H^2}{R_H} = -G(1-\Gamma)^2\frac{\mathcal{M}^2}{R_H}, \tag{14.16}$$

and according to the virial theorem its total energy is

$$E_H = \frac{\Omega}{2} = -G(1-\Gamma)^2\frac{\mathcal{M}^2}{2R_H}. \tag{14.17}$$

The total energy of the proto-halo is, under the assumptions made of equal distribution of specific angular momentum, equal to $(1 - \Gamma)$ times that of the proto-galaxy, or

$$E_H = -\frac{3}{5}(1-\Gamma)\, G\frac{\mathcal{M}^2}{R_m}. \tag{14.18}$$

Since the dark halo collapses without dissipation of energy, these two energies need to be equal; so we get

$$R_H = \frac{5}{6}(1 - \Gamma)\, R_m, \tag{14.19}$$

and the asymptotic circular speed in the rotation curve is (this is, of course, not the rotation of the halo itself)

$$V_m^2 = \frac{G\mathcal{M}}{R_H} = \frac{6}{5}\frac{G}{1-\Gamma}\frac{\mathcal{M}}{R_m}. \tag{14.20}$$

Now, let us look at the remaining material, which is gas that will eventually settle in the disk. Its mass is $\Gamma\mathcal{M}$, and its specific-angular-momentum distribution that of the Mestel sphere. Let this settle in a flat disk under conservation of specific angular momentum in the potential field of the dark halo. For convenience I will use for the potential field one that corresponds to the Fall and Efstathiou rotation curve given above. Let us first see what to expect for an exponential disk. The lower solid line in the left-hand part of Fig. 14.2 is that of an exponential disk in such a rotation curve with $\gamma = 0.1$ (this value is not important) and $R_m = 0.2h$. The specific angular momentum is expressed in units of hV_m. Then we find the best representing curve for a Mestel sphere; this is the dashed line. The only free parameter for doing this comparison is h_{max} (which determines where the distribution approaches unity), and it can be seen that we have to choose it as about 4.5 in the units used. For smaller values it will rise too steeply and for larger values too slowly. The upper solid line then shows the distribution for an exponential disk, but now with a cutoff at 4.5 scale lengths in order to reproduce the fact that the Mestel sphere has no h_s larger than h_{max}.

Then turn it around and calculate what the resulting surface-density distribution is for the angular-momentum distribution of a Mestel sphere, where I have chosen $h_{max} = 4.5hV_m = 22.5V_mR_m$, following the arguments given above. This is shown in the right-hand part of Fig. 14.2. The solid line compares this to the expected exponential disk with $h = 5R_m$ and a cutoff at $4.5h$. The resemblance is close; the vertical scale is in natural logarithms, which is close to magnitudes. Deviations of a few tenths of a magnitude are not uncommon in actual surface photometry of disks. So the conclusion from this is that if we let material in a Mestel sphere settle in the force field corresponding to a flat rotation curve under detailed conservation of angular momentum, we end up with a surface-density distribution that is exponential and truncates at 4.5 scale lengths. This is what has actually been observed in the stellar distribution in disks of galaxies, as I discussed extensively in Chapter 5.

A few further remarks need to be made before we proceed to calculate further properties of this scenario. The first is that the prediction in Fig. 14.2 shows a strong excess at small radii. This is material that was originally in the proto-galaxy close to the rotation axis. Some of this material might actually have been used to form the bulge stars, so that this is not a worrying difference. Also, the precise form of the expected distribution depends greatly on the assumed

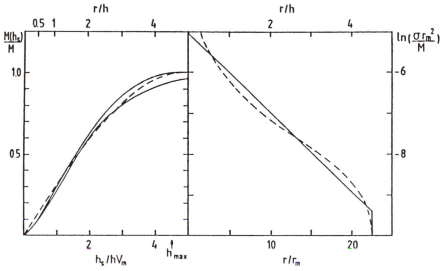

FIGURE 14.2
The left-hand panel shows the distribution of specific angular momentum in three cases. The lower solid curve is that of an infinite exponential disk in a Fall and Efstathiou (1980) rotation curve (Eq. (14.13)), using $\gamma = 0.1$ and $R_m = 0.2h$. The upper solid line is the same, but for an exponential disk with a sharp cutoff at $4.5h$. The dashed line is the distribution for a uniform, uniformly rotating sphere with maximum specific angular momentum $h_{max} = 4.5hV_m$. The right-hand panel shows the surface density of a disk with the angular-momentum distribution of a Mestel sphere in the same rotation curve and for the same h_{max}, which is also equal to $22.5V_mR_m$. The solid line is an exponential disk with scale length $h \; (= 5R_m)$ and a cutoff at $4.5h$. (From van der Kruit 1987.)

rotation curve, and the disk potential itself will modify this as well. This is also of no great consequence, because—as Freeman (1970) has shown—the self-gravitating exponential disk also has the specific-angular-momentum distribution of the Mestel sphere. So even if disk self-gravity seriously changes the potential field, we still expect the material to settle in an exponential disk.

The gas that settles in the disk will quickly start to form stars. The final process of disk formation will certainly be accompanied by cloud collisions and enhanced star formation. In this initial phase there is much turbulent gas motion and many gas concentrations, and we must expect that there will be strong effects on the orbits of these stars formed at the time of disk settling, as in the Spitzer–Schwarzschild mechanism. This initial generation of disk stars will therefore have large random motions (and intermediate metallicities) and must then settle in a thicker distribution. I suggest that this is now evident in the "thick-disk" population and maybe also in Zinn's (1985) subsystem of disk globular clusters. The subsequent star formation will form the old disk population, and will then naturally display the edges at 4 to 5 scale lengths.

This leaves the question of the H I gas that is observed beyond the optical edges. We can only assume that this is gas that fell in at a later stage and has been falling in from larger initial radii than the extent of the proto-galaxy. There is then no requirement for it to settle in the same plane as the disk. This would explain the larger angular momentum than present in the stellar disk, and also why the observed warps usually start at about the optical edges. Actually, there is no reason why this process would not also provide gas at smaller radii, which would fit in with the observation that the H_I surface density shows no feature at the optical edge.

We have thus seen that the gas with mass ΓM of the proto-galaxy will settle dissipationally in a flat disk with a scale length $h = h_{max}/\beta V_m$ and β about 4.5. In the following I will for simplicity ignore self-gravity, but this does not affect the conclusion concerning the formation of an exponential disk. The assumption roughly corresponds to the situation where the dark halo everywhere dominates the rotation curve. Otherwise we will have to assume that some unknown mechanism provides the "disk-halo conspiracy." In any case, the resulting rotation curve will be approximately flat at the level of the asymptotic value V_m for the dark halo. With the equations above for h_{max} and V_m, we can then write

$$h = \frac{25}{6} \frac{1}{2^{1/2}} \frac{\lambda}{\beta} \frac{1}{(1-\Gamma)^{1/2}} R_m. \tag{14.21}$$

The central surface density of the disk is

$$\sigma_0 = \frac{36}{625} \left(\frac{4}{3}\right)^{2/3} \frac{1}{\pi^{1/3}} \left(\frac{\beta}{\lambda}\right)^2 \frac{\Gamma}{1-\Gamma} \rho_0^{2/3} M^{1/3}. \tag{14.22}$$

The initial density ρ_0 is the density of the proto-galaxy when it detaches itself from the rest of the Universe. If galaxies form at about the same time, this density might not vary by large factors. Indeed, in the discussion of Fall (1979) on density perturbations for hierarchical clustering, the density spectrum in a critical universe $\Delta\rho/\rho$ is proportional to a low negative power of M, and therefore the dependence of σ_0 on ρ_0 and M almost disappears. So we see that if Γ is constant for all galaxies, we recover Freeman's law of constant central surface brightness. This says that everywhere in the early Universe we should have had equal ratios of dark and luminous matter, and that this still applies between present-day galaxies.

Substituting $\beta = 4.5$ and $\lambda = 0.07$, we get

$$\Gamma(1-\Gamma)^{1/2} = 1.5 \frac{\sigma_0 h}{V_m^2}, \tag{14.23}$$

$$R_m = \frac{22}{(1-\Gamma)^{1/2}} h, \tag{14.24}$$

$$R_H = 18(1-\Gamma)h, \tag{14.25}$$

$$M = 4.2 \times 10^6 (1-\Gamma)^{1/2} V_m^2 h, \tag{14.26}$$

$$\rho_0 = 9.7 \times 10^{-8} (1-\Gamma)^2 \frac{V_m^2}{h^2}. \tag{14.27}$$

Here length is in kpc, velocity in km sec^{-1}, mass in $\mathcal{M}_\odot$, and density in $\mathcal{M}_\odot$ pc^{-3}. Using for the Galaxy $h = 5$ kpc, $V_m = 220$ km sec^{-1}, and $\sigma_0 = 400$ $\mathcal{M}_\odot$ pc^{-2}, we find that $\Gamma = 0.06$, $R_m = 115$ kpc, $R_H = 90$ kpc, $\mathcal{M} = 1.0 \times 10^{12}$ $\mathcal{M}_\odot$, and $\rho_0 = 2 \times 10^{-4}$ $\mathcal{M}_\odot$ pc^{-3}. Similar values obtain for other galaxies for which the relevant data are available, namely, Γ in the range 0.04 to 0.11 and ρ_0 a few times 10^{-4} $\mathcal{M}_\odot$ pc^{-3}. The collapse factor R_m/h is of order 20, and the radius of the halo is about $18h$. Of course, λ may be different for individual galaxies within the range 0.07 ± 0.03, indicated by numerical experiments, and the estimates for the properties then change accordingly. We are in a few galaxies approaching such radii for the observed extent of HI rotation curves. The large percentage of the mass in the form of dark material is essentially needed to explain the amount of observed rotation in disk galaxies by the tidal torque hypothesis. In the clustering model of Fall (1979) we expect $\Delta\rho/\rho$ at about $9\pi^2/16$; then it follows that ρ_0 is about $5.5\langle\rho\rangle$, and for the values above for ρ_0 this occurs at a redshift z of about 3.5 in a Universe with $\Omega = 1$ and $H = 75$ km sec^{-1} Mpc^{-1}.

The luminosity of the disk is

$$L_{\text{disk}} \propto (L/\mathcal{M})\Gamma^2(1 - \Gamma)\frac{V_m^4}{\mu_0}; \tag{14.28}$$

so, for constant $\mathcal{M}/L$, Γ, and central-surface brightness μ_0, we get the Tully–Fisher relation.

It should be stressed that the scenario given here is very schematic and should not be construed as more than a working hypothesis; yet it qualitatively explains the general features that we know about spiral galaxies.

REFERENCES

Albada, T. S. van. 1982. *Mon. Not. Roy. Astron. Soc.*, **201**, 939.

Brück, H. A., Coyne, G. V., and Longair, M. S., eds. 1982. *Astrophysical Cosmology.* Vatican City: Pontifica Academia Scientiarum.

Fall, S. M. 1979. *Rev. Mod. Phys.*, **51**, 21.

Fall, S. M., and Efstathiou, G. 1980. *Mon. Not. Roy. Astron. Soc.*, **193**, 189.

Freeman, K. C. 1970. *Astrophys. J.*, **160**, 811.

Garnett, D. R., and Shields, G. A. 1987. *Astrophys. J.*, **317**, 82.

Gilmore, G., and Wyse, R. F. G. 1986. *Nature*, **322**, 806.

Gunn, J. E. 1982. In Brück *et al.*, p. 233.

Harmon, R., and Gilmore, G. 1988. *Mon. Not. Roy. Astron. Soc.*, **235**, 1025.

Kruit, P. C. van der. 1987. *Astron. Astrophys.*, **173**, 59.

Kruit, P. C. van der, and Searle, L. 1982. *Astron. Astrophys.*, **110**, 79.

Larson, R. B. 1976. *Mon. Not. Roy. Astron. Soc.*, **176**, 31.

———. 1986. *Mon. Not. Roy. Astron. Soc.*, **218**, 409.

Lewis, J. R. 1986. Ph.D. thesis, Australian Natl. Univ.

Lewis, J. R., and Freeman, K. C. 1989. *Astron. J.*, **97**, 139.

Mestel, L. 1963. *Mon. Not. Roy. Astron. Soc.*, **126**, 553.

Pagel, B. E. J., Edmunds, M. G., and Smith, G. 1980. *Mon. Not. Roy. Astron. Soc.*, **193**, 219.

Peebles, P. J. E. 1969. *Astron. Astrophys.*, **11**, 377.

Searle, L. 1971. *Astrophys. J.*, **168**, 327.

Searle, L., and Sargent, W. L. W. 1972. *Astrophys. J.*, **173**, 25.

Searle, L., and Zinn, R. 1978. *Astrophys. J.*, **225**, 357.

Zinn, R. 1985. *Astrophys. J.*, **293**, 424.

CHAPTER 15

THE MILKY WAY
IN RELATION TO OTHER GALAXIES

Pieter C. van der Kruit

In this, my final chapter, I will put the Milky Way into perspective among the realm of spiral galaxies. To this end I will first briefly summarize some properties of its nearest neighbors: the other members of the Local Group of galaxies. Then I will try to answer the question of what the Hubble type of the Galaxy is; this is preceded by a review of arguments for the value of the disk scale length, because this does play a role in some of this. An attempt will be made to identify some external galaxies that bear a close resemblance to our own, followed by some concluding remarks related to our vantage point within the Milky Way. The conclusion will be that we live in a fairly large galaxy, probably of type SbI–II, at about 1.5 to 2 scale lengths from the center. Our position close to the plane is a particularly unlucky coincidence that seriously hampers Galactic astronomy, especially in the optical, but the orientation of the plane of the Milky Way in space with respect to the nearby extragalactic Universe is particularly favorable for studies of external galaxies.

15.1 PROPERTIES OF OTHER LOCAL-GROUP GALAXIES

The Local Group consists of two major spirals, the Galaxy and M31, a smaller spiral, M33, and various dwarf systems, of which the Magellanic Clouds and the four dwarf ellipticals around M31 (M32 and NGC 205, 185, and 147) are the larger ones. Further, there is a fair number of small dwarfs that I will not be concerned with in this book. I will not attempt to give a complete description of all features of these galaxies, but will rather highlight some interesting and important studies.

Surface photometry of such nearby galaxies, which consequently have a large angular size, is difficult. Although studies on photographic plates are available from the earliest days of photometry (de Vaucouleurs 1957 observed the LMC very early on in his studies and discovered the exponential nature of its disk), the prime method today is the use of CCDs behind small telescopes (or even only a camera lens and a filter/shutter assembly). This ensures a large field of view, but provides larger pixels, sometimes up to an arcmin. This is, of course, no problem for large-angular-size objects.

Kent (1987b) observed in this way the elliptical companions of the Andromeda galaxy. There is a large variety of luminosity profiles. M32 can be represented with an $R^{1/4}$ law out to 0.34 kpc with an effective radius of 0.11 kpc. The total extent observed (out to 1.0 kpc) shows a more exponential behavior beyond 0.4 kpc, without any indication of a tidal truncation. The integrated absolute magnitude is $M_V = -16.1$ (2.4×10^8 $L_\odot$). Studies of stellar kinematics by Dressler and Richstone (1988) and others show evidence for a central black hole of 8×10^6 $\mathcal{M}_\odot$. NGC 205 has a profile that cannot be fit with either an $R^{1/4}$ or an exponential law, and has a starlike nucleus. The observed extent is 1.6 kpc, and the integrated magnitude $M_V = -16.1$ (2.4×10^8 $L_\odot$). NGC 185 also has a complicated profile, but no nucleus. Its extent is 1.4 kpc, and its absolute magnitude -15.1 (9.5×10^7 $L_\odot$). Finally, NGC 147 can be fitted very well by an exponential out to the observed maximum radius of 1.1 kpc, and the scale length is 0.5 kpc. The central surface brightness corresponds to about 22.5 B-mag arcsec^{-2}, typical for irregular dwarfs (in the morphological sense, which means that there is evidence for star formation), although its appearance is that of an old population. The integrated magnitude is -14.7 (6.5×10^7 $L_\odot$). Mould *et al.* (1983, 1984) have used CCDs to measure the brighter part of the color–magnitude diagram of NGC 147 and 205. The metallicities are low and similar ([Fe/H] about -1.2 for NGC 147, and -0.85 for NGC 205), and both systems share the property that no more than 10 percent of the stars can be of intermediate age.

Of course, a very large body of data exists for the Magellanic Clouds, and I will refer here only to the distribution of stars and of star formation. Clearly, both the LMC and the SMC experience current star formation and are rich in interstellar medium. They are companions of the Galaxy, at distances of about 50 and 65 kpc, respectively. Their integrated (absolute) magnitudes are $M_V = -18.4$ and -17.1 (2.0×10^9 and 6.0×10^8 $L_\odot$), and colors $B - V = 0.52$ and 0.61. So these are more luminous (and bluer) than the companions of M31, although the color is not blue enough to indicate a major current burst of star formation. De Vaucouleurs (1957) measured the luminosity profile of the LMC and SMC, and showed these to follow the exponential law for disks. More recent photometry, including color determinations, have been done by Bothun and Thompson (1988) with a CCD behind an 85-mm lens (the "parking-lot camera") in B, V, and R. The luminosity profiles are shown in Fig. 15.1. Both can be fitted with exponentials.

The LMC has a central surface brightness of 21.3 B-mag arcsec^{-2} after correction for inclination. This is somewhat bright, even for a normal spiral, and shows the possible effect of star formation in a late-type system. The scale length is about 1.5 kpc. This is large for a galaxy that does not show obvious spiral structure (see M33 below). The bar also has a radially exponential surface-brightness distribution; most bars in spiral galaxies, however, have a fairly constant surface brightness. At $2°.8$ a faster decline than the exponential sets in in all colors, and

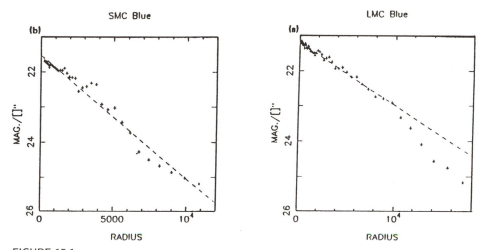

FIGURE 15.1

Radial light profiles of the Magellanic Clouds, obtained with the "parking-lot camera," a CCD behind a camera lens, with a field of view of 6.5° by 10.4°. The pixel size measured 73 arcsec. The data are shown for the B-band, and fits to exponential disks are shown by the dashed lines. The horizontal scale is in arcsec. The surface brightness is not corrected for inclination. (From Bothun and Thompson 1988.)

this may indicate the onset of a truncation of the stellar distribution caused by the tidal field of the Galaxy during a period of close proximity. The detailed color distribution shows mostly random star formation during recent times, with a small suggestion that the most recent star formation is concentrated toward the bar; if this is so, then this bar may not be a feature in the mass distribution, but rather may outline the area of most recent star-formation activity. In many respects the LMC simply is a small spiral, except that the star formation does not proceed in a manner that is organized over the disk in a pattern of spiral structure. Clearly, interaction with the Galaxy may be suspected as the cause of this. Schwering (1988) has recently produced maps of infrared IRAS brightness of both Clouds. This traces dust that is heated by hot stars, and the observed distributions correspond closely with that of tracers of recent star formation.

The SMC much more resembles dwarf irregulars, such as have been observed, for example, in the Virgo Cluster. The face-on central surface brightness is 22.3 B-mag arcsec^{-2} (assuming an inclination of 60°), which is more typical for dwarfs. The scale length of the fit in Fig. 15.1 is 0.96 kpc. Colors reveal coherent patterns of star formation, although these are not organized in spiral structure.

M33 definitely is a small Sc spiral with essentially no bulge component. Its luminosity distribution is closely exponential, although the photometry by Kent (1987b) shows a central increase in surface brightness over the exponential. This is not a bulge component, because the light is there still dominated by the spiral structure and areas of obvious star formation. The scale length is about 1.7 kpc, and the central surface brightness 21.05 B-mag arcsec^{-2}. The integrated

TABLE 15.1
Comparison of M31 and our Galaxy

	M31	*Galaxy*
Bulge:		
Luminosity ($L_\odot$)	7.7×10^9	2×10^9
Luminosity (fraction of total)	0.25	0.12
Effective radius (kpc)	2.2	2.7
Axis ratio	0.57	0.85
Velocity dispersion (km sec^{-1})	155	130
Number of globular clusters	400–500	160–200
Disk:		
Luminosity ($L_\odot$)	2.4×10^{10}	1.7×10^{10}
Scale length (kpc)	6.4	5.0
$B - V$ (errors about 0.1)	0.76	0.85
H$_I$ gas content ($\mathcal{M}_\odot$)	3×10^9	4×10^9
Rotation velocity (km sec^{-1})	260	220
IRAS infrared luminosity ($L_\odot$)	2.6×10^9	1.5×10^{10}
Gas metallicity gradient (dex kpc^{-1})	-0.04	-0.08

absolute magnitude is -18.6 ($2.4 \times 10^9\ L_\odot$), and the color $B - V = 0.50$. All these numbers are remarkably close to those of the LMC, and the only difference really is the spiral structure in M33. Also, the gas masses of M33 and the LMC are similar (9.5×10^8 and $5.4 \times 10^8\ \mathcal{M}_\odot$), as are the total infrared fluxes from IRAS (1.2×10^9 and $8.9 \times 10^8\ L_\odot$). M33 is at a projected distance of about 175 kpc from M31; so these galaxies have probably had much less interaction with each other than the LMC and the Galaxy. Deul (1988) has recently obtained maps of M33 in the IRAS bands and in H$_I$. The average dust-to-atomic-gas ratio in M33 is much lower than that in the Galaxy.

Finally, I turn to the Andromeda galaxy. Most of what follows is based on the extensive study by Walterbos (1986; see also Walterbos and Kennicutt 1987 and references therein), who observed M31 in the radio continuum, obtained multi-color optical surface photometry, and used the IRAS results. The optical data are illustrated in Fig. 15.2, where the profiles along the major axis have been plotted. Interestingly, no color gradient has been seen in the bulge, and a small but significant bluing is seen in the disk. The latter could be due to metallicity, but the contribution from young stars is also a serious possibility. The profiles can be decomposed well into an $R^{1/4}$-bulge and an exponential disk.

Andromeda is compared with the Galaxy in Table 15.1. We see that the disk of M31 is slightly bigger, but the bulge is nearly three times brighter. The H$_I$, radio-continuum, and IR luminosities are all lower in the Andromeda galaxy. M31 has not been mapped completely in CO; it is known that both M31 and the Galaxy have a ring of molecular gas. In Andromeda it peaks at about 10 kpc galactocentric radius, whereas in the Galaxy this occurs at about 4.5 kpc.

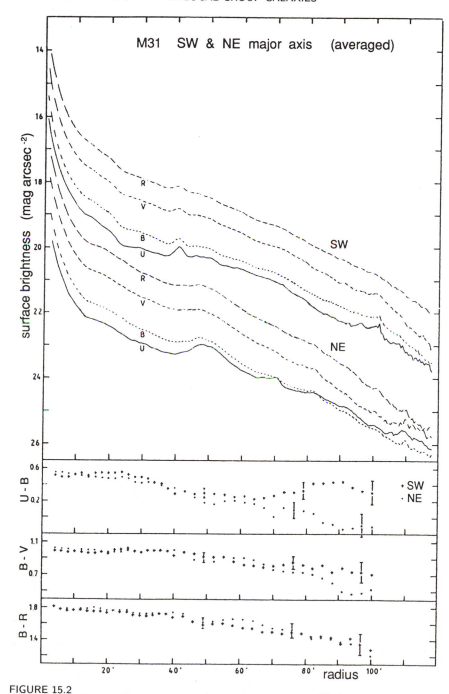

FIGURE 15.2

Radial light and color distributions along the major axis of M31 in U, B, V, and R. The profiles of the SW side have been shifted upward by 3 mag. Colors have not been corrected for foreground reddening. (From Walterbos and Kennicutt 1987.)

The surface density of molecular gas in M31 is a factor 3 to 4 lower than in the Galaxy, although the neutral-gas surface densities are somewhat higher. All information points to a lower current star-formation activity in the Andromeda galaxy.

15.2   DISK SCALE LENGTH AND THE HUBBLE TYPE OF THE GALAXY

An important question is what Hubble type applies to the Galaxy. The answer to some extent depends on what we think the disk scale length is; so I will discuss this again in somewhat more detail than I have already. As was presented in more detail earlier, the surface brightness of the Galaxy as measured with Pioneer 10 shows a disk scale length of 5.5 ± 1.0 kpc. Now we look at further evidence to derive a best value.

It is important to realize that the value of 3.5 kpc that is often seen in the literature is essentially based on a single argument by de Vaucouleurs (1979). He takes the value for the local disk surface brightness from star counts in the polar caps, assumes that the central value is Freeman's canonical one, and assumes a distance to the Galactic center. Other methods used by de Vaucouleurs are difficult to evaluate. He also uses the distribution of total gas within the solar circle; in some late-type galaxies the CO distribution follows that of the blue light. However, the Galaxy shares with M31 and NGC 891 a central depression, and there is no supporting evidence that this behavior also applies to such galaxies. Finally, he uses the surface density of H$\scriptstyle\rm I$ beyond the solar circle. There are two problems there. In the first place, the derived surface density from observations depends very sensitively on the assumed rotation curve and differs greatly between various publications. Secondly, in external galaxies the scale length of the H$\scriptstyle\rm I$ is different from that of the light. From the sample of Wevers *et al.* (1986) the ratio of H$\scriptstyle\rm I$ to blue scale length is 1.85 ± 0.35.

With the currently best value for the central surface brightness and its scatter (see Chapter 5) and the Pioneer value for the local surface brightness, it would follow that $h = 4.2 \pm 1.0$ kpc, if $R_0 = 8.5 \pm 1.0$ kpc. This already revises de Vaucouleurs' estimate. The outer H$\scriptstyle\rm I$ cannot be used well; Blitz *et al.* (1983) show an H$\scriptstyle\rm I$ distribution that is flat out to 16 or 20 kpc, and no reliable value for h can be extracted with this method.

A useful estimate comes from the observation (see Chapter 5) that young stars are seen in the anticenter direction out to at least 22 kpc. This puts R_{max} at 20 to 25 kpc. For external edge-on galaxies, van der Kruit and Searle (1982) find $h/R_{max} = 4.2 \pm 0.6$, and it would follow that $h = 5.4 \pm 1.0$ kpc.

There are also estimates possible from stellar dynamics. Lewis (1986; Lewis and Freeman 1989) has measured the velocity dispersion of old-disk giants as a function of Galactocentric radius, and finds these to show an e-folding of 8.7 ± 0.6 kpc. If, as discussed in detail earlier, this velocity dispersion falls with an e-folding twice that of the stars, then we expect $h = 4.4 \pm 0.3$ kpc.

A final method involves the asymmetric drift. The relevant equation is

$$\frac{\partial \ln\langle U^2 \rangle}{\partial R} + \frac{\partial \ln \rho}{\partial R} = \frac{V_t^2 - V_{rot}^2}{R\langle U^2 \rangle} - \frac{1}{R}\left(1 - \frac{B}{B-A}\right). \qquad (15.1)$$

Now in earlier times the first term on the left was set equal to zero in the absence of any detailed knowledge. However, we now know that it is not negligible at all (see the data of Lewis above) and is most likely equal to $-1/h$. Using this and usual values for the local Oort constants leads with Plaut's (1965) data for old-disk variables to $h = 4.3$–5.8 kpc.

From this discussion I conclude that there is ample support for a value for h as indicated by the Pioneer photometry and none for the old value of 3.5 kpc. The best value for h is 5.0 ± 0.5 kpc.

Classic ways to decide the morphological type of the Galaxy have involved arguments concerning the luminosity of the bulge, the spiral structure as determined from H I studies or the distribution of H II regions, and the local disk surface brightness. De Vaucouleurs has used over the years many such arguments, and settled on this basis on a classification of SAB(rs)bc (de Vaucouleurs and Pence 1978). In all subdivisions in this scheme the final classification is intermediate between two designations, and the suspicion is that this indicates uncertainty rather than precision.

Hodge (1983) used the scale length h_{HII} of the number distribution of H II regions to derive a type Sc, based on a relation between this parameter and the integrated magnitudes of spirals. In this plane the types Sb, Sc, and Sd form parallel sequences. Since all spirals appear to have the same central surface brightness, this segregation is only possible if the ratio between h_{HII} and h_{light} depends on type (being smaller for earlier types). Now, Hodge and Kennicutt (1983) have found a general similarity between these two scale lengths. Their figure, however, contains mostly Sc galaxies, and the Sb's may not follow this relation. Indeed, one of the two Sb's (NGC 2841) has $h_{HII} = 2.6$ kpc and $h_{light} = 5.4$ kpc. Also for M31 a fit to the number counts of H II regions gives $h_{HII} = 3.0 \pm 0.6$ kpc (van der Kruit 1987), which is also smaller than the 6 kpc for the surface brightness. Hodge's value for h_{HII} for the Galaxy is 2.35 ± 0.05 kpc, which would be consistent with the Galaxy also being of type Sb. However, using other data h_{HII} can also be estimated as 4.6 ± 1.0 kpc (van der Kruit 1987).

Color is known to correlate with Hubble type, as discussed above. For the Galaxy we unfortunately only know the color of the local disk and that with a large uncertainty, namely, $B - V = 0.85 \pm 0.15$. This is calculated from the color of the Galactic background starlight at high latitudes and is not necessarily typical for the disk as a whole. It would indicate a type of about Sb, but we are probably living in a region with relatively low current star formation.

Another property that can in principle be used is the gas content. Wevers (1984) has for this purpose defined a parameter, which is the logarithm of the ratio of surface brightness in the B band to the H I surface density, both evaluated at 3 optical scale lengths. For the Galaxy this property comes out as 0.5 ± 0.3

in solar units (van der Kruit 1986). This is similar to Sb galaxies (it is negative for Sc galaxies) and therefore points to such a classification.

One of the major criteria for the Hubble classification is the bulge-to-disk ratio. Quantitative calculations of this parameter do show a correlation as expected, but there is a large scatter within each type. The actual number gives Sbc as the most probable type, although the whole range from Sab to Scd is consistent (van der Kruit 1987).

The most compelling argument appears to come from the CO distribution. As I discussed also above, Young (1987) has reviewed the observed radial distributions. Sc galaxies show a central peak in the CO intensities, whereas at least 40 percent (improved resolution may increase this number) of the Sa and Sb galaxies have a central hole. In this respect the Galaxy is similar to Sb galaxies, and at present no Sc galaxy is known that in its CO distribution resembles the Galaxy. This observation seems to point rather decisively to Sb.

Finally, we may try to find galaxies that in their properties closely resemble the Galaxy and then look at their classification. From the section above it follows that M31 (SbI–II) is similar in many respects (including the central hole in the CO distribution), but it has a lower current rate of star formation. This may well be within the range encountered within any particular type. A galaxy that is in many respects very similar to the Galaxy is NGC 891 (van der Kruit 1984). Not only are the photometric parameters of bulge and disk of the two systems very similar, but also the CO and H I distributions and number of globular clusters are very closely the same. Both systems are seen edge-on. The appearance of the Milky Way from the southern hemisphere in the winter, when the Galactic Center culminates near the zenith in the middle of the night, is strikingly similar to pictures of NGC 891. Indeed, many illustrations comparing photographs of NGC 891 and of the Milky Way with superwide-angle cameras have been published, and the strong resemblance has been stressed often (e.g., King 1976). NGC 891 is invariably classified Sb.

On this basis I conclude that the most likely Hubble type of the Galaxy is Sb, where the strongest evidence comes from the CO distribution, the gas content, and the strong resemblance to NGC 891. It is doubtful that we will ever be able to do better, because no quantitative observable correlates so strongly with morphological type that it excludes all but one classification bin. Sc is almost certainly ruled out, but Sbc is not.

The luminosity class is more difficult to decide, but a comparison with the Virgo-Cluster galaxies can be made. For this we look at the photometry of Watanabe (1983) in the 6° core of the cluster, combined with Sandage–Tammann classifications. For Sc and Sb galaxies it turns out that all six with class I or I–II have disk scale lengths larger than 4 kpc, but for the 25 of class II this is true only for five of these. The six of class I and I–II have central surface brightness 20.6 ± 0.7 B-mag arcsec^{-2} and the five of class II 21.2 ± 1.0. For the Galaxy this parameter is 21.1 ± 0.3. In view of these limited statistics of the two photometric disk parameters, I conclude that the Galaxy is probably not of luminosity class

II, but most likely I–II. We probably live in an SbI–II galaxy, and these systems are among the largest spiral galaxies in the Universe.

15.3 THE MILKY WAY COMPARED WITH OTHER GALAXIES

We have seen that we live in a relatively large spiral galaxy of intermediate Hubble type. I have already indicated above that 10 percent of all disk stars in the Universe live in spirals with a disk scale length larger than 4 kpc. In this final section I will make some further remarks on this, and on the location and orientation of the Galaxy in space. First I will discuss the question of its appearance in relation to known nearby galaxies. In the semipopular literature and fundamental textbooks, we often see M31 or M81 described as galaxies looking much like our own.

De Vaucouleurs (see, *e.g.*, de Vaucouleurs and Pence 1978) has through the years used various arguments to identify systems that would bear a close resemblance. De Vaucouleurs and Pence have concluded on this basis that the morphological classification is SAB(rs)bcII and have published a sketch of what this morphology implies for the outsider's view. This is partly based on known morphological features, mainly from H I distributions and the distribution of H II regions, such as the definition and pitch angle of the spiral arms and the possible presence of a central bar and ring. They further identify galaxies of similar morphological type, for which the photometric parameters are also similar. These parameters are, however, integrated parameters, such as absolute magnitude, colors, diameter, mean surface brightness, and hydrogen index (H I mass to total light ratio). The galaxies are NGC 1073, 4303, 5921, and 6744. Grosbøl's (1985) photometry gives for the first three galaxies disk scale lengths of 5.0, 2.6, and 2.9 kpc, respectively, on a distance scale with a Hubble constant of 75 km sec^{-1} Mpc^{-1}. NGC 1073 has an H I profile with a width of only about 95 km sec^{-1}; its inclination is small and is in the range 20–30°, so that the rotation velocity is at most 130 km sec^{-1}, and much smaller than that of the Galaxy. NGC 6744 is a southern galaxy for which no surface photometry is available; its rotation velocity estimated from the H I profile is about 210 km sec^{-1}. Also the H I content is similar to that of the Galaxy.

I have searched the available data for galaxies that are similar to our own in some parameters that are not governed by the morphology of the spiral structure. These are a disk photometric scale length (which is Hubble-constant-dependent) of 4 to 6 kpc, a bulge luminosity of about 10 or 20 percent of the total, a bulge effective radius of 2 to 3 kpc, a disk color $B - V$ of 0.6 to 0.8, a rotation velocity (in the flat part) of 210 to 230 km sec^{-1}, and an H I content of a few times 10^9 $\mathcal{M}_\odot$. Note that the disk luminosity is not a parameter on which restrictions are put, since the galaxies are supposed to conform to Freeman's law, and therefore this is set by the scale length. Also the integrated luminosity enters through the selection on rotation velocity via the Tully–Fisher relation. Finally, the Hubble type is somewhat restricted through the ratio of bulge-to-total luminosity and disk color. Of course, the number of galaxies for which all this information is available is not very large; it is mainly set by the requirement that full surface photometry is available, so that disk-bulge separation can be performed. This does eliminate most of the photographic work, because often this extends to the brighter isophotes only when special short exposures have been taken. The rotation velocity and HI content are known already from single-dish 21-cm line observations.

I have already indicated the resemblance of the Galaxy to NGC 891. This similarity cannot come from a survey as described, since for edge-on galaxies all but the integrated disk luminosity and color are known. However, leaving these aside, NGC 891 does still come out as a close twin. Some galaxies suggested before that are often seen as such in the literature do not conform to these strict limits, if applied all at the same time. *E.g.*, M31 (as already seen above) has too large a disk scale length, too bright a bulge, and too large a rotation velocity. M81 also rotates too fast (about 255 km sec^{-1}) and has much too small a scale length (2.5 kpc according to Kent 1987a). NGC 4565 has too bright a bulge and too high a rotation velocity (about 260 km sec^{-1}). Finally, the SbI galaxy NGC 3200, which was proposed by Rubin (1983) to be similar to the Galaxy on the basis of its rotation curve, does not qualify on the basis of the disk scale length of about 8 kpc on the distance scale used here (Kent 1986).

The only galaxy that conforms to all these criteria is the Sbc(s)I–II spiral NGC 5033. Photometry of this galaxy has been given by Wevers *et al.* (1986) and Kent (1987a). The assumed distance is 14 Mpc. The disk is not entirely exponential, but the profiles from these studies can be represented reasonably well with the parameters $R_e = 2.9$ kpc and, for a color $B - V = 0.9$ and a flattening of 0.5, luminosity $4 \times 10^9 \, L_{B,\odot}$ for the bulge and for the disk $\mu_0 = 22.0$ B-mag arcsec^{-2} and $h = 5$ kpc. An optical picture is shown in Fig. 15.3. Some parameters for the three galaxies are collected in Table 15.2.

For NGC 891 the disk parameters apply only to the old-disk population, and for the bulge-to-total luminosity ratio, it has been assumed that the younger populations provide as much B light as the old disk.

It is interesting to note that NGC 5033 has a somewhat brighter bulge and lower HI mass than the Galaxy; on this basis the latter would be of somewhat earlier Hubble type than Sbc. This offsets the comparison to M31, which would make the Galaxy of slightly later type than Andromeda (Sb).

By way of concluding remarks, I will repeat some statements concerning Galactic and extragalactic research that I have made previously in the literature. The first concerns the fact that astronomy developed first in the northern hemisphere, but the Galactic center is at negative declinations. As a matter of fact, the present location on the sky has the lowest declination that the center gets during the precession cycle of 26,000 years. Half a cycle ago or in the future, the declination would be $+18°$. However, the Magellanic Clouds are always southern objects: the LMC does not get further to the north than declination $-60°$ and the SMC $-45°$. M31, which is currently $41°$ above the equator, can actually attain declinations between $10°$ and $57°$.

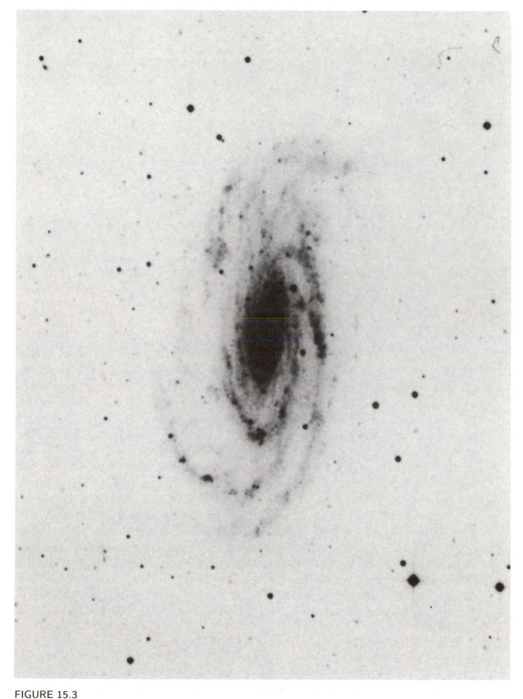

FIGURE 15.3
A IIIa-J photograph of the Sbc(s)I–II galaxy NGC 5033. This galaxy is similar to the
Galaxy in its photometric parameters of disk and bulge, disk color, rotation velocity, and
H I-mass-to-total-light ratio.

TABLE 15.2
Parameters for our Galaxy and two closely similar galaxies

	Galaxy	*NGC 891*	*NGC 5033*
Distance (Mpc)		9.5	14
Bulge:			
R_e (kpc)	2.7	2.3	2.9
b/a	0.7	0.7	?
L_{bulge} $(L_{B,\odot})$	2×10^9	1.5×10^9	4×10^9
Disk:			
μ (B-mag arcsec^{-2})	22.1	22.9 (old disk)	22.0
h (kpc)	5	4.9	5
L_{disk} $(L_{B,\odot})$	1.7×10^{10}	6.7×10^9 (old disk)	1.7×10^{10}
$B - V$	0.8	0.9 (old disk)	0.6
R_{max} (kpc)	20–25	21	22
L_{bulge}/L_{tot}	0.12	0.07	0.19
V_{rot} (km sec^{-1})	220	225	215
$\mathcal{M}_{HI}$ $(\mathcal{M}_\odot)$	8×10^9	4×10^9	4×10^9

The location of the Sun in the Galactic plane is also amusing to consider. Its velocity relative to the local standard of rest is such that in the plane the epicycle excursion (period 1.7×10^8 years compared to a rotation around the center in 2.4×10^8 years) takes it in toward the Center by 0.11 kpc in 2.0×10^7 years and out by 0.57 kpc from the present radius after 1.0×10^8 years from now. More interesting is the vertical motion. At present the location is about 12 pc from the plane as estimated from the curvature of the ridge-line in the Pioneer 10 surface-brightness distribution of the Milky Way. The vertical period is 7×10^7 years, and in about 1.5×10^7 years it will be about 85 pc from the plane. This is interesting, because then it will be situated above much of the dust lane. Had the evolution of life on Earth proceeded slower by 0.5 percent, we would have been allowed a much less obscured view of the Galactic center. We are even more unlucky, since stars of the solar age have a velocity dispersion of 20–25 km sec^{-1} in the z-direction, and stars with such a vertical velocity in the plane move out to z about 300 pc. Only 2 percent of the old-disk stars are 12 pc or less from the plane. If the dust lane has a thickness of 100 pc, no less than 90 percent or so of all old-disk stars are above or below this layer.

There is one aspect in which we are probably extremely fortunate, and that is the orientation of the Galaxy with respect to the Local Supercluster. Our Galaxy is situated somewhere toward its outer boundaries, and the Virgo Cluster appears to be the center. Now, the plane of the supercluster is oriented with an angle of 73° relative to the plane of the Galaxy, and the Virgo Cluster is located only 15° from the Galactic pole. So, as seen from Virgo ours would be one of the most face-on oriented spirals in the nearby Universe, since an inclination of 15° or less occurs in only 4 percent of all cases for random orientations. This gives us a rather unobscured view of the many galaxies in the central areas of the Local Supercluster. We are also fortunate to have a large Sb as M31 nearby (0.7 Mpc), because even in the core of the Virgo Cluster, which for a cluster is rather rich in spirals, the mean separation between two spiral galaxies with scale lengths larger than 4 kpc is 0.8 Mpc.

The favorable location of the Virgo Cluster does not apply for such galaxies as M31 or NGC 891. To illustrate this I calculated the extent on the sky of the zone of avoidance for these two galaxies (defined by latitudes ±15° as seen from their disks) and superposed these on the distribution of bright galaxies as illustrated in the original Shapley-Ames catalog (Fig. 15.4). The Virgo Cluster is at about R.A. 12^h, Dec +5°, and the plane of the Local Supercluster crosses the equator at about right angles at about R.A. 12^h and 0^h. This is, of course, only the approximate view from NGC 891, but for M31 the change would in general not be perceptible, except for the Local-Group galaxies. From M31 the view is much worse than from our Galaxy. The Virgo Cluster is located at a latitude of about 16°, which would not rule out studies of this cluster, but would make them much more difficult. Studies of our Galaxy from M31 are as hampered by large inclination as we are when observing M31. Our Galaxy would have an inclination of about 70°. Of course, most bright spirals will be seen from similar distances and at about the same inclination as from our position. Generally for most the latitude is lower and less favorable.

NGC 891 is a much more outlying member of the Local Supercluster, at a distance of about 26 Mpc from the Virgo Cluster (assuming that it is 20 Mpc from us). From NGC 891 our Galaxy will have an inclination of 72°, but it will of course be seen at latitude almost zero. The zone of avoidance actually almost coincides with the plane of the Local Supercluster. The Virgo Cluster appears at a latitude of only a few degrees, and extragalactic astronomy from NGC 891 would be very seriously hampered. The relevant parameters for the appearance of some large, major galaxies as seen from NGC 891 are given in Table 15.3.

It is evident that many spirals near NGC 891 cannot be studied because of the low latitude. Especially the large Sb galaxies M31 and NGC 4258 are cases of bad luck: these are seen almost exactly edge-on and would offer a unique opportunity to study the distribution of the stellar populations, but are only 3° from the plane.

The question arises, of course, whether we in the Galaxy are also missing some unique chances because of the orientation of the plane and therefore of the zone of avoidance. Such a case has recently appeared in the literature. Lynden-Bell et al. (1988) have suggested the presence of a "Great Attractor" at $\ell = 307°$, $b = 9°$. This is right in our zone of avoidance and would, if indeed present, have revealed itself had we been able to look in that direction unhampered by obscuration. The equatorial coordinates of the Great Attractor are R.A. $= 13^h$ 15^m, Dec $= -53°$. It can be seen from Fig. 15.4 that this position also happens

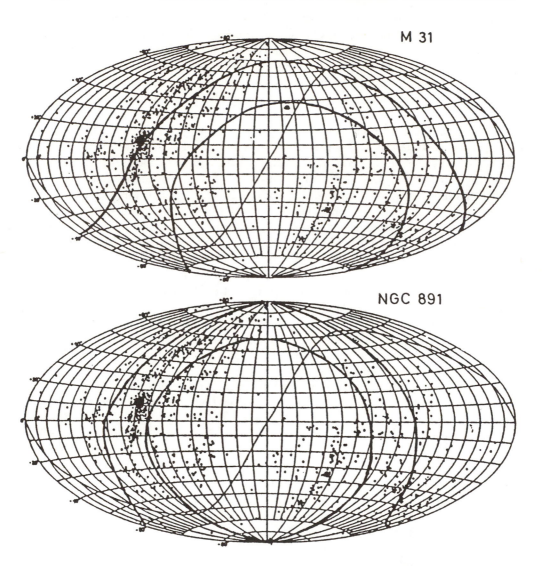

FIGURE 15.4
The zones of avoidance, defined by galactic latitude less than 15°, as seen from M31 and
NGC 891, shown by the thick lines. The coordinates are R.A. and Dec, and the center of
the figure is at R.A. 6^h 40^m, Dec 0°. The underlying distribution is that of bright galaxies
from the original Shapley–Ames catalog, and the thin line indicates the Galactic plane.
From M31 this distribution would be different only in detail, but for NGC 891 it only
shows the general features. From both galaxies much of the Local Supercluster appears at
low latitudes, especially from NGC 891, and many galaxies would be heavily obscured.
(From van der Kruit 1987.)

TABLE 15.3
The appearance of some major galaxies as seen from NGC 891

Galaxy	D(Mpc)	$|b|(°)$	$i(°)$
Ellipticals:			
M87	26.1	2	—
NGC 5128 (Cen A)	13.5	1	86 (dust)
Sb galaxies:			
M81	8.3	2	62
M31	8.8	3	90
GALAXY	9.5	2	72
NGC 7331	10.5	74	38
NGC 4258	11.4	3	88
Sc galaxies:			
NGC 925	1.7	57	56
IC 342	5.9	2	62
NGC 628	4.8	1	53
NGC 694 6	8.7	55	68
M101	11.1	14	71
M51	13.6	10	54

to be in the zones of avoidance of M31 and NGC 891. We might in this respect not be the only unfortunate astronomers in the nearby Universe.

REFERENCES

Blaauw, A., and Schmidt, M., eds. 1965. *Stars and Stellar Systems.* Chicago, IL: Univ. of Chicago Press.

Blitz, L., Fich. M., and Kulkarni, R. M. 1983. *Science*, **46**, 177.

Bothun, G. D., and Thompson, I. B. 1988. *Astron. J.*, **96**, 877.

Deul, E. 1988. Ph.D. thesis, Univ. of Leiden.

Dressler, A., and Richstone, D. O. 1988. *Astrophys. J.*, **324**, 701.

Gilmore, G., and Carswell, B., eds. 1987. *The Galaxy.* Dordrecht: Reidel.

Grosbøl, P. J. 1985. *Astron. Astrophys. Supp.*, **60**, 261.

Hodge, P. 1983. *Publ. Astron. Soc. Pacific*, **95**, 721.

Hodge, P., and Kennicutt, R. C. 1983. *Astrophys. J.*, **267**, 563.

Kent, S. M. 1986. *Astron. J.*, **91**, 1301.

———. 1987a. *Astron. J.*, **93**, 816.

———. 1987b. *Astron. J.*, **94**, 306.

King, I. R. 1976. *The Universe Unfolding.* San Francisco: Freeman, p. 388.

Kruit, P. C. van der. 1984. *Astron. Astrophys.*, **140**, 470.

———. 1986. *Astron. Astrophys.*, **157**, 230.

———. 1987. In Gilmore and Carswell, p. 27.

Kruit, P. C. van der, and Searle, L. 1982. *Astron. Astrophys.*, **110**, 61.

Lewis, J. R. 1986. Ph.D. thesis, Australian Natl. Univ.

Lewis, J. R., and Freeman, K. C. 1989. *Astron. J.*, **97**, 139.

Lynden-Bell, D., Faber, S. M., Burstein, D., Davies, R. L., Dressler, A., Terlevich, R. J., and Wegner, G. 1988. *Astrophys. J.*, **326**, 19.

Mould, J. R., Kristian, J., and Da Costa, G. S. 1983. *Astrophys. J.*, **270**, 471.

———. 1984. *Astrophys. J.*, **278**, 575.

Plaut, L. 1965. In Blaauw and Schmidt, p. 267.

Rubin, V. C. 1983. In Shuter, p. 379.

Shuter, W. L. H., ed. 1983. *Kinematics, Dynamics, and Structure of the Mily Way.* Dordrecht: Reidel.

Schwering, P. 1988. Ph.D. thesis, Univ. of Leiden.

Vaucouleurs, G. de. 1957. *Astron. J.*, **62**, 69.

———. 1979. *Observatory*, **99**, 128.

Vaucouleurs, G. de., and Pence, W. D. 1978. *Astron. J.*, **83**, 1163.

Walterbos, R. A. M. 1986. Ph.D. thesis, Univ. of Leiden.

Walterbos, R. A. M., and Kennicutt, R. C. 1987. *Astron. Astrophys.*, **198**, 61.

Watanabe, M. 1983. *Ann. Tokyo Astron. Obs., 2d Ser.*, **19**, 121.

Wevers, B. M. H. R. 1984. Ph.D. thesis, Univ. of Groningen.

Wevers, B. M. H. R., Kruit, P. C. van der, and Allen, R. J. 1986. *Astron. Astrophys. Supp.*, **66**, 505.

Young, J. S. 1987. *Star Formation in Galaxies*, NASA Conf. Proc. **2466**, p. 197.

CHAPTER 16

SOME INTERESTING
AND TRACTABLE PROBLEMS

Gerard Gilmore

Harold Bluetooth, King of Denmark (936–966), required the Viking Toko to show his proficiency with the bow by shooting an apple from his own son's head. Saxo Grammaticus' *Gesta Danorum* recorded this event around 1200, three-quarters of a century before William Tell is reputed to have suffered similar treatment at the hands of the *bailli* Gessler, and some 275 years before the first surviving written reports of the latter event. This efficient use of the historical record illustrates a useful lesson that can be applied in astronomical research: we can often learn a great deal from familiarity with the literature. It is also true that we can understand a great deal by thinking about the explanations devised to explain the first few observations in a field of research. Sometimes the lesson is derived from studying why some model now known to be erroneous was adopted, sometimes from seeing how a field was unified and made coherent from only very general data.

A useful example of the former situation is presented in Chapter 2, which discusses how we can use the failure of the Kapteyn Universe to provide a realistic picture of the stellar distribution in the Milky Way to appreciate the limitations in modeling modern star-count data, even though these data are of extremely high precision compared to those available to Kapteyn. The most relevant example of the success of a simple concept in creating order from apparently disparate data is Baade's concept of stellar populations (Baade 1944), which led to an enhanced understanding of the structure and evolution of all galaxies, not just those spirals from which it was deduced. It would be much more difficult to derive or to learn from such a concept were it first discovered today, because a large body of disparate data now exists, showing the many complexities, exceptions, and complications in the real world. Nevertheless, the concept of stellar populations is of continuing utility.

Another example of a good idea, though one which seems not to have been adequately appreciated at the time, is Strömberg's explanation of angular momentum in galaxies as originating in tidal torques applied between proto-galactic clouds at their time of maximum expansion before they collapsed to form galaxies (Strömberg 1934). Although developed specifically to explain the (Strömberg) asymmetric drift apparent in high-velocity stars near the Sun, it provides a fundamental insight into the formation of galaxies. After development through many papers in the 1950's, it became the basis for the analysis of Eggen *et al.* (1962) of the structure of the Milky Way, and for modern analyses of galaxy formation in a cosmological setting (see Ryden 1988 for a recent example).

In addition to enhancing our understanding by becoming familiar with available information, we can test our current understanding by detailed study of those situations in which available models appear inconsistent with available data, or where there is no available model. The remainder of this chapter presents some examples of such projects, all of which have been selected as being likely to provide improved understanding of important aspects of Galactic structure and evolution and also as being feasible with available techniques. This list contains only a few of the many relevant projects, and is not intended to be either comprehensive or exclusive. It is illustrative of the types of projects which are of current interest, and was chosen idiosyncratically to exemplify those areas of current Galactic research in which available understanding is least satisfactory. When considering such projects, we must remember that galaxies are neither simple nor identical. Thus detailed modeling of a specific set of data in a specific situation need not necessarily provide much information of general applicability to all galaxies.

16.1   IS THE DARK HALO TRIAXIAL?

It is highly likely that spiral galaxies have triaxial bulges. We may then ask to what extent such triaxiality is due to physical and stellar-dynamical processes in galaxy formation, and to what extent it reflects triaxiality in the underlying dark-matter distribution. One of the several ways in which this could be tested is by detailed mapping of the orientation of the velocity ellipsoid for old and high-velocity stars near the Sun.

The stellar velocity ellipsoid will be aligned with the coordinate system in which the Hamilton–Jacobi equation (which provides the equations of motion) is separable, if such a coordinate frame exists. If the potential is triaxial, there is no need for this coordinate frame to be precisely aligned with the Galactic minor axis at all points, and in general it will not be aligned with the Galactic center. The orientation of the potential can thus be discovered by careful study of the velocity ellipsoid of stars moving in equilibrium in the large-scale potential field. Old and high-velocity stars must be studied, since young stars need not be in a steady state in equilibrium with the potential, and low-velocity stars will have their orbits affected significantly by the local deviations from axisymmetry evident (as spiral arms, and so on) in the potential of the Galactic disk near the plane.

A detailed study of the space motions of high-velocity stars, with due allowance for the asymmetric biases induced by proper-motion selection techniques and with careful consideration of the effects of distance uncertainties on the space

motions, has the capability of providing important information on the shape of
the distribution of dark matter in the Galaxy. Analysis of systematic effects in
the kinematics of stars as a function of stellar age led to the discovery of Parenago's discontinuity: a change in the velocity dispersion and orientation of the
velocity ellipsoid in disk stars evident near spectral type F5. Later-spectral-type
stars can live long enough to come into equilibrium with the large-scale Galactic
Potential, and thus have kinematics which are no longer dominated by local non-
axisymmetric effects in the disk potential, but earlier-spectral-type stars cannot.
A good review of stellar age–kinematic analyses is provided by Delhaye (1965).
Extension of such studies to high-velocity stars allows study of the potential on
large scales.

16.2 DOES THE DARK HALO HAVE A HOLLOW CORE?

Association of a length scale with the dark mass would provide the most direct
indication of its nature currently available. In spite of very considerable efforts, no
such length scale has yet been identified unambiguously. The most likely places in
which a scale length might have been identified were in globular clusters (scales
$\lesssim 10$ pc), the solar neighborhood (scales $\lesssim 1$ kpc), and in dwarf galaxies (scales
~ 1 kpc). With the marginally significant exception of a few dwarf galaxies (the
available data on which are too meager for a serious analysis), no dark mass has
been identified unambiguously as being associated with any of these scales. The
next larger scale on which we might detect structure is the few kpc characteristic
of the central regions of the Galaxy.

The mass distribution in the central few kpc of the Galaxy can be derived
from the kinematics of gas and stars. Acquisition of the relevant stellar data
would be straightforward with current multi-object spectroscopic systems, and
excellent 21-cm H I data already exist. Interpretation of the kinematic data will
require detailed modeling to include the effects of noncircular motions, particularly those associated with a triaxial (stellar) mass distribution. The 21-cm
data have been interpreted to provide an *apparent* rotation curve, derived on
the assumption of no self-absorption and no noncircular motions, and are shown
in Fig. 16.1), together with a decomposition of the several identified contributions to the mass distribution. Recent careful analysis of the available 21-cm
data (Burton 1985) shows substantial self-absorption effects, and clear evidence
for noncircular motions within a few degrees of the Galactic plane. At higher
latitudes the kinematics are less confused, and suggest that the maximum in
the rotation curve of Fig. 16.1 near 500 pc is entirely an artifact of noncircular
motions and self-absorption. Thus the rotation curve which corresponds to the
Galactic potential differs substantially from the apparent rotation curve, somewhat complicating the analysis of the central mass distribution. Nevertheless, in
principle it is possible to calculate the integral mass distribution in the central
few kpc of the Galaxy to quite high precision. Comparison of that distribution

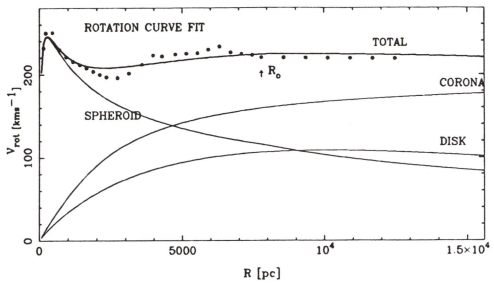

FIGURE 16.1

The Galactic apparent rotation curve, with mass models fitted including the inner maximum in the apparent rotation curve as a real dynamical feature. The dynamical reality of this inner maximum is highly suspect, so that the mass associated with the luminous component labeled "spheroid" in the figure remains uncertain. (Royal Astronomical Society)

with the luminosity distribution then reveals the spatial distribution of the dark mass, and measures any characteristic associated scale length.

The mass distribution associated with the stellar distribution in the central regions of the Galaxy has not yet been calculated reliably. Preliminary attempts have involved calculating the mass distribution from the light distribution associated with the upper giant branch. Such calculations are almost entirely indeterminate, since the range of stellar masses which occupies the giant branch in an old stellar population is extremely narrow. Thus conversion of a giant-branch luminosity to an integrated stellar mass is in effect the inversion of a delta function. There is no reason why the luminosity (and hence mass) function within perhaps 1 kpc of the Galactic center could not be calculated reliably by direct counting of low-luminosity main-sequence stars, to improve substantially this determination. The effective distance modulus of the Galactic bulge out of the plane (including absorption) is $\lesssim 15^m$, whereas good CCD data can be obtained to at least $R = 22$ in good seeing. The technical problems are similar to those in globular-cluster studies, and are routinely mastered. Thus stars as faint as $M_R \sim +7$, $\mathcal{M} \sim 0.5\,\mathcal{M}_\odot$ can be observed directly. Besides testing the environmental dependence of the stellar initial mass function, such data can be compared with the dynamical mass measures to test for structure in the distribution of dark mass.

16.3 DOES THE DARK HALO HAVE AN EDGE?

The relationship between the large-scale distributions of dark and luminous mass would be clarified if we knew if galactic dark halos had edges, beyond which their volume mass density dropped to very low values, or if they simply fade gently into a background with structure on much larger scales. Although the latter option is more consistent with currently fashionable cosmological theories, there is some evidence that the former is more correct. This evidence comes from analysis of the kinematics of the few distant globular clusters and satellites of the Milky Way for which useful data exist. Little and Tremaine (1987) have provided the most recent analysis of this type, and suggest, from analysis of data for various subsets of the available sample of 15 tracer objects, that the Galactic mass distribution effectively truncates at about 60 kpc from the Galactic center.

The validity of this conclusion, which necessarily is uncertain because the sample of tracers available is very small, can be tested by studies of distant stars in the Galaxy. Luminous carbon stars and blue-horizontal-branch stars have been identified at very large distances ($\gtrsim 50$ kpc) from the Galactic center, and very distant Miras and RR Lyraes could also be found with relative ease if they were present in significant numbers. At least the first two of these classes of stars are sufficiently common that both a density profile and a velocity distribution function can be measured reliably, with some hundreds of tracers potentially being available to distances of perhaps 60 kpc. If field stars are indeed distributed like the globular clusters, then even greater distances will be feasible.

Although several groups are attempting surveys of very distant halo stars, usually as a way to use those blue objects which "contaminate" quasar surveys, relatively little effort has as yet been devoted to the data analysis. Such analysis is complicated by the possibility that the outer halo may possess considerable angular momentum, so that much of the energy of a stellar orbit is in the unobservable transverse motion. Detailed models of such distribution functions are, however, feasible with available dynamical methods, and deserve considerable efforts. We can in principle calculate what is effectively the Galactic rotation curve to distances of up to 15 disk-luminosity scale lengths, vastly further than is possible in any other galaxy.

16.4 DOES THE DARK HALO HAVE RIPPLES?

The remaining length scale on which we might hope to detect some signature of the nature of the dark matter is that on which we see structure in galactic rotation curves. Bumps and ripples in rotation curves often correspond to features in the luminosity distribution of the galactic disk (*e.g.*, spiral arms) and hence are sometimes interpreted as evidence for dominance of the galactic potential by the mass of the disk. Since in the Milky Way, where the mass of the disk is best known independently of the form of the rotation curve, the disk does not dominate the potential at any radius (see Fig. 16.1 above), it is important to examine

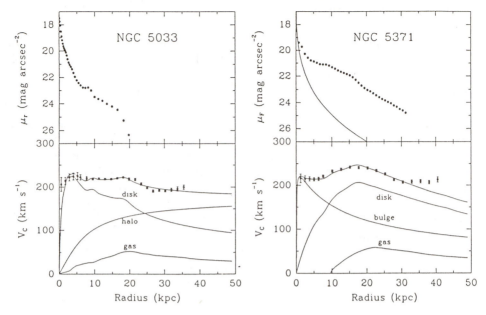

FIGURE 16.2

Luminosity profiles (top panels) and H I rotation curves (bottom panels: data are points with error bars; models for the identified components are the smooth curves) for the two identified galaxies. The detailed correspondence between the rotation curve predicted from the luminosity distribution and that observed is a test both of the validity of the mass-to-light ratios adopted in the model fits and of the amplitude of noncircular motions in the galaxies. (From Begeman 1987.)

carefully the degree of correspondence between the luminosity distribution and the kinematics.

Fig. 16.2 illustrates two unusually high-quality rotation curves, each of which is fitted somewhat less well than is typical by the luminosity distribution. The rotation-curve data and modeling are from the excellent work by Begeman (1987), and were chosen here to illustrate the information which can be derived from the residuals from such modeling when the data are of extremely high quality. They were also chosen as having almost the same velocity amplitude as does the rotation curve of the Milky Way, and to have rather similar luminosity distributions. Thus we might expect some similarity in the interpretation of the rotation curves of these galaxies and that of the Milky Way Galaxy.

The important information for our purposes here is the residual velocity between that observed and that of the model. For NGC 5033 the observed velocity is substantially in excess of that consistent with the model within ~ 2 kpc, is below that of the model for $\sim 2\text{–}5$ kpc, and is poorly fitted in slope by the model beyond ~ 20 kpc. The difficulties in interpreting rotation curves in the central regions of the Milky Way are mentioned above. Presumably, similar strictures apply in other galaxies, because of noncircular motions, which may be caused by the hydrodynamical effects of an active nuclear region, or more likely by the dynamical effects of a nonaxisymmetric (triaxial) mass distribution. If this is the explanation of the inner peak in the rotation curve of NGC 5033 (though remember that the effects of a finite beam size can be important in the inner and outer point), then noncircular motions of up to ~ 100 km sec^{-1} are evident, in agreement with those apparently seen in the central ~ 1 kpc in our Galaxy. Thus interpretation of the inner regions of rotation curves in terms of mass distributions is at least sometimes unreliable, and certainly not straightforward. If the apparent rotation curve is a fair representation of the mass distribution for NGC 5033, then a highly centrally concentrated dark-mass component exists in that galaxy.

The poor correspondence between the fine structure in the luminosity distribution of NGC 5033 and that in the rotation curve is also of interest. Detailed agreement between luminosity and velocity structure is a strong argument in favor of "maximal-disk" models of rotation curves, in which we maximize the mass-to-light ratio of the disk contribution to the rotation curve, and minimize that of the dark halo. Although apparently conservative, this approach in fact presumes the existence of two different types of dark matter (extended dark matter in a halo, and dissipational dark matter in a disk) and therefore requires careful examination. Structure is seen in the luminosity profile of NGC 5033, both in the inner regions and as an edge to the optically luminous galaxy. However, in neither case is this structure well reproduced in the velocity field. The outer drop in the rotation curve, before a later rise, has both a different slope and occurs at a different radius than does that expected from the luminosity. Again, it may well be more sensible to interpret structure in velocity fields which occurs near features in luminosity profiles as evidence for noncircular motions, associated with streaming motions in spiral arms or shearing velocities at the edge of a disk, rather than as evidence for perturbations in the potential. It would be valuable to attempt a model of the full two-dimensional velocity and luminosity distribution of a galaxy (M31 and M33 might be suitable choices), while treating as the free parameter in fitting the disk contribution the amplitude of noncircular motions associated with spiral arms and regions of star formation, rather than the disk mass-to-light ratio. This would provide an important test of the extent to which local deviations from axisymmetry in the light correspond to deviations in the potential rather than the kinematics, and would test the disk/halo mass ratio.

NGC 5371 is one of many galaxies well modeled to the edge of the luminous galaxy without the need for a dark halo. Does this mean that some galaxies do not have significant dark mass in their optical boundaries? That is, can the dark mass have a hollow core, and therefore necessarily have very high angular momentum? It would be interesting to attempt models of such galaxies which fix the disk mass-to-light ratio at a value consistent with dynamical studies of the mass distribution near the Sun and in globular clusters (*i.e.*, $\mathcal{M}/L_V \sim 2$) and treat the distribution of dark mass as the unknown parameter, to see if any structure can unambiguously be associated with the dark mass.

Using the stellar mass-to-light ratio as a free parameter in the fitting of rotation is convenient, and seems conservative, but in fact has very considerable implications. Allowing a variable mass-to-light ratio in effect requires large variations in the stellar initial mass function from galaxy to galaxy, and between the halo and disk stellar populations in a single galaxy. Making such changes consistent with the well-established similarities in color and metallicity between galaxies of similar absolute magnitude and Hubble type is far from an easy exercise, and really requires very large changes in the number of substellar-mass objects formed from galaxy to galaxy. We might wonder then how a local process like star formation could know, as a function of position, how many low-mass stars it should form to provide a flat rotation curve.

16.5   IS THE "BULGE" PART OF THE "HALO"?

The luminosity distribution of the central few kpc of the Galaxy out of the plane is dominated by a component—the "bulge" for our purposes here—which has a half-light scale length of about 0.5 kpc. This is particularly apparent in near-infrared data, but is also evident in careful optical studies. The relationship of the extended stellar halo, represented by the subdwarfs near the Sun, to the bulge is quite unknown. It is important to remember that the bulge corresponds to the only nondisk contribution to the luminosity of external spiral galaxies which has a surface bright enough to be studied.

Available data provide a somewhat confusing picture of the properties of the Galactic bulge. Analysis of those stars dominant at $12\,\mu m$ in the IRAS survey suggests that they are intermediate-age long-period variables (Harmon and Gilmore 1988). The majority of the bulge population at low Galactic latitudes must be older than the Sun, and may be as old as the metal-rich globular clusters (Terndrup 1988). Chemical-abundance data for a sample of K giants in Baade's Window show them to be metal-rich, with modal abundance perhaps twice solar (Rich 1988). The distribution of abundances for these stars is consistent with that expected from the simple model of chemical evolution with a closed box (no inflow or outflow; see Chapter 13), but with effective yield significantly higher than that derived from observations in the solar neighborhood. Similar abundance data for planetary nebulae and RR Lyrae stars (Gratton *et al.* 1986), however, provide a modal abundance of one-half solar, consistent with the same effective yield as is seen near the Sun. A point which is related to this is why the metal-rich stars in the bulge should be appropriate as templates in stellar-population studies of giant elliptical galaxies. The absolute magnitude of the bulge is about -18, whereas big ellipticals are several magnitudes brighter. Given the well-established relations between absolute magnitude and color and metallicity (or at least line strength), it is really very unlikely that the integrated spectrum of the bulge can look like that of a much bigger galaxy. It would be interesting to investigate this further, because it suggests that the very metal-rich stars must provide a minor component of the integrated light of the Galactic bulge.

Currently available kinematic data are inadequate to provide any useful information on the structure of the bulge. By analogy with studies of the bulges of external spirals, we might expect the Galactic bulge to be rotating, with both velocity dispersion and mean rotation about half the asymptotic amplitude of the HI rotation curve, or about 100 km sec^{-1}. Straightforward application of the radial Jeans equation to the kinematics of stars near the Sun also predicts a value for the velocity dispersion of both the disk and the halo stellar populations of $\sim$ 125 km sec^{-1} near the Galactic center. Systematic rotation will reduce this value, whereas triaxiality of the central-bulge potential may either increase or decrease it substantially, depending on the geometry of the situation.

There are several fundamental properties of the bulge which could be established from straightforward observations. These include estimation of the abundance distribution for stars of such low luminosity that dredgeup has not affected their atmospheric abundances, thereby providing a distribution function of abundances which is representative of that at the time of stellar formation. A sufficiently large sample of stars must be observed to clarify the following points.

1. Are the very metal-rich stars a tail of a distribution which is represented by the abundance distribution seen in the old planetary nebulae and RR Lyrae stars, or *vice versa*?

2. Where are the very metal-rich old-disk (and thick-disk) stars which are expected in significant numbers if there really is a radial abundance gradient in the disk?

3. Where are the subdwarfs, like those near the Sun, whose density distribution must also peak in the center of the Galaxy?

If these several groups of stars are intimately related, we might expect a smooth distribution function to be found. If, however, they represent discrete phases of the evolution of the Galaxy, a multi-modal distribution can be expected, like that found for the globular-cluster system. Kinematic and spatial distribution data for these same classes of stars are necessary for a serious analysis, and could readily be obtained.

One of the most important properties of the bulge which is amenable to test is the age range of the metal-rich stars. Although it is often assumed that these stars are as old as the globular clusters, there is in fact no strong evidence in favor of this hypothesis. It would indeed be mildly surprising if that population of stars which is the youngest in chemical terms, in that the greatest number of generations of massive stars must have had time to evolve and explode before its formation, and which is young in dynamical terms, in that a substantial amount of dissipation of binding energy occurred before star formation, was at the same time among the oldest in a chronological sense.

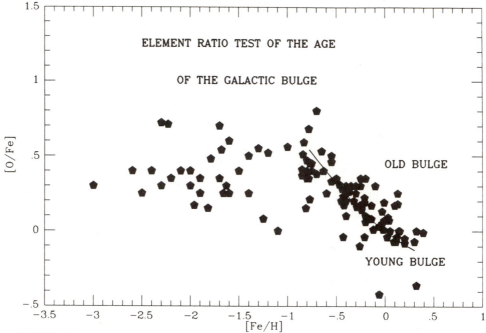

FIGURE 16.3

The relationship between the oxygen-to-iron ratio and stellar metallicity. The systematic decrease in this ratio occurs about 1 Gyr after the onset of significant star formation. If the Galactic bulge really is as old as the metal-poor globular-cluster system, as is often assumed, it will also show a relative excess of oxygen in spite of its high value of [Fe/H].

There is a simple test of the hypothesis that the very metal-rich stars in the central bulge are indeed as old as the globular clusters, and that is to measure the ratios of oxygen (preferably) or the α elements (Mg, Ca, Si, Ti). Old subdwarfs near the Sun and giants in globular clusters show a systematic overabundance of oxygen and the α elements, with [O/Fe] and [α/Fe] $\sim$ +0.4 for [Fe/H] $\lesssim$ −1 (Fig. 16.3). The explanation of this behavior is discussed in Chapter 13, and is simply that these elements are produced only in supernovae from high-mass progenitors, and tend to be dominant in the very early stages of the evolution of a stellar system. After about 1 Gyr (Type I) supernovae from lower-mass progenitors dominate the element production in the Galaxy, so that all element ratios tend toward the solar values. This age corresponds to a metallicity of about −1 dex. If the central bulge really is as old as the globular clusters, it must also have element ratios consistent with dominance of its chemical evolution by high-mass short-lived supernova progenitors. Thus measurement of the ratios [O/Fe] and [α/Fe] for the bulge stars with [Fe/H] $\approx$ +0.3 is a unique test of the hypothesis that the bulge is as old as the globular clusters. If these ratios are significantly less than +0.4, then the bulge is younger than the clusters.

16.6 ARE GLOBULAR CLUSTERS PART OF THE STELLAR HALO?

The relationship between field subdwarf stars and the globular clusters remains poorly understood. It is clear that the globular clusters form a two-component system, with a metal-poor majority having a metal-abundance distribution and kinematics like those of the halo subdwarfs, and a more metal-rich minority being part of the thick disk. The number of clusters is, however, too small to allow those distributions to be observed to high precision, regardless of the precision of the data for an individual cluster. One parameter of fundamental importance whose evaluation could be improved considerably is the age distribution. The primary source of uncertainty, given that high-precision CCD color–magnitude data can now be routinely obtained, is the appropriate evolutionary sequence to fit to the data. The problem here involves adopting the correct set of element ratios, in part to choose the appropriate evolutionary track, and in part to check for differences between clusters and field stars. Are clusters formed from gas with the same element patterns as field stars, or are they significantly self-enriched? The requirements are identical to those discussed above for determining the age of the Galactic bulge.

The observational situation is both easier and harder, however. Most clusters are farther from us than is the bulge; so those stars which cannot be affected by dredgeup are fainter; yet all stars are indistinguishable at a given color on the main sequence. The observed width of the main sequence in 47 Tuc corresponds to an upper limit of only 40 K in the corresponding temperature range, or a range in [Fe/H] of only 0.04 dex. Thus we may proceed by obtaining low-signal-to-noise spectra of a large sample of faint stars, and add them to form a high-signal-to-noise composite. With current multi-object spectroscopic methods, all the (noisy) spectra can be obtained at the same time, so that good spectra of cluster main-sequence stars could be obtained for analysis.

The two-component nature of the cluster system in our Galaxy is well established. Similar halo/thick-disk structure in the M31 cluster system is also probably present, though it is less well established as yet (Elson and Walterbos 1988). Detailed correspondence between both the spatial and the chemical-abundance distributions of field stars and of globular clusters would be strong support for an intimate evolutionary history, and suggest that the formation of clusters is an integral part of the formation of stellar systems. Thus it is important to calculate in detail the dynamics of the M31 cluster system, and also to discover if some or all of the disk population of globular clusters in our Galaxy is really associated with the Galactic old disk. Only the brightest 100 or so M31 clusters have as yet been studied. More could be. Space motions have been derived for a few Galactic globular clusters, in one case (M71) providing a total space motion of only 50 km sec^{-1}, suggestive but not conclusive evidence that it is associated with the thin disk (Cudworth 1985). Space motions for a larger sample of disk clusters would be of considerable importance in discovering the relationship between the field and globular-cluster stellar systems.

16.7   WHAT ARE THE SHAPE AND DENSITY PROFILES OF THE STELLAR HALO?

The shape of a distribution of stars is related to the gravitational potential in which the stars orbit by stellar kinematics. Thus knowing any two of a stellar spatial distribution, stellar kinematics, and the gravitational potential will give us the third to some level. In practice the first two of these can be observed with relative ease, and used to constrain the potential and hence the mass distribution.

Halo stars near the Sun have a noticeably anisotropic velocity dispersion, which requires a flattening of $c/a \approx 0.5$ in the axis ratio of the isodensity contours, even if the gravitational potential is round (see Chapter 11). Recent starcount data showing that the spatial distribution is indeed flattened that much are also discussed in Chapter 11. Mapping the shape of the stellar halo uses a very simple method: we count stars in a fixed range of apparent magnitude and absolute magnitude at several latitudes in the $\ell = 90°, 270°$ plane. Such stars will then all be at the same distance from both the Sun and from the Galactic center, so that the relative number of halo stars as a function of Galactic latitude depends solely on the axis ratio of the stellar distribution. This has been done for main-sequence stars to $V = 20$, but it seems that available fainter data are not reliable enough to provide useful information. Since halo stars at $V = 20$ are only a few kpc distant, only a very local mapping of the Galactic halo shape is available. It would be straightforward to extend this mapping to much larger distance (perhaps 10–20 kpc) by using halo K giants. These stars are bright, so that data to only 15–16 mag are required, but are sufficiently rare that large areas must be surveyed, and good photometry is required to provide reliable distances. Nevertheless, objective-prism surveys can find halo giants efficiently, and well-developed photometric systems (DDO) exist for the subsequent high-precision work, so that the first reliable model of the shape of the stellar halo at large distances from the Sun and from the Galactic center could be completed.

16.8   IS THERE A REAL AGE RANGE IN THE STELLAR HALO?

The range in age evident among halo stars is one of the most important observational clues to the formation and early evolution of the Galaxy. The available indirect evidence, primarily from modeling the distribution of element ratios shown in Fig. 16.3 suggests that at least most of the metal-poor subdwarf halo and most of the metal-poor globular clusters formed in less than about 1 Gyr (Chapter 13), with the thick disk and the metal-rich globular clusters being one to a few Gyr younger than this. However, some direct evidence is inconsistent with this old age for the halo stars.

Schuster and Nissen (1989) have recently completed $uvby\beta$ photometry of a large sample of high-velocity F stars, the results of which are summarized in Fig. 16.4. A real age scatter of several Gyr is evident in these data. If this age range is real, then a significant change will be required in our understanding of

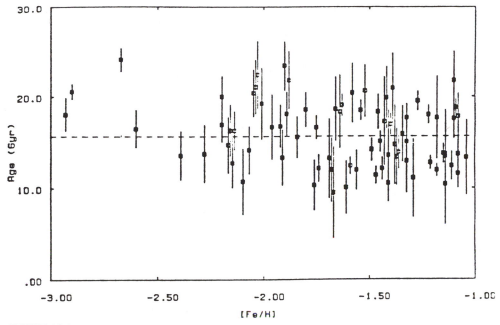

FIGURE 16.4
The distribution of ages and metallicities of a sample of high-velocity F stars near the Sun. (From Schuster and Nissen 1989.) The age spread evident in these data will require a major revision in the conventional picture of the formation of the Galaxy if real.

the timescales of proto-galactic collapse and dynamical evolution, the lifetimes of supernova-Type-I progenitors, and the chemical evolution of the Galaxy. Confirmation of this result is of considerable importance. Some complications which will need to be dealt with include deconvolving the metal-poor tail of the thick disk, which might well be significantly younger than the true halo in some formation models; allowing for the effect of real scatter in the element ratios (if there is real scatter) on the choice of the appropriate age to ascribe to each star; and the precision of the calibration of the theoretical and observational photometric systems. The first of these complications can be resolved only by aquisition of a substantially larger sample. The second and third require high-resolution and high-signal-to-noise spectroscopy and high-precision spectrophotometry, respectively. The spectroscopy can be used to measure the element ratios for each star of interest, thereby removing the uncertainty in the choice of the isochrone from the uncertainty in the age dating. The spectrophotometry can be used to bypass the calibration of the photometry, by using directly the several temperature, gravity, and abundance-sensitive features in the spectra which the $uvby\beta$ photometric system seeks to measure. After all, narrow-band photometry is nothing but an observationally efficient method to obtain low-resolution spectrophotometric data. If the calibration of the supposedly more convenient observational method is limiting our scientific goals, then a more direct method must be employed. In this way, a large sample of relatively bright metal-poor stars could be studied both more reliably and in more detail than heretofore.

16.9   HOW WELL KNOWN ARE ELEMENT RATIOS?

The chemical-element-ratio data shown in Fig. 16.3 are of considerable importance for reconstructing the early evolutionary history of the Galaxy. Most of the information is derived from the amount of scatter evident in the data and from the change in slope of the relationship at [Fe/H]≈ -1. The scatter in the data at low abundances can be interpreted to measure the level of inhomogeneity in the proto-galaxy, and so can be used to discriminate between models of galaxy formation involving a small number of discrete long-lived structures, and those in which the proto-galactic gas is efficiently virialized before substantial star formation. The scatter in the data at higher abundances arises from any variations in the star-formation rate, which may therefore also be constrained (Chapter 13). The sharpness of the change in slope of the [*element*/Fe] *vs.* [Fe/H] relationship is a measure of the extent to which the nucleosynthetic products of Type I supernovae dominate over those of Type II supernovae. This dominance in turn depends on the mass range of supernova progenitors, the nucleosynthetic yields of supernovae of different types, and the history of the star-formation rate. A second break in the [α/Fe] *vs.* [Fe/H] relationship at ~ -0.4 dex further constrains these parameters.

The changes in slope of the various element-ratio relationships seem to occur at those abundances (-1 dex and -0.4 dex) which correspond to the halo–thick disk and the thick disk–thin disk transitions, which therefore relates the Galactic chemical-evolutionary history to its dissipational history, providing further information. Hence finding out the true scatter, as well as the location, sharpness, and number of discontinuities present in the various element-ratio relationships, is probably the most important feasible test of the early evolution of the Milky Way. Present observational methods are adequate, though a very substantially enlarged sample of stars must be studied before appropriate statistical analyses can be undertaken.

16.10   WHAT IS THE SECOND PARAMETER?

The morphology of the horizontal branch (HB) in the HR diagram depends on age and metal abundance, and on at least one other parameter, which may well be age again. *Bona fide* HB stars are old (none is known less than about 12 Gyr old), but otherwise their distribution is not well understood. Most metal-poor globular clusters have blue HB's, though some have red HB's. Are these clusters perhaps younger? Most metal-rich clusters have HB's which are so red that RR Lyraes are not found in them. However, RR Lyraes are common in the field

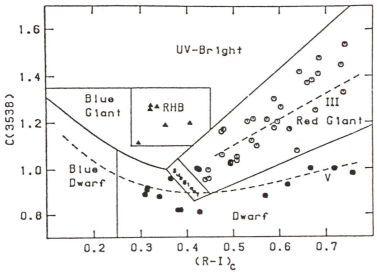

FIGURE 16.5

A two-color diagram illustrating the potential of DDO and $R - I$ photometry to identify field red-horizontal-branch stars. (From Norris *et al.* 1985.)

at even higher abundances than those of the metal-rich clusters. Perhaps the metal-rich field RR Lyraes are older than the metal-rich globular clusters?

Whatever the explanation, mapping the morphology of the horizontal-branch stars in the field is of considerable interest. BHB stars can be found extremely easily from photometric and objective-prism surveys, and RR Lyraes can be found from photometric surveys for variables. RHB stars can also be found, though as yet no systematic survey for field RHB stars has been undertaken. Fig. 16.5 illustrates the way in which such stars could be found. The discrimination could be carried out to first order from photographic or drift-scan CCD data, since the RHB stars lie so far in $C(3538)$ from other (subgiants and main-sequence) stars of similar $R - I$ color. With such a survey we could map the morphology of the horizontal branch in the field stars for the first time.

16.11 IS THERE AN ABUNDANCE GRADIENT IN THE HALO?

The existence of an abundance gradient in the stellar halo is a sensitive test of the amount of dissipation which occurred during halo formation, *modulo* the potentially complicating effects if there were later violent relaxation of the potential. The present observational status of an abundance gradient is confused. The inner regions of the Galaxy are certainly much more metal-rich than are the outer parts, which might be interpreted as a gradient. However, there is a severe semantic problem here, in that it is far from clear that the inner "bulge" is in any way related to the outer "halo." The abundance distribution of those stars which might be considered in some definition to belong to the halo but which are presently in the central regions of the Galaxy is unknown, and quite likely unknowable. Given that all stars have roughly the same velocity dispersion in the central 1 kpc or so, and no abundance-dependent definition of the halo has any meaning when we are searching for possible changes in abundance, no reliable sample of *bona fide* halo stars could be found. We must proceed by disentangling the abundance and kinematic distribution functions of all those stellar populations which coexist in the bottom of the Galactic potential well. Thus measurement of an abundance gradient in the halo interior to the Sun is an extremely subtle problem.

At large distances from the center a gradient could be found. The easiest way to do this is from spectra or appropriate photometry (*e.g.*, DDO for K giants) of distant stars *in situ*. Again we must be careful to avoid an abundance bias when selecting a sample of stars. The color of the giant branch is a strong function of abundance, with a very metal-poor giant branch being too blue to be included in most surveys to date. Similarly, the main-sequence turnoff of an old metal-rich population is very red, so that selection of (say) F/G stars as turnoff tracers will inevitably provide a severely biased result. RR Lyraes and BHB stars are even more obviously biased in this way. A careful selection is, however, possible, provided that some efficient method of locating a sample of stars in an evolutionary state which is not abundance-dependent can be found. Objective-prism and multicolor photometric surveys of G/K stars to identify the distant giants is the most promising prospect here, though a considerable extension to the blue of available surveys will be needed to avoid a bias against just those very metal-poor stars we most wish to find.

Another way to find the amplitude of any halo abundance gradient is to search for a correlation between space motion and metallicity in a sample of halo stars near the Sun. Conversion of the result derived from some sample of stars observed near the Sun to the state of the Galaxy, however, requires careful modeling of the selection effects in local samples of proper-motion stars. Although a gradient might or might not be detected with relative ease, how do we relate this to the larger volume of the Galaxy? We could readily imagine a situation in which there was a strong correlation between orbital properties and metallicity which prevented stars in some range of abundances from ever reaching the Sun, and thereby excluded them from the local sample. For example, there might be a steep radial decrease in metallicity *and* a steep radial increase in orbital angular momentum. Thus there would be a large population of very metal-poor stars on orbits well outside the solar radius, and on sufficiently circular orbits that they never cross the solar circle. These stars would never be counted in a sample defined near the Sun, but might well dominate in an *in situ* sample. Many other correlations between abundance and orbital properties can be imagined that would complicate the interpretation of local data. All could be modeled to some extent, however, and tested by relatively limited samples of distant objects.

We can deduce the most useful form of the analysis required from dynamical consideration. The total orbital energy of a star provides the support against the gravitational force required to maintain some density distribution. This orbital energy may be shared between "peculiar motions," which provide pressure support, and streaming motion, which provides angular-momentum support. The interrelation between pressure and angular-momentum support is described by the asymmetric-drift equation discussed earlier (Eq. 11.1). This equation shows that stars whose orbits have greater total energy, and also have zero net angular momentum, must correspondingly have a larger scale length associated with their spatial distribution than do stars on lower-total-energy orbits. Stars on high-energy orbits must have formed from gas which had dissipated less of its energy than had the gas which formed stars with lower-total-energy orbits. Thus analysis of the relationship between pressure (effectively, radial velocity dispersion) and metallicity for samples of stars with zero net angular momentum can enable us to decide if significant dissipation occurred during formation of the stellar population represented by these stars. In purely observational terms, is a radial abundance gradient indicated by very-low-abundance stars near the Sun?

The relevant observational data are shown in Fig. 16.6, where the important feature for our purposes here is the tendency of groups of stars with no net angular momentum to cross the constant-energy lines at low abundances. If real, this result means that the oldest stars in the Galaxy formed during dissipational collapse of the Galactic halo. The number of stars in the bins farthest to the right in Fig. 16.6 is, however, small. Since these stars have the most extreme abundances and velocities in the available data, it is highly likely that a disproportionate number of errors will be included. The effect of random errors in the distances can be calculated; they tend to move the data points vertically in the figure, but not to the right. Still, the data set needs to be enlarged and made as precise as is possible. We need space motions and abundances for a large sample of stars with $[Fe/H] \leq -1.5$; we could then test if the first star formation in the Galaxy occurred during a period of dissipational collapse.

16.12   IS THERE A STELLAR POPULATION WITH INTERMEDIATE ANGULAR MOMENTUM?

There are two points in Fig. 16.6 with $V_{rot} \approx -100$ km sec^{-1} and with small radial velocity dispersion. These correspond to the c-type (small amplitude of variation) RR Lyraes and the long-period variables (LPVs) with periods in the range 150 to 200 days. This period range for the LPVs corresponds to that seen in the thick-disk globular clusters, so that both of these groups of stars are old.

It is interesting that two groups of stars which are defined on criteria specified by their internal properties should have space motions which are so similar, and which are almost perfectly intermediate between the well-established halo and thick-disk populations in the Galaxy. Does this mean that there really is a continuum of stellar populations between the extreme halo subdwarfs and the high-angular-momentum disk (as has been argued extensively by Sandage 1987), or does it mean that there are many discrete subpopulations of stars in

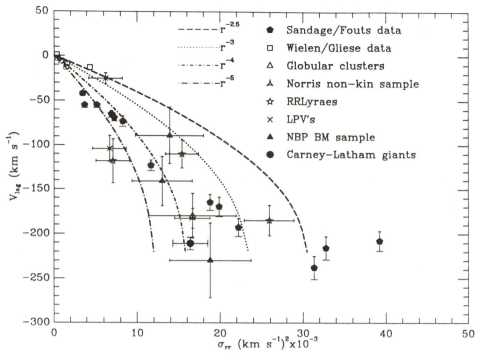

FIGURE 16.6

The relation between the radial velocity dispersion σ_{RR} and the asymmetric drift V_{rot} of samples of old stars in the Galaxy. The data for field stars are binned by metallicity. The key identifies data from sources identified in Fig. 11.1. The model lines correspond to different solutions of the equation relating the allowed amount of random and organized motion in orbits of given total energy (Eq. (11.1)). The tendency for the data to cross the model lines at low V_{rot} (highly negative V_{lag})—that is, there is a radial abundance gradient for low-abundance stars, shows that some star formation took place during dissipational collapse of the Galaxy. Confirmation of the reality of this result is highly desirable.

the Galaxy? This last prospect is particularly important, since it relates directly to the merger history of the Galaxy.

The most likely way to make progress in this field is to enlarge the sample of high-latitude LPVs with intermediate periods (say, 100 to 300 days) for which radial velocities are known. Solutions for the space motions as a function of period could then be obtained, thereby testing the discrete-populations $vs.$ continuous-relation theories. The LPVs can be found readily in the IRAS catalogue (though this catalogue is biased against the lowest-abundance and hence shortest-period LPVs) and are also of such large amplitude of variability in the visual that high-latitude variability surveys should find large numbers of more distant LPVs.

16.13   IS THERE STRUCTURE IN PHASE SPACE FOR OLD STARS?

As stars orbit in the Galactic potential, small initial differences in velocities become apparent as large differences in their positions at later times. A group of stars with similar though not identical initial positions and velocities will slowly lose their uniform phase, and become phase-mixed around their orbits. Such stars occupy diffusion orbits, which are perhaps best imagined as a doughnut, with stars scattered around the ring. Any such doughnut will be destroyed by subsequent violent relaxation of the potential (at least in the inner parts of the Galaxy), so that no clear record of the dynamical structure of the initial Galaxy remains. Should the Galaxy have merged with or swallowed any stellar system subsequent to (the last) violent relaxation, a dynamical record of that merger may, however, still exist in phase space. Similarly, any group of stars which has been disrupted from a (nearly) bound system will also be apparent as structure in phase space. Obvious examples of this latter situation include the remnants of disrupted globular clusters and of large star-formation regions. In the star-forming regions, phase-space structure can be so prominent that recognizable structure (*e.g.*, Gould's Belt, open clusters) can often still be seen in coordinate space.

The amount of structure which exists in phase space is a useful record of the merger history of the Galaxy, and is worthy of considerable study. The two most obvious ways to pursue this study are by detailed study of moving groups containing many stars, and by extension of such analyses to groups of only two or more stars.

The study of moving groups is a difficult observational problem, primarily since data of very high precision would be required to assign individual stars to some subset of phase space on a star-by-star basis. Since such data are usually not available, we must proceed by statistical methods. A major advance in the study of moving groups came with the discovery by Eggen (1987) that almost all stars assigned a high probability of membership in the HR 1614 moving group showed a similar, and intrinsically very rare, CN excess. The stars assigned to the HR 1614 moving group on astrometric criteria were later shown to form a chemically homogeneous group as well. This major result opens the possibility of detailed study of the age and chemical-evolutionary history of moving groups, using the element-ratio techniques discussed above for the Galactic bulge. Most stars currently assigned to moving groups are apparently bright, so that a combination of high-resolution spectral analysis to define the detailed elemental ratios, together with precise space motions from parallax and radial-velocity measurements, could provide unique information on the merger history of the Galaxy, and the extent to which star formation is dominated by a few large complexes.

16.14   WHAT IS THE TIDAL FIELD IN THE GALACTIC DISK?

There is a well-established relationship between age and velocity dispersion for thin-disk stars, in that older stars have larger random velocities (and correspondingly smaller mean rotational velocity) than do younger stars. The detailed form of this age–velocity dispersion relationship has been investigated by several authors (see Fuchs and Wielen 1987), and the theoretical analysis of the manner in which various secular scattering mechanisms act to increase the velocity dispersion of a group of stars is also well detailed. The velocity diffusion is due to large masses concentrated to the plane of the Galaxy, with giant molecular clouds and spiral arms being the most likely and most important scatterers. However, detailed analysis (Binney and Lacey 1988) shows that even the combination of giant molecular clouds and spiral arms is barely able to explain the amplitude of the age–velocity dispersion relation, with an upper limit of about 20 km sec^{-1} in the vertical velocity dispersion possibly being explained in this way. (For a more detailed discussion, see Chapter 9.)

The aim of analyses of this type is to quantify the amplitude of structure in the gravitational field of the Galaxy. A direct measure of that structure would of course be invaluable, and is potentially available from the study of wide binary stars. Very wide binary stars are weakly bound (the velocity of escape in the widest binaries known is much less than 1 km sec^{-1}), and hence are sensitive probes of the level of structure in the Galactic tidal field.

Detailed analyses of the distribution of wide binaries have been undertaken (Wasserman and Weinberg 1987), but have been frustrated by difficulties with the available data, in that the absolute number of wide binaries toward the north Galactic pole (the best studied field) is implausibly large. A recent larger-scale observational study of the distribution of wide binaries (Saarinen and Gilmore 1989) shows very large variations in their surface density, with the binaries themselves being weakly clustered. These data are summarized in Fig. 16.7, which illustrates the nonuniform surface distribution of the optical binaries, and illustrates the extent to which the data toward the north Galactic pole are not representative of those on larger scales. Besides providing the means to map the small-scale tidal field in the Galactic disk, these stars also contain a wealth of information regarding the initial distribution of angular momentum and binding energy in star formation, and are suitable as probes of small-length-scale variations in the uniformity of the interstellar medium during star formation. Precise radial velocities to isolate the physical binaries, and detailed spectroscopic analysis of their elemental abundance distributions, would provide a substantial improvement in our understanding of the structure and evolution of the Galactic disk.

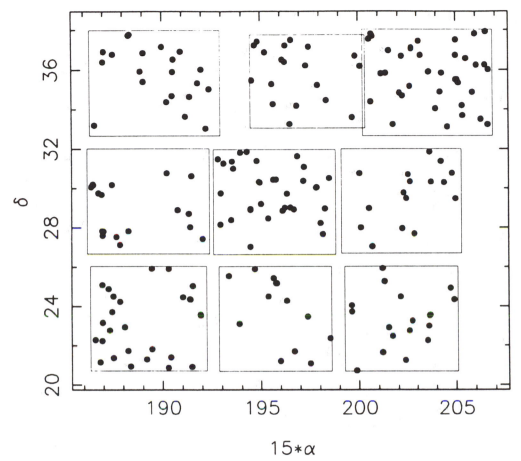

FIGURE 16.7
The distribution of the centers of optical binaries toward the north Galactic pole. In all cases both components are brighter than $B = 12$. Each point represents one optical binary, and the squares indicate the area measured. The distribution of binaries is nonuniform, with a significant local excess in the central field, closely aligned with the north Galactic pole, and with considerable clustering of the binaries themselves.

16.15   WHY ARE GALACTIC HALOS BLUE?

Many elliptical galaxies have a short-wavelength ultraviolet ($\lambda \leq 2000\,\text{Å}$) flux considerably in excess of that expected by fitting stellar-population models to the visual light. The explanation of this excess remains controversial, with the most-likely models basically involving either massive hot young stars or a short-lived luminous hot evolutionary state associated with normal old low-mass stars. An important constraint on these models follows from IUE observations showing the UV excess to be extended, and to be distributed (roughly) like the galaxy's optical light. It is interesting to compare these results with those of studies of the UV-bright sources in the Galactic halo, where the individual sources of UV radiation can be identified and studied in detail.

Besides hot horizontal-branch and blue-straggler stars, which may well provide a large part of the excess flux, there is one other source of UV radiation in the Galactic halo which might be relevant for deciding between models involving old stars and those involving young stars. This source is a small population of apparently young high-mass stars currently farther into the halo than they have any right to be. Extremely careful study of some of these stars shows them to be apparently normal main-sequence B stars, with element ratios similar to those in the Galactic disk, but with radial velocities far too low to allow the star to have formed in the Galactic disk and to have traveled to its present position in its main-sequence lifetime (Keenan *et al.* 1986).

Regardless of whether or not these massive young stars resolve the problem of the extended short-wavelength excess in other galaxies, their existence is an intriguing puzzle. Is there some rare phase of stellar evolution in old low-mass stars which is indistinguishable from main-sequence stars? If the B stars are what they seem, and they cannot have formed in the Galactic disk, where were they formed?

There are currently no regions which we recognize as sites of formation of high-mass stars at high Galactic latitudes, giant molecular clouds being conspicuously absent. Thus, either high-mass stars can form in a very-low-density environment, and the origin of the UV excess in ellipticals is explained, or these stars are subluminous in an ill-understood way. Further study is evidently required. Relevant considerations for these stars include the consistency of their number and distribution with the Type II supernova rate in Galactic halos, the fraction which may be binary, the number of similar A and F stars which make up a mass function, the similarity of this to mass functions in OB associations in the Galactic disk, and the spatial distribution of the B stars. To help test the possibility that they really are massive young stars, it would be interesting to search for possible counterparts in the Magellanic Clouds and particularly in M31. If they do exist in the halos of other Local Group galaxies, then they are certainly luminous and almost certainly the resolution of the puzzle concerning the UV excess in other galaxies. We then merely have the problem of explaining how sufficient mass gets together to form high-mass stars in a low-density environment.

16.16　IS THE THICK DISK A TAIL OF THE THIN DISK?

The relationship between the thin and thick disks provides important information on the dynamical evolution of the Galaxy, and on the formation mechanism for the thick disk. Several of these possibilities are discussed in Chapter 13. One of the most interesting ways to discriminate between plausible models would be to find out if there is a smooth relationship between the thick and the thin disks. The scattering mechanisms identified to explain the age–velocity dispersion

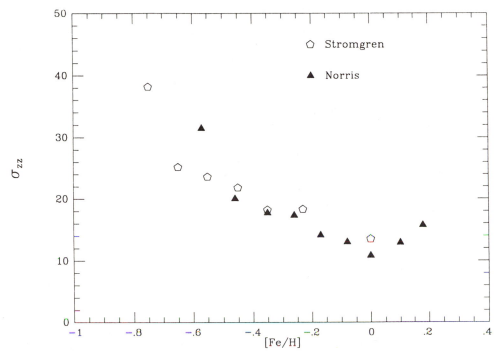

FIGURE 16.8

The relationship between vertical velocity dispersion and metallicity for samples of stars near the Sun. Models of the thick disk which require continuity with the thin disk explain these data as a smooth correlation, whereas discrete-origin models invoke overlapping distribution functions. Distinction between these models is clearly not possible with these data.

relationship observed in old disk stars near the Sun involve a combination of giant molecular clouds and spiral arms. It is only just possible with this mechanism to explain the observed vertical velocity dispersion for the old disk of ~ 20 km sec^{-1}. The vertical dispersion of the thick disk is perhaps 40–50 km sec^{-1}. Thus, since the difficulty in explaining a vertical velocity dispersion increases steeply with increasing velocity (since the high-velocity stars spend all their time far from the plane, and never feel the gravitational perturbations of the disk scatterers), a fundamentally new scale of massive structure must be invoked. That is, some very major mass perturbation (a bar?) must have existed in the Galactic disk at some time if the thick disk is a tail of the thin disk. Study of the relationship between the thick and thin disks is one of the few ways in which the existence of a massive bar in the Galactic disk in the distant past could be tested.

The present state of the data to test the kinematic relationship of the thick and thin disks is shown in Fig. 16.8. It is clear that available or even feasible local kinematic data are unlikely to be able to resolve the uncertainty. It is therefore necessary to consider the relationship of the thick and thin disks in all possible parameters.

The other ways in which the thick disk might be a tail of the thin disk are in metallicity and in age. Calculation of the abundance and kinematic distribution functions for a very large sample of stars with abundances below perhaps -0.3 dex will be necessary for this. Care must be taken not to bias the result by selection of blue (F) stars, since old metal-rich stars, which might well exist, would not be found in such a survey. The most suitable data set might have been the Hipparcos catalogue, and narrow-band photometry (to provide ages and abundances) and radial velocities (to provide space motions) will be the necessary observations. Further considerations involve explaining the thick-disk long-period variables and RR Lyrae stars. These populations both seem metal-rich and are certainly old, yet do not seem to be associated with the old thin disk. The age range in the thick disk is of particular importance in improving understanding of its origins.

16.17   WHAT IS THE GALACTIC WARP?

Many galaxies, including our own, show large-amplitude warps in the outer parts of their disks. The most likely explanation of these warps has recently been provided by Sparke and Casertano (1988), who show that warps are likely to be vertical oscillatory modes of the disk. The type of mode and the amplitude of the resulting warp are sensitive functions of the core radius of the dark halo, and any cutoff in the outer disk. Thus detailed study of galactic warps can provide information on the dynamics of the warps themselves, and also on the distribution of the dark mass.

The best available models of this type are shown in Fig. 16.9, and could be extended and tested from studies of the warp in the Galaxy and in M31. The Milky Way Galaxy shows a large-amplitude warp in H I, and an associated stellar warp also exists, as observed in IRAS sources by Djorgovski and Sosin (1989). A stellar and H I warp is evident in M31. The K giants in a Galactic stellar warp would have $V \approx 20$, and be readily observable. If a sufficiently large sample of associated stars can be found, then the mean motion of the warp could be measured to high precision, and provide an important test of the model.

Observations of warp stars would also establish the mean abundance in the outermost disk, thereby deciding if the gas at the edge of the star-forming disk has primordial abundance, and so is likely to be relatively recently accreted, or is processed. This is important to help understanding of the almost invariably seen (and almost invariably ignored) asymmetry in the H I distribution in warps. The velocity dispersion of the stars in the Galactic warp should be very small. A self-gravitating exponential disk in its outer parts should have a velocity dispersion similar to the $\sim 6\text{--}8$ km sec^{-1} of the interstellar medium. Measurements of this dispersion would be useful to help find the reason for the apparent sharp edges to galactic stellar disks.

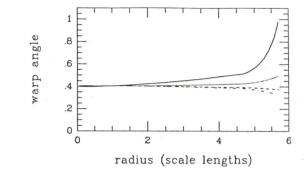

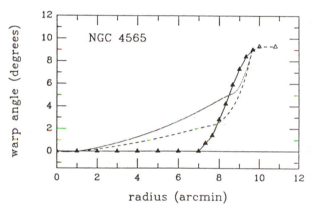

FIGURE 16.9
Models and observations of galactic warps, showing the sensitivity of the properties of the warp to the parameters of the dark halo. The top figure illustrates the decrease in the expected warp angle as the core radius of the dark mass increases, so that large core radii do not allow large-angle warps. The bottom figure shows the observed warp in NGC 4565 (triangles) with illustrative models. The general features of the observations are reproduced, and require a core radius of ~ 4 kpc for the dark matter in NGC 4565. (From Sparke and Casertano 1988.)

16.18   WHAT ARE THE KINEMATICS OF SPIRAL BULGES?

Kinematic data are available for only two bulge components for late-type spiral galaxies. The data are limited to surface brightnesses greater than about $\mu = 24.5$ mag/square arcsec, and were obtained up to 1 kpc from the galaxies' major axes. Surface-brightness distributions for these galaxies are shown in Fig. 16.10, and illustrate the complexity in understanding these kinematic data. Decomposition of the data into standard photometric components, using least-squares methods, provides the components shown. Both the form of the luminosity distribution and the number of photometric components required to match the data are different in the two galaxies, illustrating the difficulty in comparing data from other galaxies either with each other or with data for our Galaxy. Most of the available kinematic data for these two galaxies, including those data thought to be measuring the bulge, contain a substantial contribution from thin-disk light, so that the available evidence for high systemic rotation velocities far from the plane and from the minor axes of spiral bulges is at best uncertain. It is extremely difficult to calculate the fraction of thin-disk light which is contributing to the extant data. The amount by which the thin disk contaminates the bulge

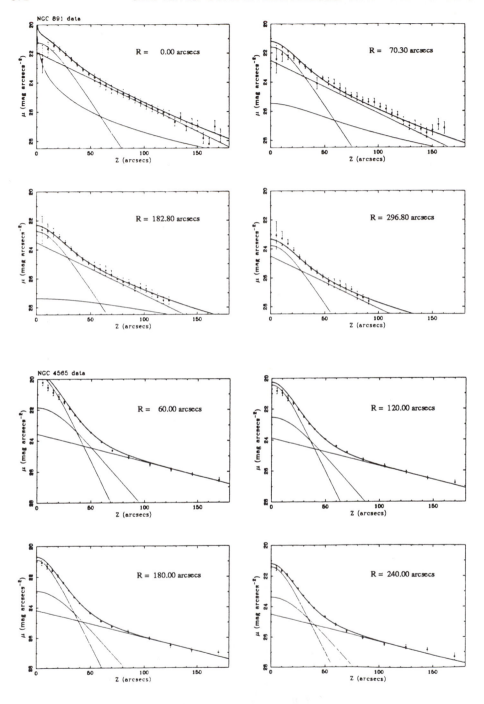

FIGURE 16.10 (Facing page)
Luminosity distributions for NGC 891 (top four panels) and NGC 4565 (lower four panels) fit by three-component models typical of those used to model galaxy surface photometry. The very different contributions of the three components in each galaxy to the surface-brightness distributions, and the considerable contribution by thin-disk light several disk scale heights from the plane, illustrate the difficulty in isolating the kinematics of the bulge component. (From Shaw and Gilmore 1989.)

light is a sensitive function of the way in which we model the surface-brightness distribution, and hence cannot be calculated reliably.

Data at much greater distances from the major axis are essential to minimize this problem. Such data could be obtained, using standard long-slit methods and CCD detectors (all the published data used lower-quantum-efficiency detectors than are currently available) provided that regular chopping to measure the sky reliably is undertaken. It would also be interesting to examine the extent to which we could proceed using fiber-optic methods. A number of fibers could be positioned on the galaxy (along some isophote), and other fibers could be positioned well off the Galaxy to provide simultaneous sky values. It should in principle be possible to observe reliably the kinematics of a spiral bulge well away from the thin disk to perhaps 0.5 R_e or even 1.0 R_e, where R_e is the bulge half-light radius, using this method.

16.19   WHAT IS $K_z(z)$ FAR FROM THE SUN?

There is no significant missing mass associated with the Galactic disk near the Sun. Is that true in other disks and in other places in the Galactic disk?

Photometric studies of edge-on spiral galaxies show them to have disks whose thickness is remarkably constant with galactocentric distance. Since the mass in the disk is decreasing (roughly) exponentially with radius, this implies that the vertical stellar velocity dispersion is also decreasing (roughly) exponentially with radius, provided that the disk mass-to-light ratio is roughly constant with radius. Thus we can test the constancy of the disk mass-to-light ratio by verifying the predicted radial dependence of the vertical velocity dispersion for the old disk. This apparently simple experiment is complicated a little by the requirement that we use the old-disk velocity dispersion, since it is that component which has constant thickness. Thus the observationally more straightforward HI observations do not really test the relevant thing. However, planetary nebulae are reasonably common, and should make ideal old-disk probes. A further minor complication is that we cannot measure a vertical velocity dispersion and a vertical thickness at the same time. Some indirect analyses have been attempted using inclined galaxies (see Chapter 10) but are necessarily more model-dependent. Perhaps the best way would be to measure the radial-velocity profile for several face-on galaxies. If all showed similar kinematics, we could be confident that they were representative of all galaxies, and the radial profile of the disk mass-to-light ratio would be derivable.

Kinematic analyses of other galaxies can, however, measure only changes in the disk mass-to-light ratio, not its absolute value. For that measurement we need both spatial distribution and kinematic data, and hence are restricted to our Galaxy. Although it is possible in principle to calculate the $K_z(z)$ law at other radii in our Galaxy from precise proper-motion data (see Kuijken and Gilmore 1989 for a discussion of one of several such methods), a more direct probe is the interstellar H I gas (Knapp 1988). The thickness of the H I distribution can be measured directly, as can its kinematics. Although this method suffers from all the usual difficulties associated with distance measurements, it has considerable promise. Particularly in the anticenter it could prove very powerful. A careful study must allow not only for the distance uncertainties, but also for the radial change in the contribution of the H I gas to the total mass of the disk. This contribution rises rapidly with increasing radius, and quickly becomes dynamically very important. Nevertheless, there are no difficulties in principle with this method, which holds considerable promise for calculation of the disk mass-to-light ratio far from the Sun.

16.20   ARE THERE METAL-POOR YOUNG CLUSTERS NEAR THE SUN?

The distribution of abundances for stars in the Galactic disk changes with both time and position. The most important variables determining the form of this distribution include the initial abundance of the disk gas, the fraction of the available gas which has ever formed into stars, the distribution of masses of newly formed stars (the stellar initial mass function), and the history of gas flows into and out of the star-forming regions. All these variables may themselves be functions of both position and of time. The distribution of abundances of different elements can also be a function of the star-formation rate. Both the star-formation rate and the rate of accretion of gas into the Galactic disk are certainly functions of time and of position, and are of particular importance because they determine the luminosity evolution of the Galactic disk as well as its chemical evolution. Thus detailed understanding of the abundance distribution of disk stars is of fundamental importance in understanding the evolution of disk galaxies.

The data available for study of the chemical evolution of the Galactic disk include the age–metallicity relation for stars near the Sun, the fraction of the disk mass which is gas and that fraction which is stellar, and the radial variations of these properties. The fractions of the local Galactic disk which are stellar and gaseous are derived from calculation of the Galactic $K_z(z)$ law, and are discussed in Chapter 8. The available local data for the age–metallicity relation are summarized in Fig. 16.11, which also includes some data for extreme open clusters. The radial gradient of abundances in the disk can be measured for the present interstellar medium from H II regions, and for stars from studies of

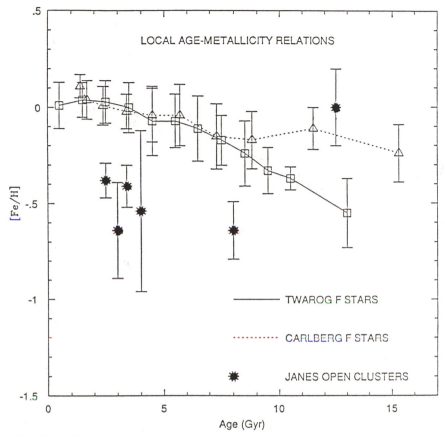

FIGURE 16.11
Age–metallicity relations for local F stars from Twarog (1980) and from Carlberg *et al.* (1985), together with data for some extreme open clusters. The cluster data imply that open clusters near the Sun exist with solar abundance and ages of 12 Gyr, and with less than one-half solar abundance and very young ages. It is very difficult to reconcile these data with plausible age–metallicity relations and with plausible radial abundance gradients.

K giants. These studies are in reasonable (though far from perfect) agreement, and suggest a negative radial abundance gradient in the disk of ~ 0.05–0.10 dex kpc^{-1}. The dependence of this gradient on age is not reliably established as yet.

Because understanding the luminosity evolution of galaxies requires an understanding of the local abundance distribution, we must look carefully at the reliability of the local data. Although the F-star trends in Fig. 16.11 look in reasonable agreement for ages less than ~ 10 Gyr, in fact the agreement is much less encouraging. The two analyses shown in Fig. 16.11 are analyses of the same data set. Thus the substantial differences at old ages reflect entirely the quality of the available photometry, and not the reliability of the determination. A further very serious limitation on all extant studies is the selection bias used in derivation of the sample. The stars chosen for analysis to date are F stars, primarily since well-calibrated photometric systems for their study exist. However, metal-rich ($[Fe/H] \gtrsim 0$) stars older than about 10 Gyr have a main-sequence turnoff which is much redder than the F-star spectral type. Thus there are two fundamental problems with the result shown in Fig. 16.11. First, why are any old metal-rich "F stars" present in the sample at all, since such stars should have evolved into giants long ago? Second, what would the abundance distribution look like if the sample were not heavily biased against metal-rich old stars? It is obviously not a total coincidence that the slope of the Twarog age–metallicity relationship follows very closely the isochrone for late F stars.

Some evidence that the true age–metallicity relationship is in fact quite unknown, and may not even exist as a well-defined correlation, comes from the open-cluster data also shown in Figure 16.11. These data points have been deliberately selected to illustrate the extreme points known, and so will contain a disproportionate number of errors. Calculation of the true age–metallicity relation for open clusters, with due allowance for the important effects of radial and possibly time-dependent abundance gradients, is of very considerable importance and is a straightforward observational project for a small telescope.

However, assuming for the present that the data are reliable, what do they mean, and are they consistent with other data? The metal-poor young open clusters all lie ~ 1 kpc exterior to the solar circle, and so might be interpreted as a steep abundance gradient exterior to the Sun. This is, however, unlikely for three reasons. The first is the awkward coincidence that such a steep change in the abundance gradient should start so near to where we happen to live; the second is that such a steep gradient is not seen in other samples of field stars or in the interstellar medium; the third is that in general galactic disks show small and smooth radial color gradients. To make these color gradients consistent with abrupt steepening of an abundance gradient requires considerable fine tuning, and is unlikely to be a general phenomenon.

It is more likely that the apparent steep abundance gradient in open clusters is in part another selection effect and in part a measure of the (apparently considerable) scatter in the age–metallicity relationship. Open clusters are most easily found in the Galactic anticenter, since both interstellar reddening and the background star density are lower there than toward the Galactic center. Thus a larger population of clusters will be found beyond the Sun, and hence a larger number of extreme objects. A real large scatter in the abundance distribution at any age and position will further populate the wings of the distribution. Additionally, some clusters (though not the youngest) have quite high space velocities, and so really sample the intrinsic abundance distribution several kpc from the Sun. A smooth abundance gradient similar to that seen in field stars, together with some intrinsic scatter, then explains the large number of low-abundance clusters. The radial density profile in the Galactic disk in fact acts against this effect, since the number of objects decreases with increasing radius. However, the number of open clusters seen is a convolution of their creation rate and their destruction rate. Since the surface density of giant molecular clouds, which are efficient at destroying open clusters, decreases very rapidly exterior to the Sun, we also expect a radially increasing fraction of all open clusters which are ever formed to survive. This effect will act to counteract, and may even overwhelm, the radially decreasing formation rate of clusters.

It would be interesting to improve both the quantity and the reliability of the available data, and to carry out a detailed study of the processes noted above. One of many important general conclusions, in addition to the intrinsic interest in understanding the dynamics of the clusters, is in obtaining reliable measures of the true scatter in the abundance distribution as a function of time and of position. This scatter is related both to the efficiency of mixing in the interstellar medium (abundance-ratio studies of the young open clusters will be of considerable importance here) and to the accretion history of the Galactic disk. Any plausible model of Galaxy formation in a cosmological context requires accretion of primordial gas onto the disk for an extended period, with this accretion occurring at systematically increasing radii. The outermost parts of the disk should still be accreting primordial material today. Tests of these predictions from detailed abundance studies would be of considerable interest.

REFERENCES

Baade, W. 1944. *Astrophys. J.*, **100**, 137.

Begeman, K. 1987. Ph.D. thesis, Univ. of Groningen.

Binney, J., and Lacey, C. 1988. *Mon. Not. Roy. Astron. Soc.*, **230**, 597.

Blaauw, A., and Schmidt, M., eds. 1965. *Galactic Structure*. Chicago, IL: Univ. of Chicago Press.

Burton, W. B. 1985. In van Woerden *et al.*, p. 83.

Carlberg, R., Dawson, P., Hsu, T., and VandenBerg, D. 1985. *Astrophys. J.*, **294**, 674.

Cudworth, K. 1985. *Astron. J.*, **90**, 65.

Delhaye, J. 1965. In Blaauw and Schmidt, p. 61.

Djorgovski, S., and Sosin, C. 1989. *Astrophys. J. (Letters)*, **341**, L13.

Eggen, O. J. 1987. In Gilmore and Carswell, p. 211.

Eggen, O. J., Lynden-Bell, D., and Sandage, A. 1962. *Astrophys. J.*, **136**, 748.

Elson, R., and Walterbos, R. 1988. *Astrophys.J.*, **333**, 594.

Fich, M., ed. 1987. *The Mass of the Galaxy*. Toronto: CITA.

Fuchs, B., and Wielen, R. 1987. In Gilmore and Carswell, p. 375.

Gilmore, G., and Carswell, R., eds. 1987. *The Galaxy*. Dordrecht: Reidel.

Gratton, R. G., Tornambè, A., and Ortolani, S. 1986. *Astron. Astrophys.*, **169**, 111.

Harmon, R., and Gilmore, G. 1988. *Mon. Not. Roy. Astron. Soc.*, **235**, 1025.

Keenan, F. P., Lennon, D. J., Brown, P. J. F., and Dufton, P. L. 1986. *Astrophys. J.*, **307**, 694.

Knapp, G. 1987. In Fich, p. 35.

Kuijken, K., and Gilmore, G. 1990. *Mon. Not. Roy. Astron. Soc.*, to be submitted.

Little, B., and Tremaine, S. 1987. *Astrophys. J.*, **320**, 493.

Norris, J., Bessell, M., and Pickles, A. 1985. *Astrophys. J. Supp.*, **58**, 463.

Rich, M. 1988. *Astron.J.*, **95**, 828.

Ryden, B. 1988. *Astrophys. J.*, **329**, 589.

Saarinen, S., and Gilmore, G. 1989. *Mon. Not. Roy. Astron. Soc.*, **237**, 311.

Sandage, A. 1987. In Gilmore and Carswell, p. 321.

Schuster, W., and Nissen, P. 1989. *Astron. Astrophys.*,, **222**, 69.

Shaw, M. A., and Gilmore, G. 1989. *Mon. Not. Roy. Astron. Soc.*, **237**, 903.

Sparke, L., and Casertano, S. 1988. *Mon. Not. Roy. Astron. Soc.*, **234**, 873.

Strömberg, G., 1934. *Astrophys. J.*, **79**, 460.

Terndrup, D. 1988. *Astron. J.*, **96**, 884.

Twarog, B. 1980. *Astrophys. J.*, **242**, 242.

Wasserman, I., and Weinberg, M. D. 1987. *Astrophys. J.*, **312**, 390.

Woerden, H. van, Allen, R. J., and Burton, W. B., eds. 1985. *The Milky Way Galaxy.*
    Dordrecht: Reidel.

ACKNOWLEDGMENTS

The authors and publisher would like to thank the following organizations and individuals for the use of copyrighted material.

Annual Review of Astronomy and Astrophysics. The following figures are reproduced with permission:

> Figures 2.3, 8.2, 8.3, 11.1, 11.2, 11.3, 11.4, 11.6, 13.4, 16.6, 16.8, Volume 27, ©1989 by Annual Reviews Inc.
> Figure 2.4, Volume 24, ©1986 by Annual Reviews Inc.
> Figure 3.1, Volume 25, ©1987 by Annual Reviews Inc.
> Figure 3.14, Volume 26, ©1988 by Annual Reviews Inc.
> Figure 10.7, Volume 16, ©1978 by Annual Reviews Inc.

Astronomy and Astrophysics, Figures 5.4, 5.5, 5.6, 5.9, 5.10, 5.11, 5.12, 5.13, 5.14, 7.2, 10.3, 10.9, 10.10, 10.11, 10.12, 10.14, 10.15, 10.16, 12.1, 14.2, 15.2, 16.4.

W.H. Freeman and Company

> Mihalas and Binney, *Galactic Astronomy*, 1981, Figure 4.3.
> King, *The Universe Unfolding*, 1976, Figure 4.4.

Gordon and Breach Science Publishers, *Fundamentals of Cosmic Physics*, Volume 5, Figure 12.6.

Kitt Peak National Observatory, Figure 3.5.

Kluwer Academic Publishers, Figures 3.2, 3.3, 3.4, 3.6, 10.8, 15.4.

National Radio Astronomy Observatory, Figures 3.7 and 6.6.

Observatories of the Carnegie Institution of Washington, Figures 4.5, 5.10, 9.4, 9.5, 10.8, and 15.3.

Theo Schmidt-Kaler, University of Bochum, Figure 3.8.

Springer-Verlag, Verschuur and Kellerman, *Galactic and Extragalactic Radio Astronomy*, 2nd edition, 1988, Figures 4.2 and 4.6.

University of Chicago Press, Blaauw and Schmidt, eds., *Galactic Structure*. ©1965 University of Chicago Press. Figure 7.1.

INDEX